Advanced Thermoelectric Materials - Theory, Development, and Applications

Edited by Uday M. Basheer Al-Naib

Published in London, United Kingdom

Advanced Thermoelectric Materials - Theory, Development, and Applications
http://dx.doi.org/10.5772/intechopen.1006434
Edited by Uday M. Basheer Al-Naib

Contributors
Anita Ioana Visan, Claudiu Hapenciuc, Di Wang, Hong Liu, Irina Negut, Mark Auslender, Mohammad Ali Nasiri, Ronald Edgar Pirela La Cruz, Rony Shneck, Sergey Nemov, Sergio Rafael Velásquez Guzmán, Uday M. Basheer Al-Naib, Xiang Li, Xinli Cheng, Xin Peng, Yahui Xue, Yao Lu, Zinovi Dashevsky

First published in London, United Kingdom, 2025 by IntechOpen
IntechOpen is the global imprint of INTECHOPEN LIMITED, registered in England and Wales, registration number: 11086078, 167-169 Great Portland Street, London, W1W 5PF, United Kingdom

For EU product safety concerns: IN TECH d.o.o., Prolaz Marije Krucifikse Kozulić 3, 51000 Rijeka, Croatia, info@intechopen.com or visit our website at intechopen.com.

British Library Cataloguing-in-Publication Data
A catalogue record for this book is available from the British Library

Advanced Thermoelectric Materials - Theory, Development, and Applications
Edited by Uday M. Basheer Al-Naib
p. cm.
Print ISBN 978-1-83634-824-5
Online ISBN 978-1-83634-823-8
eBook (PDF) ISBN 978-1-83634-825-2

If disposing of this product, please recycle the paper responsibly.

Meet the editor

Dr. Uday M. Basheer Al-Naib is an Assistant Professor in Materials Engineering with over a decade of academic and research experience in advanced ceramics, metal-ceramic joining, and thermoelectric materials. He holds a Ph.D. in Materials Engineering from Universiti Sains Malaysia and has served as a faculty member at Universiti Teknologi Malaysia. His research interests include solid slip casting, friction welding, thermal barrier coatings, and energy materials. Dr. Basheer has published extensively in peer-reviewed journals and edited several scientific volumes in the field of materials science. He is actively involved in collaborative research and mentoring, making significant contributions to the advancement of sustainable materials and innovative fabrication techniques.

Contents

Preface **IX**

Chapter 1 **1**
Advanced Thermoelectric Materials: A Snapshot of Theory, Development, and Applications
by Mohammad Ali Nasiri

Chapter 2 **31**
Thermoelectric Complete Response
by Ronald Edgar Pirela La Cruz and Sergio Rafael Velásquez Guzmán

Chapter 3 **51**
Electron and Phonon Transport Mechanisms in Thermoelectric Materials
by Hong Liu and Xinli Cheng

Chapter 4 **85**
High-Perfection Bulk and Film Thermoelectrics Based on Indium-Doped Lead Telluride
by Sergey Nemov, Mark Auslender, Rony Shneck and Zinovi Dashevsky

Chapter 5 **123**
Nanofluidic Thermoelectric Materials
by Di Wang, Xin Peng and Yahui Xue

Chapter 6 **139**
Flexible Thermoelectric Films and Devices
by Xiang Li and Yao Lu

Chapter 7 **179**
High-Temperature Ceramic Thermoelectric Generators: Potential Challenges and Future Prospects in Automotive Applications
by Uday M. Basheer Al-Naib

Chapter 8 **199**
Recent Advances in Thermoelectric Materials for Biomedical Applications: Energy Harvesting and Wearables
by Anita Ioana Visan, Irina Negut and Claudiu Hapenciuc

Preface

Thermoelectric materials have emerged as a vital research focus in the quest for sustainable energy technologies. Their unique ability to convert thermal energy into electrical energy and vice versa opens a wide range of power generation, cooling, and energy harvesting applications. This edited book, *Advanced Thermoelectric Materials – Theory, Development, and Applications*, brings together contemporary research covering the theoretical foundations, material innovations, fabrication techniques, characterization methods, and real-world applications of thermoelectric materials. The book begins with Chapter 1 and a comprehensive overview of the field, offering a historical perspective and mapping out the current landscape of thermoelectric materials research. It introduces readers to key themes explored throughout the volume, including the role of material structure, performance metrics, and application domains.

Chapter 2 provides a focused summary of the fundamental physical effects that govern thermoelectric behavior, such as the Seebeck, Peltier, and Thomson effects. It outlines essential performance indicators like the thermoelectric figure of merit (ZT) and presents the theoretical basis for understanding and optimizing thermoelectric systems. Building upon this, Chapter 3 examines the underlying electron and phonon transport mechanisms in thermoelectric materials. This chapter highlights how microscopic transport processes influence macroscopic material properties and discusses engineering strategies for enhancing thermoelectric performance by controlling scattering events and carrier mobility. Chapter 4 focuses on advanced bulk and thin-film materials, particularly those based on doped lead telluride systems. The chapter addresses synthesis routes and structural perfection techniques, demonstrating their influence on charge transport and thermal conductivity.

Chapter 5 explores novel nanofluidic and confined-phase thermoelectric materials. These systems introduce new opportunities for manipulating heat and charge transport through interface effects, nanoconfinement, and fluid dynamics, representing a frontier in thermoelectric material design. Chapter 6 presents the development of flexible thermoelectric films and devices. This chapter discusses fabrication strategies and challenges involved in creating mechanically compliant thermoelectric systems suitable for integration into wearable and portable electronics. The application of ceramic-based thermoelectric materials in high-temperature environments is addressed in Chapter 7. This contribution examines the performance of ceramic thermoelectric generators in automotive and industrial contexts, focusing on thermal stability, material compatibility, and engineering design. Finally, Chapter 8 investigates thermoelectric materials developed for biomedical and wearable applications. The chapter showcases recent advances in soft, lightweight materials capable of harvesting body heat for powering low-energy medical and electronic devices.

This book is intended for a broad readership, including materials scientists, physicists, engineers, and graduate students who are interested in the theoretical principles, material development, and engineering applications of thermoelectrics. By combining foundational knowledge with state-of-the-art research, the book aims to serve both as a reference and as a source of inspiration for future innovations in the field. We extend our sincere appreciation to the contributing authors for their valuable insights and to the reviewers for their detailed and constructive feedback. Special thanks are due to the editorial and production teams for their consistent support throughout the preparation of this book. We hope that this book will provide a meaningful contribution to the ongoing development of advanced thermoelectric materials and will stimulate further interdisciplinary research and collaboration.

Dr. Uday M. Basheer Al-Naib
Centre for Advanced Composite Materials (CACM),
Universiti Teknologi Malaysia (UTM),
Malaysia

Chapter 1

Advanced Thermoelectric Materials: A Snapshot of Theory, Development, and Applications

Mohammad Ali Nasiri

Abstract

This chapter provides a comprehensive overview of high-performance thermoelectric materials and their transformative role in energy conversion technologies. Beginning with the fundamentals of thermoelectricity, the chapter traces the historical development of materials like bismuth telluride and explores the advancements in organic, hybrid, and nanostructured thermoelectric systems. Special focus is given to cutting-edge materials, including topological insulators, 2D materials, and ionic thermoelectric materials, which exploit ionic conductivity and thermodiffusion for efficient low-grade heat harvesting. Key strategies such as alloying, nanostructuring, and doping are discussed alongside computational methods that drive material innovation. The chapter also highlights diverse applications, from waste heat recovery to wearable devices and green energy systems, emphasizing their role in sustainable energy solutions. Advanced phenomena, including the Soret effect, and state-of-the-art modeling techniques are analyzed to provide insights into optimizing performance. By synthesizing historical perspectives, breakthroughs, and future potentials, this chapter serves as a resource for researchers and engineers aiming to advance thermoelectric materials and systems. It addresses current challenges and outlines future directions to illuminate the path toward more efficient and sustainable energy conversion technologies.

Keywords: high-performance materials, ionic thermoelectric, Soret effect, low-grade thermal energy harvesting, Seebeck effect

1. Introduction

The search for efficient technologies of energy conversion has become paramount in the quest for sustainable energy solutions. One such avenue can be found in thermoelectric materials, a class of materials possessing the capability to directly transform heat energy into electrical power via the Seebeck effect [1]. Considering global warming, which is an environmental challenge today, devices made of thermoelectric materials can play a role in reducing the harm of global warming by directly converting heat into electrical energy [2, 3]. The conventional sources of energy are per se connected with limited resource supply and environmental impact, making the satisfaction of the fast-growing demand in power very challenging. Fossil fuel combustion not only causes the release of greenhouse gases but also the consumption of limited reserves; hence, alternative and sustainable energy conversion technologies have to be found. In such a scenario, thermoelectric materials emerge as a very promising solution that will capture waste heat from various sources and convert it into useful electricity [4, 5]. On the other hand, ensuring the sustainability of thermoelectric materials remains a challenge. A sustainable material such as lignin, a complex natural polymer, can be combined with other thermoelectric materials [6, 7]. Recent research has shown that thermoelectric systems can be built using this material. In these systems, positive and negative ions, rather than electrons, serve as electrical carriers in lignin-based microchannels, enabling the observation of the Seebeck effect in this structure [8, 9].

In thermoelectric materials, charge transfer typically requires a medium such as a metal or semiconductor, where the differential potential is governed by the movement of electrons. In contrast, ionic thermoelectric systems rely on the movement of ions (both positively and negatively charged) within an electrolyte or fluid. These fluids exhibit electrical conductivity due to the presence of mobile ions. In such systems, the driving force for ion movement is associated with thermodiffusion phenomena, where a temperature gradient induces ionic motion. Therefore, the movement of particles, such as ions, under the influence of a temperature gradient is referred to as the Soret effect in the context of ionic systems [8, 10]. Qiao et al. introduced a linearized approach to solving the Poisson-Nernst-Planck equations, employing a gradient-flow framework and simplifying the logarithmic function [11]. Fish provided a theoretical model to analyze nanocomposite electrolytes, describing the space charge layer and its influence on the conductivity of insulating spheres within ionically conductive materials [12]. Fleharty examined nanochannels with electric double layers comparable to the channel width, solving the Poisson-Boltzmann equation using charge regulation boundary conditions [13]. Moya et al. investigated cylindrical nanopores containing ternary electrolyte solutions, analyzing the effects of wall charge polarity. They applied the network simulation method to solve the Poisson-Boltzmann equation and used modified Navier-Stokes and Nernst-Planck equations to determine velocity and conductivity [14]. Jing et al. evaluated the electroviscous effect on electric double layers, revealing its insignificance at high Z potential [15]. Luo and Keh derived expressions for electrophoretic mobility and conductivity in salt-free suspensions, highlighting their dependence on surface charge density [16]. Varner studied dilution effects in ionic liquid supercapacitors to improve performance [17]. Huang explored hydrodynamic slippage and its impact on nanoscale power generation [18]. Levy et al. modeled the dielectric response of ionic liquids, accounting for ion-dipole interactions and deriving a closed-form dielectric constant [19]. Stout et al. summarized ionic transport principles [20], and Qiao et al. further refined numerical schemes for solving ionic transport equations [11].

In other hand, Topological materials, characterized by their distinct band structures, have garnered significant interest in condensed-matter physics, thermoelectrics, spintronics, and other related fields. Thermoelectric technology, which enables the direct conversion of heat to electricity or vice versa, holds considerable promise in addressing the global energy crisis and advancing solid-state cooling solutions. In recent research systematically investigates the magneto-thermoelectric transport properties of a high-quality single-crystalline $Bi_{88}Sb_{12}$ topological insulator. It observes a large magneto-Seebeck effect due to the material's ultrahigh mobility and linear band dispersion, which results in an enhanced magneto-zT under magnetic fields as low as 1 T. A high zT value of ~1.7 ± 0.2 is achieved at 180 K and 0.7 T, particularly significant below 300 K. The study highlights the importance of magneto-thermoelectric correlations as an effective strategy for optimizing thermoelectric performance. It suggests future exploration of p-type $Bi_{1-x}Sb_x$ with hole doping to further improve the material's response [21].

In conclusion, thermoelectric materials represent a promising avenue for sustainable energy conversion, addressing both global energy demands and environmental concerns. Advances in ionic thermoelectric systems, particularly those utilizing sustainable materials like lignin, offer new pathways for efficient waste heat recovery. Additionally, the exploration of topological materials, such as $Bi_{1-x}Sb_x$, demonstrates the potential for enhanced thermoelectric performance through magneto-thermoelectric effects. By integrating novel material systems and theoretical models, future research can further optimize thermoelectric technologies for widespread energy applications. Here, we do not categorize thermoelectric materials based on high or low-temperature gradients; instead, we classify them according to material types. Our focus is on sustainable materials and enhancing the performance of thermoelectric materials in light of recent research. Additionally, we aim to identify integrated systems that strive to develop compact electrical or sensor modules.

2. Theoretical foundations

Thermoelectric materials represent the research field where the groundbreaking application of the Seebeck effect, a basic concept that permits the direct conversion of temperature gradients into electrical voltage, is put into practice. Such a characteristic makes up the very core of harvesting lost heat and converting it directly into precious electrical output. But before introducing different classes of material and their uses, understanding the principal laws with which the basics of thermoelectricity are embodied would first be appropriate [22]. The very essence of thermoelectricity is dependent upon the behavior of the charge carriers in a material across a temperature gradient. Keeping this view in mind, if two materials with different conductivity to electricity are brought in contact, then electrical charge will start moving from the high-conductivity material, like from the hot side to the low-conductivity material, the cold side. This migration of charge carriers creates the potential difference, hence establishing an electric current. Hence, this would allow conversion of thermal energy into electrical energy on the principle of thermoelectricity [23, 24].

The Seebeck effect is the cornerstone of thermoelectric phenomena. Discovered by Thomas Johann Seebeck in the early nineteenth century, this effect illustrates the generation of an electromotive force (EMF) across a temperature gradient. The magnitude of the generated voltage is proportional to the temperature difference across the material. Harnessing the Seebeck effect allows for the creation of thermoelectric generators capable of converting waste heat from various sources, such as industrial processes or car exhaust, into electricity [25–27]. On the other hand, named after Jean Charles Athanase Peltier, the Peltier effect is the reverse of the Seebeck effect. It describes the absorption or release of heat at an electrical junction, resulting in a temperature gradient. This effect is fundamental in thermoelectric cooling applications, where an electric current is applied to transfer heat from one side of the material to the other. Peltier devices are found to be used in electronic cooling systems and temperature control applications [28]. Also, the Thomson effect, also known as the Kelvin or Joule-Thomson effect, describes the temperature change that occurs when a current flows through a homogeneous conductor subjected to a thermal gradient. Though less commonly employed in practical applications than the Seebeck and Peltier effects, the Thomson effect plays a role in understanding the overall behavior of thermoelectric materials under varying conditions [29, 30]. To quantify the efficiency of a thermoelectric material, researchers commonly refer to the figure of merit, denoted as zT. The zT value amalgamates various material properties, including electrical conductivity, thermal conductivity, and temperature, into a single metric. Higher zT values signify superior thermoelectric performance. The quest for materials with elevated zT values is a driving force in the development of high-performance thermoelectric materials, as it directly correlates with their ability to convert heat into electrical power effectively. In subsequent sections, we will explore how different classes of materials strive to optimize zT values for enhanced energy conversion efficiency [31, 32].

Figure 1 Considering that the sign of α_n for electrons is negative, it depicts the production of an electric field that opposes the temperature gradient $\vec{\nabla}T$ [33]:

$$\vec{E} = \alpha_n \vec{\nabla} T \tag{1}$$

The electronic contribution to the Seebeck coefficient can be expressed generally as

$$S = \frac{k_B}{e} \int \frac{E_F - E}{K_B T} \cdot \frac{\sigma(E)}{\sigma} dE \tag{2}$$

In this equation, e stands for elementary charge, kB for Boltzmann, and constant σ(E) for electric conductivity, which takes energy-dependent scattering mechanisms and density of states into consideration. By integrating electric conductivity over all energies ($\sigma = \int \sigma(E) d(E)$), one may determine the total conductivity, or σ, where T is the temperature. Furthermore, we take into consideration the mobility edge, where $E_F - E_v$ denotes the Fermi level location with respect to the valence band edge.

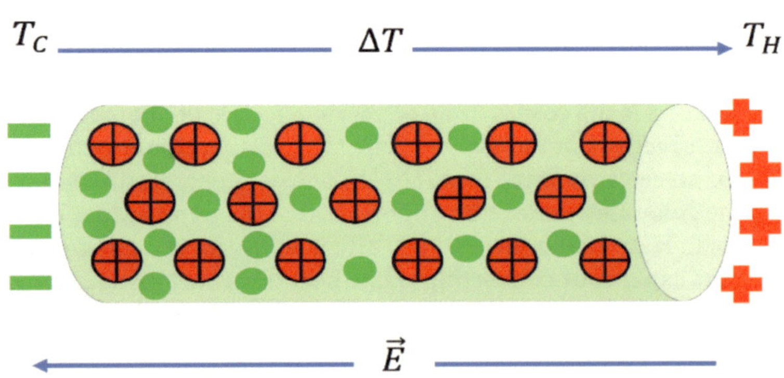

Figure 1.
An illustration of the thermoelectric effect in an equilibrium system, when a material is exposed to an external temperature gradient. The mean free path of the more energetic electrons is longer. Green dots represent these higher energetic electrons, which then diffuse to the cool side until an electric field (E) is created to prevent any further diffusion, reproduced with permission.

The figure of merit for thermoelectric materials is represented by zT, according to the expression

$$zT = \frac{\sigma S^2}{\kappa} T \tag{3}$$

In this case, T is the absolute temperature, σ is the electrical conductivity, S is the Seebeck coefficient, and κ is the thermal conductivity. As a result, zT uses in-verse temperature units. The value of zT is roughly 1 for thermoelectric materials that are efficient. The higher the zT number, the better thermoelectric generators and refrigerators work. Furthermore, the power factor ($PF = S^2\sigma$) is typically employed to compare materials' thermoelectric efficiencies with similar thermal conductivities. The thermal conductivity can be expressed as $\kappa = \kappa_L + \kappa_E$, which is the sum of the lattice contribution (κ_L) and the electronic contribution (κ_E). Because charge carriers also contribute to heat transfer, the electronic contribution to thermal conductivity rises with carrier concentration:

$$\kappa_E = L\sigma T = ne\mu_c LT \tag{4}$$

where L is the Lorenz factor, 2.44×10^{-8} ($J^2\ C^{-2}\ K^{-2}$) for free electrons, n charge carrier density, e is Elementary charge, μ_c is charge carrier mobility and T is absolute temperature, and has dimension of K (kelvin).

The efficiency of a thermoelectric device is fundamentally defined

$$\eta = \frac{P_0}{q_h} \tag{5}$$

According to the first law of thermodynamics, q_h is the rate of heat flow from the hot to the cool section, and P0 is the electrical power production:

$$\eta_C = \frac{T_H - T_C}{T_H} \tag{6}$$

The efficiency of thermoelectric devices (η) must be calculated using zT and the Carnot efficiency (η_C), which is derived from the temperature differential (ΔT) and the hot-side temperature (T_H). The formula for this is $\eta_C = \Delta T/T_H$. The word "this relationship" accurately describes

$$\eta = \eta_C \frac{\sqrt{1+zT} - 1}{\sqrt{1+zT} + \frac{T_C}{T_H}} \tag{7}$$

On the other hand, the charge on ions plays a crucial role in the mechanism of voltage generation in ionic thermoelectric materials. This process is driven by the interplay between ion transport, electrostatic interactions, and thermal gradients (Soret effect), allowing for the efficient conversion of heat into electricity. Ionic thermoelectric materials leverage the interplay between ion transport, electrostatic interactions, and thermal gradients to convert heat into electricity efficiently. In lignin-derived ionic membranes infused with KOH electrolyte, the formation of an electric double layer (EDL) plays a crucial role in facilitating selective ion diffusion. Upon immersion in KOH, the lignin surface transforms into negatively charged alkoxide ($C–O^-$) groups, attracting K^+ cations and establishing a structured ionic environment. This arrangement enables efficient charge transport along aligned channels. Under a thermal gradient, ion migration occurs due to the Soret effect, where differences in temperature drive the redistribution of cations and anions. The resulting imbalance in ionic concentration, coupled with electrostatic interactions within the charged channels, enhances thermally driven ionic conductivity and the ionic Seebeck effect. The synergy between selective ion diffusion and thermophoresis highlights the potential of lignin-based membranes for sustainable energy harvesting, offering a pathway toward efficient utilization of low-grade thermal energy [8, 34].

3. Material classes

Thermoelectric Materials encompass diverse categories, with notable examples including inorganic materials like bismuth telluride and lead telluride, organic and organic–inorganic hybrids such as conducting polymers [5] and metal–organic frameworks [35], nanostructured materials like silicon nanowires [36] and quantum dots [37], and advanced materials like topological insulators [38] (e.g., bismuth selenide) and 2D materials like graphene [39]. Ongoing research endeavors aim to optimize the thermoelectric efficiency of these materials by tailoring their properties, exploring novel synthesis methods, and advancing their performance in diverse operating conditions. The collective goal is to harness these materials for converting waste heat into valuable electrical energy, contributing significantly to the development of sustainable energy technologies.

3.1 Inorganic materials

In the area of thermoelectric materials, inorganic compounds have historically played a very significant role, with typical examples being bismuth telluride and lead telluride. These materials have received a lot of attention due to their favorable thermoelectric properties, making them very basic in thermoelectric device development.

Development of Traditional Inorganic Thermoelectric Materials:

Among the inorganic thermoelectric materials, bismuth telluride (Bi_2Te_3) and lead telluride (PbTe) share significant importance. Known since the middle of the twentieth century, bismuth telluride has a high Seebeck coefficient and low thermal conductivity, so it was considered the material of choice for cooling applications. On the other hand, good electrical properties are hosted by the lead telluride, which makes it particularly effective in power generation. Bi_2Te_3 has been under investigation since 1954 and is still one of the most commonly applied TE materials. Nevertheless, Bi_2Te_3 remains a material with a high potential of interest and has recently been subject to remarkable improvements. The thermoelectric figure of merit zT of material limits the efficiency of thermoelectric energy converters. Recent progress in zT, particularly nanostructure-based approaches aiming to limit phonon heat conduction, is reaching the fundamental limit. The thermal conductivity cannot be reduced below the amorphous limit. Joseph P. Heremans and colleagues explored the enhancement of the Seebeck coefficient by a distortion of the electronic density of states and showed thallium impurity levels that gave high Seebeck coefficients in lead telluride, PbTe. This engineering of the band structure resulted in an improvement in zT in p-type PbTe by a factor of two to values above 1.5 at 773 Kelvin. The combination of this new physical principle with nanostructuring for lowering thermal conductivity can further improve zT and broaden the uses for thermoelectric systems [40]. Another intriguing class of materials is the silver-copper chalcogenide family, which has garnered significant attention for its exceptional thermoelectric properties. Remarkably, studies have reported zT values ranging from 0.3 to 1.6, showcasing their potential for high-efficiency thermoelectric applications. These materials exhibit a unique combination of high electrical conductivity and low thermal conductivity, which is often attributed to their complex crystal structures and effective phonon scattering mechanisms [41]. Nanostructured Bi_2Te_3-based materials significantly outperform their bulk counterparts in thermoelectric efficiency, as reflected in their higher zT ~ 1.4 values. Bulk Bi_2Te_3 typically achieves zT ~ 0.8–1.0 at room temperature (300 K), which serves as a benchmark for comparison [42]. In contrast, nanostructured variants, such as nanocomposites (e.g., Bi_2Te_3/Sb_2Te_3) and superlattices [43], achieve zT values ranging from 1.2 to 2.4, benefiting from reduced lattice thermal conductivity and enhanced power factors due to quantum confinement and optimized carrier concentrations [44, 45]. Specifically, superlattices hold the record with zT values up to 2.4 at 300 K, while doped nanostructures (e.g., Pb-doped Bi_2Te_3) [46] and alloyed systems (e.g., p-type $Bi_{0.5}Sb_{1.5}Te_3$) [47] achieve zT ~ 1.6–1.9 making them highly promising for thermoelectric applications. Overall, the advancements in nanostructuring and alloying have pushed zT values to new heights, highlighting their potential in high-performance thermoelectric devices [48].

Moving on to $CoSb_3$-based materials, binary skutterudites with a $CoAs_3$-type structure have been investigated [49], showing enhanced zT values as high as 1.1 at ~550°C in n-type nanostructured $CoSb_{3-x}Te_x$ skutterudite compounds. Additional substitution of IVB-group elements (Si, Ge, Sn, and Pb) for Sb, particularly Sn, was found to be the most effective in enhancing zT by extending the solubility limit of Te in $CoSb_{3-x}Te_x$, reducing thermal conductivity more significantly than electrical conductivity [50].

In another research study, a comprehensive exploration of the Mn doping effect on $CoSb_3$-skutterudite was reported, encompassing both theoretical and experimental investigations. An innovative approach was adopted, focusing initially on single Mn-doped $Mn_xCo_{1-x}Sb_3$ samples before delving into co-doping with Te to prepare $Mn_xCo_{1-x}Sb_{2.85}Te_{0.15}$ samples, evaluating their impact on both conduction types. Theoretical density functional theory (DFT) results unveiled the feasibility of single Mn doping, a finding successfully validated experimentally with the production of pure samples featuring a low solubility of Mn ($x \leq 0.05$ in $Mn_xCo_{1-x}Sb_3$) positioned at the Co site. Up to a certain level of Mn substitution, a minor amount of the binary MnSb phase was formed. In terms of transport properties, it was demonstrated for the first time that Mn stabilizes p-type conduction with a substantial Seebeck coefficient, resulting from a slight reduction of the valence electron concentration (VEC) per formula due to Co substitution. This led to an improvement in the power factor (PF) and the resulting figure of merit zT in the near-room temperature range (T < 450 K). Co-doping with Te was subsequently undertaken, revealing an unexpectedly prevalent formation of the additional MnTe phase over Co substitution. Structural investigations highlighted that Te substitutes for Sb without filling the icosahedral voids present in the $CoSb_3$ skutterudite. Moreover, a significant enhancement of the PF was achieved, reaching an ultrahigh value of 4.7 mW m^{-1} K^{-2}at 725 K, rarely reported for rare-earth-free skutterudites. This was attributed to a remarkable increase in the Seebeck coefficient in the co-doped $CoSb_3$ skutterudite. The combined experimental and theoretical findings indicated that various factors contribute to the substantial enhancement of the Seebeck coefficient, allowing for a maximum zT value of 1 at 725 K. The controlled formation of the MnTe secondary phase was identified as a key factor, slightly reducing thermal conductivity and maximizing an unexpected PF to a level comparable to that reported for expensive rare-earth-filled skutterudites. This achievement contributes to the development of a new generation of sustainable thermoelectric (TE) materials by exploring the magnetic and composite effects, aiming to provide stable, cost-efficient, and high-performance materials for thermoelectricity based on unfilled skutterudites [51]. Ongoing research in $CoSb_3$-based thermoelectric nanocomposites involves innovative cold sintering processes, such as CSP and post-annealing techniques. Aida Serrano et al. reported successful sintering of CoSb3-based thermoelectric nanocomposites by CSP with suitable mechanical integrity and similar morphological and structural properties as the starting powders. The thermoelectrical properties are shown to be dependent on the sintering process, with the highest power factor (1000(200) $\mu W\ m^{-1}\ K^{-2}$) and zT (0.12(3) at RT) achieved for nanocomposites sintered by CSP followed by a subsequent post-annealing at 500°C, presenting the highest $CoSb_3$ phase content with adequate grain and crystallite size [52].

3.1.1 Recent advancements and improvements

Recent years have shown renewed interest in the performance enhancement of traditional inorganic thermoelectric materials. Scientists have introduced a number of new synthesis methods and alloying and nanostructuring strategies to improve such material's figure of merit. The advances in material designs and fabrication have resulted in the optimization of thermoelectric properties, thus showing continuing efforts to improve the efficiency and applicability of inorganic thermoelectric materials.

This collection portrays a comprehensive view of diverse inorganic materials for thermoelectricity, from state-of-the-art thermoelectric materials to much less conventional ones but with huge potential for future applications. Although being concerned, like many authors, with the new figure of merit, zT max, we also emphasize the significance of the average zT, especially in the temperature region of interest, along with a few other significant thermoelectric parameters. Potential fulfillment of this important checklist has entered an era of exceptional materials advancement and discovery through a combination of data mining, machine learning, and high-throughput calculations, providing active research in modeling and experimental instrumenting. The pursuit of Earth-abundant, non-toxic starting materials remains a cardinal constraint for driving down cost and environmental impact. Coupling of atomistic and density of states calculations with band structure engineering will be instrumental in deciding upon effective additives to either enhance electrical conductivity or reduce thermal conductivity. In spite of these advances, the way is one of dealing with a vaster field: leading the material to the same level of performance as p-type, leading to good mechanical properties and stability, degrading mechanisms controlling, interfacing of thermoelectric modules, processing costs to levels where broad market competitiveness is attainable. A detailed assessment of the production costs, cost-effective material processing, and manufacturing routes has been considered long overdue for the purpose of wider adoption of thermoelectrics [53].

3.2 Organic and organic-inorganic hybrid materials

A paradigm change in the sector is shown by the introduction of organic and organic–inorganic hybrid thermoelectric materials. These materials provide new possibilities for thermoelectric applications because of their tunable electronic characteristics, cheap production cost, and flexibility.

Creating polymer–inorganic thermoelectric composites offers a viable strategy to address the poor electrical conductivity, low Seebeck coefficient, and inadequate power factor (PF) of polymers, as well as the expensive nature of inorganic materials. Multiple techniques have been utilized to produce these hybrid polymer–inorganic substances, including physical mixing, solution mixing, and in situ polymerization. It is important to note that the thermal conductivity of polymers is approximately an order of magnitude lower than that of conventional inorganic thermoelectric materials. This characteristic positions polymers like polyaniline (PANI), polythiophene (PTH), and poly(3,4-ethylenedioxythiophene) (PEDOT) as excellent candidates for the advancement of new thermoelectric materials [54]. One way to enhance the performance of this class of thermoelectric materials is through doping. This process, referred to as chemical or primary doping, can be accomplished via two methods: (1) A charge transfer (CT) mechanism that involves the incorporation of a redox-active molecule such as tetrakis (dimethylamino) ethylene (TDAE) into the conducting polymers, where the doping mechanism is founded on a CT interaction between the donor and the acceptor. 2) A comprehensive investigation was conducted to examine the complex effects of primary doping processes, which utilize acid-base ($A^+ - B^-$) and charge transfer (CT) mechanisms via H_2SO_4 and TDAE, respectively. Additionally, the study explored the secondary effects of ethylene glycol (EG) on the thermoelectric (TE) performance of both p-type Te/PEDOT:PSS and n-type Ag_2Te/PEDOT:PSS hybrids. In particular, the results have shown pronounced differences between EG post-treatment of pure PEDOT:PSS versus Ag_xTe/PEDOT:PSS hybrid systems. Unlike improved conformation exhibited in pure PEDOT:PSS systems, EG post-treatment mostly disturbs morphological conformations of PEDOT chains in hybrid systems. The upset negatively impacts on the lamellar domains, which raises the energy conversion layers (ECLs) full width at half maximum. This generally degrades the whole TE performance by increasing carrier mobility, μ, and conductivity, σ. Besides, deep insight is given into how these numerous post-treatment processes create structural and electrical changes in hybrid organic–inorganic materials, underlining complex results realized in Ag_xTe/PEDOT:PSS hybrid systems in comparison with pure polymer systems. This work puts these strategies in the broader perspective of hybrid electronic material research and focuses on PEDOT-based hybrid materials for thin-film TEs [55].

3.3 Nanostructured organic-inorganic hybrid thermoelectric materials

Organic thermoelectric materials exploit the intrinsic features of carbon-based molecular structures and polymers. Tuning their electronic structure is a highly promising approach for designing materials with improved thermoelectric performance. Moreover, hybrid materials have been developed in which organic and inorganic components interact synergistically to provide flexibility without sacrificing stability, thereby expanding their potential for diverse applications.

3.3.1 Nanostructured materials

Nanostructured materials represent a cutting-edge approach in the pursuit of high-performance thermoelectric materials. Gervino-Solona and colleagues claim a novel interconnected nanostructure of Bi_2Te_3, which achieves impressive thermoelectric performance despite using only 10% of the material typically required for thin films. The addition of sodium lignosulfonate during electrochemical deposition significantly improves the material's morphology and output performance. The fabrication method is scalable, offering the potential for flexible thermoelectric devices, particularly in energy-harvesting applications for wearable technologies [56]. Utilizing phonon scattering and quantum confinement processes, nanostructured thermoelectric materials such as nanowires and nanoparticles can dramatically change the thermal and electrical transport characteristics [57]. By overcoming the conventional trade-off between electrical and thermal conductivity [58], these materials seek to further the limits of thermoelectric efficiency [59]. Recent research shows that the addition of single-wall and multi-wall carbon nanotubes (SWCNTs and MWCNTs) to the conductive polymer base on PEDOT can significantly enhance its thermal conductivity. Researchers have demonstrated that these organic materials can increase the thermoelectric performance of flexible polymers. The authors claim that films with MWCNTs exhibit slightly higher electrical conductivity than those with SWCNTs. Using the LbL technique, the primary electrode made of MWCNTs achieved a conductivity of 31.26 Scm^{-1}, playing a critical role in subsequent polymerization. For SWCNTs, the conductivity was recorded at 3.49 Scm^{-1} under the same conditions [5].

For the development of thermoelectric generators that have nanometer thickness, increasing the electrical sheet resistance and thermal resistance is one of the challenges ahead. As a result, methods such as manufacturing thermoelectric micrometric strips with nanometer thickness and creating an interconnected network of these generators can be a way to overcome this problem. However, creating more costs for using the micrometric dimension manufacturing system and the need for more professional operators leads to an increase in the price of the final product. To find the optimal dimensions, theoretical methods and simulation software can help in finding the optimal dimensions quickly [59].

3.4 Advanced materials

Advanced materials like topological insulators have a good talent for thermoelectric applications. Due to their special boundary states that are topologically shielded from backscattering at non-magnetic impurities and defects, topological insulators show enormous promise in the domains of electronics and magnetism. Interestingly, most topological insulators are also very good thermoelectric materials since they have comparable properties to thermoelectric compounds [60]. Quantized energy quanta linked to lattice vibrations, phonons are essential for a number of important physical processes, including heat capacity, electron–phonon/magnon-phonon interactions, thermal conduction, and lattice thermal expansion. Gaining better control over phonons, one of the main heat carriers in thermal transport, has important applications in a number of energy materials domains, such as waste heat recovery, phononic devices, and thermoelectric conversion. THz phonons in natural crystalline materials have gained interest recently because of the importance of atomic lattice vibrations in physics. Topological phonons in crystalline materials have been theoretically classified into a number of different forms, such as Dirac/Weyl point phonons and [61] nodal-line phonons, using a framework akin to the topological classifications in electronic systems [62]. Wearable technology with maintenance-free batteries is necessary for the "Internet of Things" [63, 64] rapid development, and thermoelectric energy conversion based on large-area flexible materials has garnered a lot of interest. Recently, a fully printed origami thermoelectric generator (TEG) module has demonstrated significant advancements in power output, power density, and scalable manufacturing. By employing screen printing for the thermoelectric and contact materials, researchers developed a high-performance TEG module using synthesized n-type Ag_2Se and p-type Bi-Sb-Te materials. The design incorporates an origami-inspired folding process to enhance thermal impedance matching, with heat flow directed in-plane along the module's height. The printed module achieves a low resistance of 56 Ω and demonstrates substantial power output, indicating its potential for effectively powering IoT sensors [65]. Because of their superior transport characteristics and high-power factors, 2D materials—such as graphene and similar materials—are among the most promising large-area flexible materials for thermoelectric applications. The experimental reports on thermoelectric devices of graphene, black phosphorus, transition metal dichalcogenides, and other 2D materials are used in this review to survey both single-crystalline and polycrystalline 2D materials. Particular attention is paid to their carrier density-dependent thermoelectric characteristics and power factors that are optimized by Fermi level tuning methods. When the relevant performances of 2D materials and widely utilized thermoelectric materials are compared, it becomes clear that 2D materials have much higher power factors [66].

4. Methods for enhancing thermoelectric performance

In the pursuit of more efficient thermoelectric materials, various strategies have been employed to enhance the figure of merit (zT), a critical parameter for assessing thermoelectric performance. This section explores key methodologies, including alloying, nanostructuring, doping, and the increasingly influential role of computational methods in material design.

4.1 Alloying

This is the classical strategy where alloying several different elements in a material can be deliberately combined to improve thermoelectric performance [67]. A proper choice of alloying elements could modify the electronic structure and phonon scattering mechanisms in a material [68]. That is to say, the optimization for increasing electrical conductivity and decreasing lattice thermal conductivity will contribute to elevating zT. Much success in alloying has already been achieved with traditional inorganic materials; for example, bismuth telluride alloys [69] have been prepared in tailored compositions and show considerable enhancement of thermoelectric efficiency. Bismuth telluride has long dominated TE technology, the scarcity of tellurium has shifted focus to abundant alternatives like $Mg_3(Sb,Bi)_2$-based materials, which exhibit a high thermoelectric figure of merit (zT > 1) over a broad temperature range (300–773 K) [70].

4.2 Nanostructuring

Manipulating materials at the nanoscale, or "nanostructuring," has been proposed as a novel technique to improve thermoelectric qualities. Based on quantum confinement, material design at the nanoscale modifies electrical and thermal transport properties through the use of nanowires, nanoparticles, and even hierarchical nanostructures. However, increased phonon scattering at the interfaces has been found to significantly inhibit thermal conductivity, and in some situations, quantum confinement effects in such nanostructured materials may even enhance electrical conductivity. zT values are enhanced when these two elements are combined. The inherent trade-offs between electrical and thermal conductivity have been particularly promisingly overcome by nanostructuring, which has led to unprecedented levels of thermoelectric efficiency.

Our group has achieved a significant advancement in the power factor augmentation of a conductive polymer by employing a sustainable technology aimed at improving the efficiency of a flexible thermoelectric module built on films based on carbon nanotubes infused with PEDOT. With an emphasis on conductivity and doping control through the incorporation of CNTs, a strategic combination of the Layer-by-Layer method for forming the layer of CNTs and electrochemical polymerization for the conducting polymer was utilized in the synthesis of effective thermoelectric nanocomposites. After measuring the Seebeck coefficient and assessing the samples' electrical conductivity, the samples were fully characterized. The power factor could be further estimated from these data. The synergistic combination of PEDOT and SWCNTs produced a competitive power factor of 131.1 $\mu Wm^{-1} K^{-2}$, which is comparable to the performance of conventional inorganic thermoelectric materials. This method has the potential to be a very successful way to recycle heat waste because it is simple, affordable, and environmentally benign. Bright futures lie ahead for the advancement of flexible and effective thermoelectric module technology, thanks to green methods [5].

Antonios and coworkers have carefully investigated the effect of increasing the degree of modulation on the thermoelectric efficiency of width-modulated nanowires. More specifically, this work systemically elongates the modulation by adding modulation units and explains the effects of multiple modulation units on transport, keeping an eye on the optimization of the modulation profile toward maximum TE efficiency. The investigations took into consideration two-quantum dot and multiple-QD modulation scenarios, couplings between modulation units, and periodic and aperiodic sequences interacting with each other. Among the main observations made was that the electron transport property and thermoelectric power factor are augmented with an increase in modulation units and the maximum increase being for periodic modulation. The study went ahead to establish that phonon thermal conductance decreases with rising modulation, attaining the SL value for periodic modulation. Most importantly, the researchers uncovered the fact that a drastic increment in zT accompanies modulation, specifically for aperiodic modulation profiles where quasi-localized states for electrons are created. It will open new routes to optimize width modulation in order to achieve maximum TE efficiency, mostly in scenarios dominated by quantum effects in transport [71].

4.3 Doping

Doping is the deliberate introduction of foreign components into a material with the goal of altering its electrical structure. It is a very commonly used strategy. Researchers can manipulate the charge carrier concentration and mobility, which impacts the material's thermoelectric properties, by carefully selecting dopants [72]. Doping is a flexible approach that may be used with a wide range of thermoelectric materials, both inorganic and organic. Optimizing electrical conductivity and the Seebeck coefficient by precise control of the doping process results in improved thermoelectric performance [73]. Zhibin and coworkers try to enhance the thermoelectric performance of BiCuSeO for its practical application in thermoelectric power generation and thermoelectric catalysis. According to the authors, the practical application of BiCuSeO is hindered by the fact that it shows lower thermoelectric efficiency. Bulk materials of BiCuSeO were prepared using the solid-phase reaction method combined with the ball milling method and spark plasma sintering. In this respect, the authors attempted to optimize thermoelectric performance through the synergistic enhancement of both carrier concentration and mobility. Enhancement was achieved by Al doping into a BiCuSeO matrix. Here, adjustments in carrier mobility were to be attained by an energy band adjustment. The results indicated that Al doping enlarged the bandgap and increased the carrier mobility of BiCuSeO; hence, it is a useful way to improve the thermoelectric properties at middle and high temperatures. Besides, Pb was used as a doping element to adjust the carrier concentration of BiCuSeO. Such Al and Pb dual-doping exhibited a great effect on improving carrier concentration without losing high carrier mobility. This method could improve electrical conductivity without sacrificing a large Seebeck coefficient. The power factor for the Al/Pb-doped BiCuSeO reached ~7.67 $\mu Wcm^{-1} K^{-2}$ at 873 K. The thermal conductivity for all doped samples was found to be low in the measured temperature range. Finally, the zT of Al/Pb-doped BiCuSeO was ~1.14 at 873 K, which is far greater compared to the pure phase. This successful dual-doping strategy eventually improved the thermoelectric properties of the BiCuSeO matrix [74]. Mg_3Bi_2-based thermoelectric materials have also been recognized as highly promising materials for thermoelectric applications. The sample preparation of Mg_3Bi_2 was carried out by a high-temperature synthesis. This study has focused on the influence of Mg/Sb content on the electrical transport properties and properties related to semiconductor-semimetal. The efficiency in terms of introducing electrons from excess Mg by high-temperature synthesis is relatively lower in comparison with that achieved by ball milling. This is mainly attributed to the high vapor pressure of Mg during synthesis. Herein, we present the substitution of Sb/Te at the Bi site, and its effect in bringing about a transition from p-type to n-type conduction. The increased content of the stronger Mg content weakens the material's behavior on the semiconducting side and strengthens the material's semi-metallic tendency with a corresponding enhancement of the material's conductivity. To this particular goal, the present results are of vital value regarding the optimization of the performance in the thermoelectric material series containing Mg_3Bi_2 and help to shed light on factors like synthesis methods and elemental doping used for the tailoring of their electrical properties [75].

4.4 Computational methods

Advances in computational methods have revolutionized the field of thermoelectric materials by enabling researchers to predict and design new materials with targeted properties. Quantum mechanical calculations, density functional theory (DFT), and high-throughput screening methodologies facilitate the exploration of vast material databases, accelerating the discovery of novel thermoelectric candidates. Computational tools It is now possible to predict and design new thermoelectric materials, thanks to advances in computational methods. Such efforts include quantum mechanical calculations, density functional theory, and high-throughput screening methodologies that allow for huge material databases to be mined quickly and easily for a variety of thermoelectric candidates. The computationally determined electronic structure and bandgap engineering are helpful in developing phonon properties of materials that will turn out to be guiding experiments toward the synthesis of the best possible materials. What has happened, in view of this synergy of experimental and computational approaches, is a radical contraction of the cycle of development for materials and an expanded range of their application in the domain of thermoelectricity. Furthermore, these CFD tools prove exceedingly useful for researchers, designers, engineers, and analysts since they provide a view that assists in early decision-making, thereby saving much valuable engineering design process time [76]. On the other hand, Rodrigo et al.'s study explores the potential of novel silicide-tetrahedrite modules for energy generation. In the research, performance studies were carefully made on thermoelectric generators using computer simulations via the finite element method and the implicit finite difference method. For the verification of their computational models, the team uses validation against data measured on a specifically designed system working with commercial TE devices. It is found that high predictive capability is obtained with the developed models, with deviations of $\leq$10%. The implicit finite difference method (IFDM) analyzes the power performance developed by the silicide-tetrahedrite TEGs under different temperature differentials ΔT, while the finite element method (FEM) [77] gives a detailed investigation of temperature distributions within the test system. This work thus shows the role computational modeling can play in understanding the behavior of new thermoelectric materials and proves the reliability of developed models in predicting the performance of TEGs [78]. Except for this, the search for identifying new high-performance TE materials that are economical and environmentally friendly remains a pressing task for the Thermoelectric society. However, the advancement in this task has been considerably exacerbated by the conventional trial-and-error approach being tedious and expensive. In fact, the field of TE materials research is now in the midst of a transformative change due to remarkable progress in the hardware of computers, efficient methods of computation, advanced artificial intelligence algorithms, and exponential growth in material data. Upon this paradigm shift, innovative strategies are built to transcend these limitations. A number of electrical and thermal performance descriptors for new materials have been advocated, accompanied by the creation of efficient high-throughput (HTP) calculation methods. The high-throughput methods, hence, allow the rapid screening of potential TE materials from large material databases. Moreover, a few HTP experiments using doped systems have been devised to gain a larger density of information in a single experiment, thus reducing the time and cost involved. In addition, the inclusion of machine learning (ML) methods in thermoelectrics has added to newer dimensions. The previous sections in this review present a description of the HTP strategies used in discovery processes in general, performance descriptors applications, HTP calculations, HTP experiments, and ML. Moreover, it highlights the challenges of using this area and explores the possible directions of future research [79].

On the other hand, Tiryaki and coworkers have examined novel techniques for characterizing high-performance thermoelectric materials, specifically p-type $Sb_xBi_yBa_zB_tYb_wTe_3$ and n-type $Bi_xBa_yB_zYb_tTe_3$-based thermoelectric alloys. Two different cases of n-type $Bi_{2-x}Te_3$ and p-type $Sb_{1.5}Bi_{0.5-x}Te_3$ will be presented, where in all cases, x varies from 0 to 0.5 for Ba, B, and Yb, respectively, when tools for measuring electrical and thermal conductivity are not available. First, the experimental data at different temperatures for different compositions of these alloys were obtained. Then, training was performed using 26 machine-learning algorithms. More importantly, some compounds' transport properties, like the p-type $Sb_{1.5}Bi_{0.2}B_{0.3}Te_3$ and n-type $Bi_{1.9}Ba_{0.1}Te_3$, have been predicted only using the best ML algorithm. The predictive analysis is divided into two cases: case 1, in which only the Seebeck coefficient and electrical resistivity are forecasted, and case 2, for which heat capacity values Cp and thermal diffusivity were predicted. Both cases are under identical experimental conditions, classified just due to the availability of particular measuring equipment for thermoelectricity. Case 1 refers to situations in which the Seebeck coefficient and electrical resistivity-measuring device is not available and need an ML-driven prediction. Similarly, case 2 corresponds to the situation wherein the device measuring heat capacity and thermal diffusivity is not available and is in need of an ML-driven prediction. More importantly, as shown in this work, the challenge of the variation in the scales of the data during the training and test phases is addressed by normalization techniques [80].

Put another way, a rapidly expanding influence from computational approaches unites traditional methods of alloying and nanostructuring with doping, offering a potent toolkit for improving thermoelectric performance. When combined, these approaches are pushing thermoelectric materials toward increased efficiency and broader applications. Recently, researchers have demonstrated that integrating theoretical approaches with numerical computations for the Matrix Transfer Method can lead to the development of metamaterial structures. Ultimately, they successfully combined this structure with a thermoelectric conductive polymer to create a hybrid energy harvesting device [77].

5. Applications

In this section, we try to show recent applications of thermoelectric materials. Here, we show in an ionic system, lignin-like sustainable material can be used to make a conductive membrane that can generate valuable Seebeck voltage. Also, it is interesting to discuss hybrid energy thermoelectric devices that combine metamaterial structure and organic conductive polymer. Furthermore, we demonstrate how thermoelectric technology can be scientifically combined with electrical generator technology in space to mitigate global warming. Finally, we explore the applications of thermoelectric materials in the automotive and sensor industries.

5.1 Lignin-derived ionic conducting membranes

In recent research, the ionic thermoelectric properties of lignin-derived membranes were investigated in the low-temperature heating zone. Membranes infused with aqueous KOH electrolyte were evaluated for ionic conductivity and Seebeck coefficients. The selective ion transport mechanism, thermal ion diffusion, and intermolecular bonding were illustrated, with phenolic groups on lignin surfaces converting into anionic alkoxides upon KOH immersion. This transformation induced negatively charged channels that attracted K^+, forming an electric double layer (EDL) and facilitating selective ion diffusion. The membranes demonstrated an ionic Seebeck coefficient of up to 5.7 ± 0.4 mV K^{-1}, attributed to increased lignin content enhancing nanochannel dominance and thermoelectric diffusion efficiency [8].

According to **Figure 2**, thermal conductivity was measured in both dry and wet membranes. Dry membranes exhibited low thermal conductivity (0.10–0.14 W $m^{-1} \cdot K^{-1}$), while KOH infiltration increased conductivity to

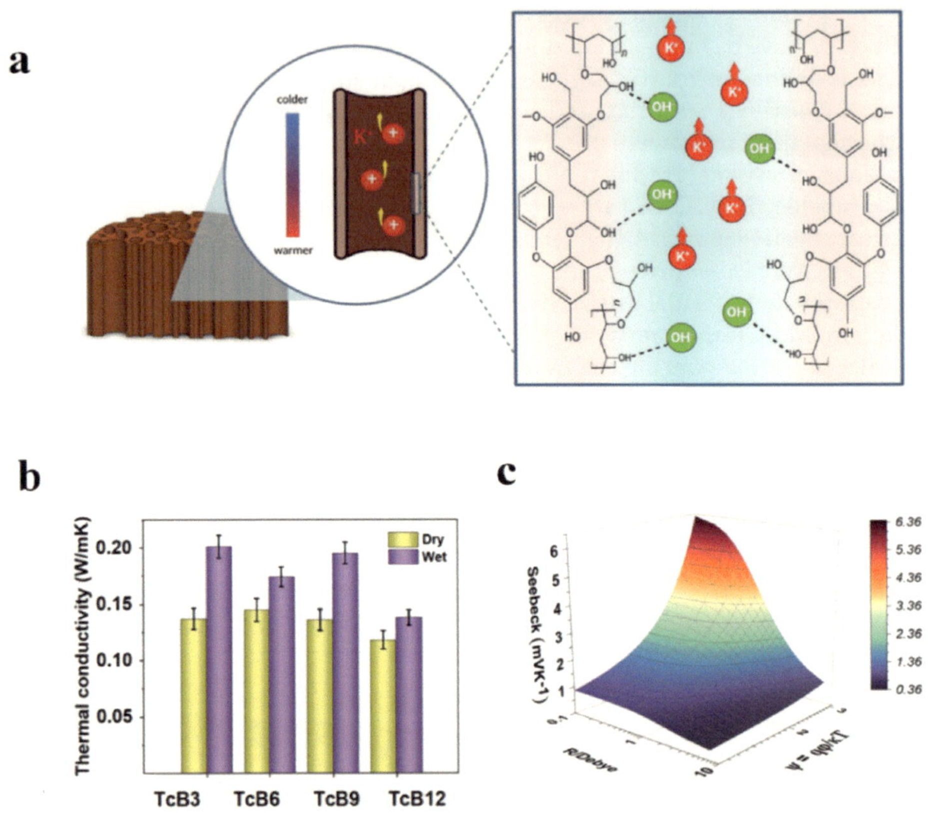

Figure 2.
(a) Schematic representation of selective ion diffusion within the channels of lignin-derived ionic conducting membranes, and ionic thermoelectric performance of synthesized membranes in terms of (b) thermal conductivity, and (c) estimated ionic Seebeck value in lignin-derived membranes (parallel) at channel radii and EDL potential. Reproduced with permission [8] (Copyright 2023, Wiley-VCH).

0.17–0.20 W $m^{-1}{\cdot}K^{-1}$ due to water uptake. Numerical modeling further confirmed that narrower channel radii and higher EDL potentials optimized the ionic Seebeck effect. Enhanced lignin content increased functional group ionization, leading to higher EDL potentials and thermopower. However, the experimental thermopower (5.71 mV K^{-1}) fell slightly below the modeled maximum (6.36 mV K^{-1}) due to channel size variation and tortuosity in real samples. These findings validate the potential of lignin membranes for efficient thermoelectric applications [8].

5.2 Integration of organic thermoelectric materials with metamaterial structures

Figure 3 shows that enhancing light absorption is crucial for advancing solar thermoelectric generators. Conventional light absorbers often rely on a back mirror, typically a thick metal film, to minimize reflectivity by facilitating interference between incident and reflected light. However, such continuous metal films pose a challenge for thermoelectric applications as they can short-circuit the Seebeck voltage. Recently, researchers introduced a back mirror-free selective light absorber tailored for thermoelectric devices. The design optimizes the combination of materials with high and low refractive indices coated with a semi-transparent electrode. Unlike a back mirror, the semi-transparent electrode can be patterned to preserve

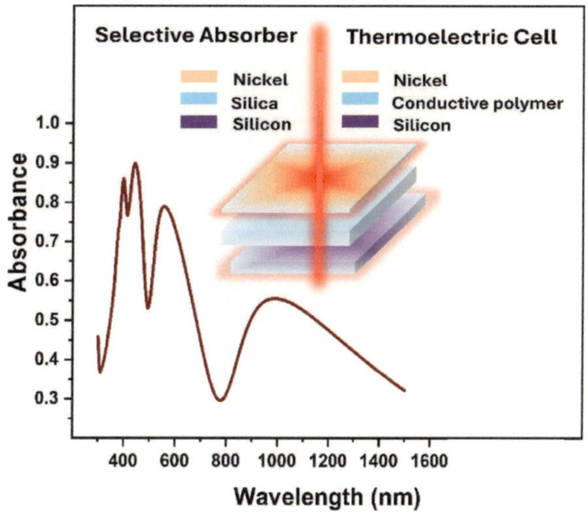

Figure 3.
An illustration of the integration of thermoelectric materials with metamaterial structures [77]. (Copyright 2024, Advanced Optical Materials).

the Seebeck voltage. This approach also allows the low-refractive-index material to be substituted with a transparent thermoelectric material, enabling efficient heat-to-energy conversion without compromising absorption performance [77]. The researchers show a back mirror-free selective absorber for thermoelectric applications has been numerically simulated and experimentally realized. Designed with layers of nickel, silica, and silicon on a transparent substrate, the structure has demonstrated scalable fabrication potential. By eliminating the back mirror, improved thermoelectric performance has been achieved, along with greater tolerance to fabrication parameter variations, such as layer thickness or refractive index. The silica layer has been replaced in some cases with a transparent thermoelectric polymer (conductive PEDOT: PSS) for light-to-electricity applications. Devices fabricated in this way have shown over 60% absorption and a Seebeck coefficient exceeding 22 μVK^{-2}, paving the way for innovative combined thermoelectric-photovoltaic.

5.3 Space exploration

For decades, thermoelectric materials have played an important role in space exploration. Inherent in these materials is the ability to power long-duration missions and deep-space probes with radioisotope thermoelectric generators at the core of space exploration. RTGs convert heat emanating from radioactive isotopes into electric power. The technology has been used since the Voyager, Cassini, and Mars rovers [82]. Xinyu Miao et al. present a coupling system integrating alkali metal thermoelectric converter (AMTEC), TEG, and microwave power transmission (MPT) subsystems to enhance thermoelectric efficiency in rovers and lunar bases by recycling AMTEC's waste heat (**Figure 4**). Thermodynamic analyses reveal that AMTEC primarily determines system performance, with output power and efficiency varying by key parameters (J, i, and heat exchange area). The combined system achieves maximum efficiency at 37% under optimal conditions, suggesting potential for efficient, low-cost, high-power spacecraft applications in space exploration [81].

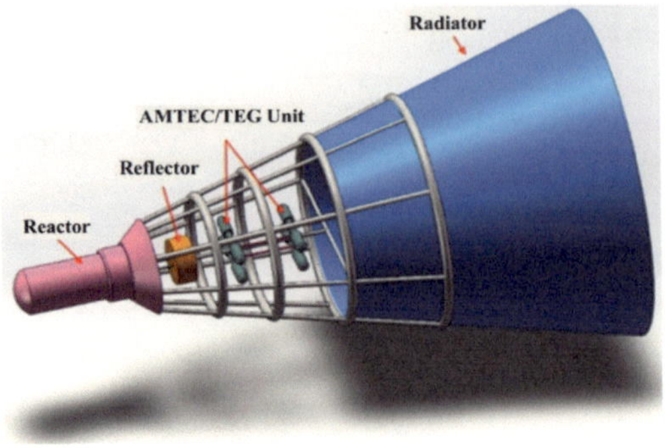

Figure 4.
Nuclear space power stations (NSPS) with unit alkali metal thermoelectric converter (AMTEC) with TEG combined unit. Reproduced with permission [81] (Copyright 2024, Journal of Annals of Nuclear Energy*).*

5.4 Automotive and sensor industry

In the automotive sector, high-efficiency thermoelectric materials apply to more than just waste heat recovery. First, they can be used in advanced cooling systems to improve the efficiency of temperature control in e-vehicles and traditional combustion engine vehicles. Second, the thermoelectric modules in hybrid vehicles capture and turn into electricity part of the energy wasted as heat through the exhaust system, thereby contributing to total energy efficiency. Kim et al. developed a fiber-shaped thermoelectric temperature sensor tailored for advanced wearable applications. This sensor utilizes a continuous graphene fiber with two halves in distinct reduction states, achieved by treating each half with hydroiodic acid (HI) at different concentrations. The differing reduction levels create a seamless junction with varied Seebeck coefficients, enabling thermoelectric functionality. This flexible graphene thermocouple demonstrates high sensitivity (12.5 μV/K), linearity, and fast response time (0.24 s), making it suitable for integration into wearable textiles, like gloves, without needing external power [83].

6. Challenges and future directions

Several challenges still remain, and future directions beckon toward more efficient, cost-effective, and far-reaching applications as high-performance thermoelectric materials continue in the process of evolution. Addressing these challenges is very critical to unlocking the full potential of thermoelectric technology.

6.1 Challenges and future

6.1.1 Challenges

In this section, it was highlighted the challenging points for researchers that need to be addressed to advance the development of thermoelectric materials and their applications. In order to advance the thermoelectric industry, factors such as cost and scalability, efficiency, operating temperature range, and environmental protection are crucial to address.

6.1.1.1 Cost and scalability

One of the primary challenges lies in the cost of manufacturing and scalability of high-performance thermoelectric materials [73]. Many of the materials with superior thermoelectric properties, such as certain rare or toxic elements, can be expensive and hinder large-scale deployment. Finding cost-effective alternatives without compromising performance is a critical challenge.

6.1.1.2 Efficiency trade-offs

Traditional thermoelectric materials often face a trade-off between electrical conductivity and thermal conductivity. Enhancing one property tends to negatively impact the other, limiting the overall efficiency of the material. Striking the right balance and overcoming this trade-off is an ongoing challenge in material design [84].

6.1.1.3 Temperature range

Many thermoelectric materials exhibit optimal performance only within specific temperature ranges. This limitation hinders their applicability in environments with fluctuating or extreme temperatures. Developing materials that maintain high performance across a broader temperature spectrum is a challenge for achieving versatility in applications [85].

6.1.1.4 Environmental impact

A few high-efficiency thermoelectric materials may include elements that are environmentally sensitive or might have a possible negative impact on the environment during mining and manufacturing. The sustainability of thermoelectric materials to the environment remains the latest concern [86].

6.1.2 Future

6.1.2.1 Future directions

Materials Discovery and Design: Improvements in the area of computational methods, using machine learning and artificial intelligence, open up exciting avenues for the accelerated discovery and design of new thermoelectric materials. High-throughput simulations and predictive modeling guide researchers toward the identification of materials with optimal properties and reduce the time and resources required for experimental exploration [87].

6.1.2.2 Multidisciplinary approaches

In this respect, if the challenges that now exist are to be surmounted, it will be incumbent upon at least some diversified fields of disciplines, such as materials science, physics, chemistry, and engineering. It is expected that multidisciplinary study may awaken creativity and facilitate an understanding of the complex interactions among multiple factors that influence thermoelectric performance [77].

6.1.2.3 Flexible and stretchable thermoelectric materials

With the rising demand for wearable devices and flexible electronics, the development of thermoelectric materials that are not only efficient but also flexible and stretchable is a promising avenue. Such materials could open up new possibilities for energy harvesting in unconventional settings [88].

6.1.2.4 Nanostructured and 2D materials

By adjusting electrical and thermal transport characteristics at the nanoscale, more studies in nanostructured materials such as nanowires and 2D materials such as graphene, could help overcome efficiency trade-offs. These materials present special chances to customize thermoelectric performance [89].

6.1.2.5 Hybrid and composite materials

Investigating hybrid and composite materials, which combine the best features of several material classes, may have a positive feedback loop and enhance overall performance. This strategy might lessen the difficulties brought on by personal material constraints [90].

6.1.2.6 Real-world integration

One of the most important future directions is to close the gap between laboratory-scale demonstrations and real-world implementations. Incorporating thermoelectric materials into useful devices, systems, and infrastructure necessitates taking technical and financial concerns into account in addition to material obstacles [91].

In conclusion, although the creation and use of high-performance thermoelectric materials have not yet been successful and are really difficult to execute, continuous efforts have opened up new ways for research and development. Potential applications of thermoelectric technology can create tremendous changes in energy harvesting, waste heat recovery, and the use of widely available, environmentally friendly materials. Thermoelectric materials present an exciting pathway for sustainable energy solutions, offering opportunities for energy harvesting and waste heat recovery. However, challenges such as cost, efficiency trade-offs, environmental impact, and scalability remain significant barriers. Addressing these hurdles requires advancements in computational design, multidisciplinary collaborations, and the exploration of nanostructured, hybrid, and flexible materials. Looking ahead, the integration of thermoelectric technology into real-world applications holds transformative potential, driving innovations in energy systems and contributing to a greener, more sustainable future.

Acknowledgements

The authors acknowledge the financial support from the Spanish MCIN(2D-HETEROS PID2020-117152RB-100, TED2021-129298B-I00, PID2019-111724GA-I00, co-financed by FEDER, and Excellence Unit "Maríade Maeztu" CEX2019-000919-M), the Generalitat Valenciana (CIDE-GENT/2018/005,

PROMETEO2020-016). This study forms part of the Quantum Communication program and was supported by MCIN with funding from the European Union NextGenerationEU and GeneralitatValenciana (COMCUANTICA/011). Thanks his support to this work through the grant PID2021-124845OA-I00 funded by MCIN/AEI/10.13039/501100011033 and by the "European UnionNextGenerationEU/PRTR."

Conflict of interest

The authors declare no conflict of interest.

Nomenclature

α_n	Seebeck coefficient for electrons	VK^{-1}
E	electric field	Vm^{-1}
∇T	temperature gradient	Km^{-1}
S	Seebeck coefficient	VK^{-1}
k_B	Boltzmann constant	1.38×10^{-23} JK^{-1}
e	elementary charge	1.602×10^{-19} C
$\sigma(E)$	energy-dependent electrical conductivity	Sm^{-1}
σ	total electrical conductivity	Sm^{-1}
T	absolute temperature	K
E_F	fermi level energy	eV
E_v	valence band edge energy	eV
zT	thermoelectric figure of merit	Dimensionless
κ	thermal conductivity	$Wm^{-1}{\cdot}K^{-1}$
κ_L	lattice thermal conductivity	$Wm^{-1}{\cdot}K^{-1}$
K_e	electronic thermal conductivity	$Wm^{-1}{\cdot}K^{-1}$
PF	power factor	$Wm^{-1}{\cdot}K^{-2}$
L	Lorenz factor	2.44×10^{-8} $W\Omega K^{-2}$
n	charge carrier density	m^{-3}
μ_c	charge carrier mobility	$m^2V^{-1}{\cdot}s^{-1}$
q_h	rate of heat flow from hot to cold section	W
P_0	electrical power output	W
η	thermoelectric conversion efficiency	Dimensionless
η_C	Carnot efficiency	Dimensionless
T_H	hot-side temperature	K
T_C	cold side temperature	K

Nomenclature: Key terms and symbols in advanced thermoelectric materials.

Author details

Mohammad Ali Nasiri[1,2]

1 Institute of Molecular Science (ICMol), University of Valencia, Valencia, Spain

2 Materials Science Institute, University of Valencia (ICMUV), Valencia, Spain

*Address all correspondence to: nasiri@alumni.uv.es

References

[1] Kim M, Park D, Su P-C, Kim J. Enhancing thermoelectric performance through acid-treated NbFeSb: A versatile approach for sustainable energy conversion. Journal of Alloys and Compounds. 2024;**976**:173243. DOI: 10.1016/j.jallcom.2023.173243

[2] Barzegar M, Goracci G, Martauz P, Dolado JS. Sustainable geopolymer concrete for thermoelectric energy harvesting. Construction and Building Materials. 2024;**411**:134398. DOI: 10.1016/j.conbuildmat.2023.134398

[3] Hooshmand Zaferani S, Ghomashchi R, Vashaee D. Strategies for engineering phonon transport in Heusler thermoelectric compounds. Renewable and Sustainable Energy Reviews. 2019;**112**:158-169. DOI: 10.1016/j.rser.2019.05.051

[4] Caballero-Calero O, Ares JR, Martín-González M. Environmentally friendly thermoelectric materials: High performance from inorganic components with low toxicity and abundance in the earth. Advanced Sustainable Systems. 2021;**5**. DOI: 10.1002/adsu.202100095

[5] Nasiri MA, Tong SY, Cho C, Gómez CM, Cantarero A, Culebras M. Synthesis of PEDOT/CNTs thermoelectric thin films with a high power factor. Materials. 2024;**17**:1121. DOI: 10.3390/ma17051121

[6] Shao Q, Li Y, Liang Z, Chen Z, Xu A, Qiu X, et al. Lignin: A multifunctional and sustainable photothermal material for solar-driven thermoelectric generation and desalination. Composites Part B: Engineering. 2024;**284**:111694. DOI: 10.1016/j.compositesb.2024.111694

[7] Zhang H, Li H, Wang W, Li P, Liu S, Yang M, et al. Biomass lignin as dispersion to substantially enhance carbon nanotubes thermoelectric converter for energy harvesting. Carbon N Y. 2024;**229**:119489. DOI: 10.1016/j.carbon.2024.119489

[8] Muddasar M, Nasiri MA, Cantarero A, Gómez C, Culebras M, Collins MN. Lignin-derived ionic conducting membranes for low-grade thermal energy harvesting. Advanced Functional Materials. 2024;**34**. DOI: 10.1002/adfm.202306427

[9] Kao S, Hsu C, Hong S, Jeng U, Wang C, Tung S, et al. Host–guest complexation of α-cyclodextrin and triiodide ions for enhanced performance of ionic thermoelectric capacitors. Advanced Energy Materials. 2025;**10**:2405502. DOI: 10.1002/aenm.202405502

[10] Jabeen N, Muddasar M, Menéndez N, Nasiri MA, Gómez CM, Collins MN, et al. Recent advances in ionic thermoelectric systems and theoretical modelling. Chemical Science. 2024;**15**:14122-14153. DOI: 10.1039/D4SC04158E

[11] Qiao T, Qiao Z, Sun S, Zhou S. An unconditionally energy stable linear scheme for Poisson–Nernst–Planck equations. Journal of Computational and Applied Mathematics. 2024;**443**:115759. DOI: 10.1016/j.cam.2024.115759

[12] Fish JS, Li C-P, Fehribach JD, Wolden CA, O'Hayre R, Bunge AL, et al. Poisson–Boltzmann model of space charge layer effects on conductivity in randomly distributed nanoionic composites. Electrochimica Acta. 2012;**83**:454-462. DOI: 10.1016/j.electacta.2012.07.122

[13] Fleharty ME, van Swol F, Petsev DN. The effect of surface charge regulation on conductivity in fluidic nanochannels. Journal of Colloid and Interface Science. 2014;**416**:105-111. DOI: 10.1016/j.jcis.2013.10.051

[14] Moya AA, Nikonenko VV. A comparative theoretical study of potential distribution and conductivity in cation- and anion-exchange nanoporous membranes filled with ternary electrolytes. Electrochimica Acta. 2015;**180**:929-938. DOI: 10.1016/j.electacta.2015.09.025

[15] Pulyala P, Jing M, Gao W, Cheng X. Solution composition dependent Soret coefficient using commercial MicroScale thermophoresis instrument. RSC Advances. 2023;**13**:15901-15909. DOI: 10.1039/D3RA00154G

[16] Luo RH, Keh HJ. Electrophoresis and electric conduction in a salt-free suspension of charged particles. Electrophoresis. 2021;**42**:2134-2142. DOI: 10.1002/elps.202100181

[17] Varner S, Wang Z-G. Effects of dilution in ionic liquid supercapacitors. Physical Chemistry Chemical Physics. 2022;**24**:27362-27374. DOI: 10.1039/D2CP03398D

[18] Huang W-H, Chang C-C, Yang R-J. Effect of hydrodynamic slip on thermoelectric response in negatively-charged nanofluidic channels. International Journal of Heat and Mass Transfer. 2023;**214**:124454. DOI: 10.1016/j.ijheatmasstransfer.2023.124454

[19] Levy A, Andelman D, Orland H. Dipolar Poisson-Boltzmann approach to ionic solutions: A mean field and loop expansion analysis. The Journal of Chemical Physics. 2013;**139**. DOI: 10.1063/1.4826103

[20] Stout RF, Khair AS. Diffuse charge dynamics in ionic thermoelectrochemical systems. Physical Review E. 2017;**96**:022604. DOI: 10.1103/PhysRevE.96.022604

[21] Pan Y, He B, Feng X, Li F, Chen D, Burkhardt U, et al. A magneto-thermoelectric with a high figure of merit in topological insulator Bi88Sb12. Nature Materials. 2025;**24**:76-82. DOI: 10.1038/s41563-024-02059-9

[22] Jia N, Cao J, Tan XY, Dong J, Liu H, Tan CKI, et al. Thermoelectric materials and transport physics. Materials Today Physics. 2021;**21**:100519. DOI: 10.1016/j.mtphys.2021.100519

[23] Yang G, Sang L, Zhang C, Ye N, Hamilton A, Fuhrer MS, et al. The role of spin in thermoelectricity. Nature Reviews Physics. 2023;**5**:466-482. DOI: 10.1038/s42254-023-00604-0

[24] Liu J, Liu Q, Lin S, Leung MY, Ma Y, Tao X. Wearable thermoelectric generators: Materials, structures, fabrications, and applications. Physica Status Solidi (RRL)—Rapid Research Letters. 2023;**17**. DOI: 10.1002/pssr.202200502

[25] Abouelregal AE, Tiwari R, Nofal TA. Modeling heat conduction in an infinite media using the thermoelastic MGT equations and the magneto-Seebeck effect under the influence of a constant stationary source. Archive of Applied Mechanics. 2023;**93**:2113-2128. DOI: 10.1007/s00419-023-02375-7

[26] Gupta S, Das S, Dutta R. Peltier and Seebeck effects on a nonlocal couple stress double porous thermoelastic diffusive material under memory-dependent Moore-Gibson-Thompson theory. Mechanics of Advanced Materials and Structures. 2023;**30**:449-472. DOI: 10.1080/15376494.2021.2017525

[27] Kjelstrup S, Kristiansen KR, Gunnarshaug AF. Seebeck, Peltier, and Soret effects: On different formalisms for transport equations in thermogalvanic cells. The Journal of Chemical Physics. 2023;**Bedeaux D, 158**. DOI: 10.1063/5.0131731

[28] Xie L, Yang J, Liu Z, Qu N, Dong X, Zhu J, et al. Highly efficient thermoelectric cooling performance of ultrafine-grained and nanoporous materials. Materials Today. 2023;**65**:5-13. DOI: 10.1016/j.mattod.2023.03.021

[29] Zhang Z, Peng Z, Jiang P, Sin Ang Y, Zhang C, Ma Z. Thomson effect in thermionic refrigeration: Enhanced performance of graphene/2D-semiconductor/graphene heterostructure cooler. Journal of Applied Physics. 2023;**133**. DOI: 10.1063/5.0153026

[30] Riss A, Garmroudi F, Parzer M, Pustogow A, Mori T, Bauer E. Thermoelectric power factor of composites. Physical Review Applied. 2024;**21**:014002. DOI: 10.1103/PhysRevApplied.21.014002

[31] Zhang M, Gao Z, Lou Q, Zhu Q, Wang J, Han Z, et al. Achieving high carrier mobility and thermal stability in Plainified rhombohedral GeTe thermoelectric materials with zT > 2. Advanced Functional Materials. 2023. DOI: 10.1002/adfm.202307864

[32] Sun Y, Wu H, Dong X, Xie L, Liu Z, Liu R, et al. High performance BiSbTe alloy for superior thermoelectric cooling. Advanced Functional Materials. 2023;**33**. DOI: 10.1002/adfm.202301423

[33] Tritt TM. Thermoelectric phenomena, materials, and applications. Annual Review of Materials Research. 2011;**41**:433-448. DOI: 10.1146/annurev-matsci-062910-100453

[34] Muddasar M, Menéndez N, Quero Á, Nasiri MA, Cantarero A, García-Cañadas J, et al. Highly-efficient sustainable ionic thermoelectric materials using lignin-derived hydrogels. Advanced Composites and Hybrid Materials. 2024;**7**:47. DOI: 10.1007/s42114-024-00863-0

[35] Raptopoulou CP. Metal-organic frameworks: Synthetic methods and potential applications. Materials. 2021;**14**:310. DOI: 10.3390/ma14020310

[36] Schierning G. Silicon nanostructures for thermoelectric devices: A review of the current state of the art. Physica Status Solidi (A). 2014;**211**:1235-1249. DOI: 10.1002/pssa.201300408

[37] Sothmann B, Sánchez R, Jordan AN. Thermoelectric energy harvesting with quantum dots. Nanotechnology. 2015;**26**:032001. DOI: 10.1088/0957-4484/26/3/032001

[38] Heremans JP, Cava RJ, Samarth N. Tetradymites as thermoelectrics and topological insulators. Nature Reviews Materials. 2017;**2**:17049. DOI: 10.1038/natrevmats.2017.49

[39] Li D, Gong Y, Chen Y, Lin J, Khan Q, Zhang Y, et al. Recent progress of two-dimensional thermoelectric materials. Nanomicro Letters. 2020;**12**:36. DOI: 10.1007/s40820-020-0374-x

[40] Heremans JP, Jovovic V, Toberer ES, Saramat A, Kurosaki K, Charoenphakdee A, et al. Enhancement of thermoelectric efficiency in PbTe by distortion of the electronic density of states. Science. 1979;**2008**(321):554-557. DOI: 10.1126/science.1159725

[41] Li N, Zhang Q, Shi X, Jiang J, Chen Z. Silver copper chalcogenide thermoelectrics: Advance, controversy,

and perspective. Advanced Materials. 2024;**36**. DOI: 10.1002/adma.202313146

[42] Luo D, Li Y, Yan Y, Hu X, Fan X, Chen W-H, et al. Realizing ultrahigh ZT value and efficiency of the Bi2Te3 thermoelectric module by periodic heating. Energy Conversion and Management. 2023;**296**:117669. DOI: 10.1016/j.enconman.2023.117669

[43] Sousa V, Goto M, Claro MS, Pyrlin S, Marques L, Modin EB, et al. PbSe quantum dot superlattice thin films for thermoelectric applications. Advanced Functional Materials. 2024;**34**. DOI: 10.1002/adfm.202409216

[44] Wredh S, Dai M, Hamada K, Rahman MA, Adanan NQ, Zamiri G, et al. Sb_2Te_3–Bi_2Te_3 direct photo–thermoelectric mid-infrared detection. Advanced Optical Materials. 2024;**12**. DOI: 10.1002/adom.202401450

[45] Nepal R, Bajracharya P, Kumar R, Kolagani R, Budhani RC. Thermoelectric response of textured Sb_2Te_3–BiSb and Sb_2Te_3–Bi_2Te_3 thin film junctions. Applied Physics Letters. 2024;**124**. DOI: 10.1063/5.0187554

[46] Park O, Park SJ, Kim H-S, Lee SW, Heo M, Kim S. Enhanced thermoelectric transport properties of Bi_2Te_3 polycrystalline alloys via carrier type change arising from slight Pb doping. Materials Science in Semiconductor Processing. 2023;**166**:107723. DOI: 10.1016/j.mssp.2023.107723

[47] Qin H, Liu Y, Zhang Z, Wang Y, Cao J, Cai W, et al. Improved thermoelectric performance of p-type $Bi_{0.5}Sb_{1.5}Te_3$ through Mn doping at elevated temperature. Materials Today Physics. 2018;**6**:31-37. DOI: 10.1016/j.mtphys.2018.07.002

[48] Yang J, Daqiqshirazi M, Ritschel T, Bahrami A, Lehmann S, Wolf D, et al. Interfacial distortion of Sb_2Te_3–Sb_2Se_3 multilayers via atomic layer deposition for enhanced thermoelectric properties. ACS Nano. 2024;**18**:17500-17508. DOI: 10.1021/acsnano.3c13152

[49] Bérardan D, Alleno E, Godart C, Puyet M, Lenoir B, Lackner R, et al. Improved thermoelectric properties in double-filled Cey/2Yby/2Fe4−x(Co/Ni)xSb12 skutterudites. Journal of Applied Physics. 2005;**98**. DOI: 10.1063/1.1999854

[50] Liu W-S, Zhang B-P, Zhao L-D, Li J-F. Improvement of thermoelectric performance of $CoSb_{3-x}Te_x$ Skutterudite compounds by additional substitution of IVB-group elements for Sb. Chemistry of Materials. 2008;**20**:7526-7531. DOI: 10.1021/cm802367f

[51] Bourgès C, Zhang W, Raut KK, Owada Y, Kawamoto N, Mitome M, et al. Investigation of Mn single and Co-doping in thermoelectric $CoSb_3$ -Skutterudite: A way toward a beneficial composite effect. ACS Applied Energy Materials. 2023;**6**:9646-9656. DOI: 10.1021/acsaem.3c01725

[52] Serrano A, Caballero-Calero O, Granados-Miralles C, Gorni G, Manzano CV, Rull-Bravo M, et al. $CoSb_3$-based skutterudite nanocomposites prepared by cold sintering process with enhanced thermoelectric properties. Journal of Alloys and Compounds. 2023;**931**:167534. DOI: 10.1016/j.jallcom.2022.167534

[53] Freer R, Ekren D, Ghosh T, Biswas K, Qiu P, Wan S, et al. Key properties of inorganic thermoelectric materials—Tables (version 1). Journal of Physics: Energy. 2022;**4**:022002. DOI: 10.1088/2515-7655/ac49dc

[54] Jin H, Li J, Iocozzia J, Zeng X, Wei P, Yang C, et al. Hybrid organic–inorganic

thermoelectric materials and devices. Angewandte Chemie International Edition. 2019;**58**:15206-15226. DOI: 10.1002/anie.201901106

[55] Rubio-Govea R, Félix R, Wilks RG, Bär M, Mazzio KA. Understanding the effects of primary and secondary doping via post-treatment of P-type and N-type hybrid organic–inorganic thin film thermoelectric materials. Advanced Electronic Materials. 2023;**9**. DOI: 10.1002/aelm.202300076

[56] Cervino-Solana P, Ramirez-Peral MJ, Martín-González MS, Caballero-Calero O. Thermoelectric bismuth telluride nanostructures fabricated by electrodeposition within flexible templates. Heliyon. 2024;**10**:e36114. DOI: 10.1016/j.heliyon.2024.e36114

[57] Mao J, Liu Z, Ren Z. Size effect in thermoelectric materials. npj Quantum Materials. 2016;**1**:16028. DOI: 10.1038/npjquantmats.2016.28

[58] Dong H, Wen B, Melnik R. Relative importance of grain boundaries and size effects in thermal conductivity of nanocrystalline materials. Scientific Reports. 2014;**4**:7037. DOI: 10.1038/srep07037

[59] Nasiri MA, Seijas-Da Silva A, Serrano Claumarchirant JF, Gómez CM, Abellán G, Cantarero A, et al. Ultrathin transparent nickel electrodes for thermoelectric applications. Advanced Materials Interfaces. 2024;**11**. DOI: 10.1002/admi.202300705

[60] Xu N, Xu Y, Zhu J. Topological insulators for thermoelectrics. npj Quantum Materials. 2017;**2**:51. DOI: 10.1038/s41535-017-0054-3

[61] Li J, Liu J, Baronett SA, Liu M, Wang L, Li R, et al. Computation and data driven discovery of topological phononic materials. Nature Communications. 2021;**12**:1204. DOI: 10.1038/s41467-021-21293-2

[62] Ding Z, Zeng Y, Liu W, Tang L, Chen K. Topological phonons and thermoelectric conversion in crystalline materials. Advanced Functional Materials. 2024;**34**. DOI: 10.1002/adfm.202401684

[63] Ding Z, Li G, Wang Y, Du C, Ye Z, Liang L, et al. Ultrafast response and threshold adjustable intelligent thermoelectric systems for next-generation self-powered remote IoT fire warning. Nanomicro Letters. 2024;**16**:242. DOI: 10.1007/s40820-024-01453-x

[64] Markiewicz M, Dziurdzia P, Skotnicki T. Randomly moving thermoelectric energy harvester for wearables and industrial internet of things. Nano Energy. 2024;**126**:109565. DOI: 10.1016/j.nanoen.2024.109565

[65] Franke L, Georg Rösch A, Khan MI, Zhang Q, Long Z, Brunetti I, et al. High power density $Ag_2Se/Sb_{1.5}Bi_{0.5}Te_3$ -based fully printed origami thermoelectric module for low-grade thermal energy harvesting. Advanced Functional Materials. 2024;**34**. DOI: 10.1002/adfm.202403646

[66] Kanahashi K, Pu J, Takenobu T. 2D materials for large-area flexible thermoelectric devices. Advanced Energy Materials. 2020;**10**. DOI: 10.1002/aenm.201902842

[67] Moshwan R, Shi X-L, Liu W-D, Liu J, Chen Z-G. Entropy engineering: An innovative strategy for designing high-performance thermoelectric materials and devices. Nano Today. 2024;**58**:102475. DOI: 10.1016/j.nantod.2024.102475

[68] Xia M, Record M-C, Boulet P. Influence of disorder in high-entropy alloys on thermoelectric properties and phase stability. The Journal of Physical Chemistry C. 2024;**128**:12010-12022. DOI: 10.1021/acs.jpcc.4c02309

[69] Bano S, Chetty R, Babu J, Mori T. $Mg_3(Sb,Bi)_2$-based materials and devices rivaling bismuth telluride for thermoelectric power generation and cooling. Device. 2024;**2**:100408. DOI: 10.1016/j.device.2024.100408

[70] Ni HL, Zhu TJ, Zhao XB. Thermoelectric properties of hydrothermally synthesized and hot pressed n-type Bi_2Te_3 alloys with different contents of Te. Materials Science and Engineering: B. 2005;**117**:119-122. DOI: 10.1016/j.mseb.2004.11.001

[71] Stefanou A-D, Zianni X. Towards optimizing width modulation for maximum thermoelectric efficiency. Micromachines (Basel). 2023;**14**:2176. DOI: 10.3390/mi14122176

[72] Usop R, Hasnan MMIM, Mohamad M, Ahmad MK, Said SM, Salleh F. Thermoelectric power-factor of Ag-doped TiO_2 thin film. Micromachines (Basel). 2022;**13**:2169. DOI: 10.3390/mi13122169

[73] Park O, Lee SW, Park SJ, Kim S. Phase formation behavior and thermoelectric transport properties of S-doped $FeSe_2$−xSx polycrystalline alloys. Micromachines (Basel). 2022;**13**:2066. DOI: 10.3390/mi13122066

[74] Wang Z, Zhao H, Luo X, Han W, Wang H, Meng L, et al. Study on the targeted improvement mechanism of the carrier concentration and mobility of BiCuSeO ceramics. Micromachines (Basel). 2023;**14**:1757. DOI: 10.3390/mi14091757

[75] Yang J, Wang Z, Zhao H, Luo X, Han W, Wang H, et al. Effect of composition adjustment on the thermoelectric properties of Mg_3Bi_2-based thermoelectric materials. Micromachines (Basel). 2023;**14**:1844. DOI: 10.3390/mi14101844

[76] Sornek K, Papis-Frączek K. Numerical and experimental analysis of a prototypical thermoelectric generator dedicated to wood-fired heating stove. Micromachines (Basel). 2023;**14**:145. DOI: 10.3390/mi14010145

[77] Nasiri MA, Serrano-Claumarchirant JF, Gómez CM, Cantarero A, Canet-Ferrer J. Back Mirror-free selective light absorbers for thermoelectric applications. Advanced Optical Materials. 2024. DOI: 10.1002/adom.202402079

[78] Coelho R, Casi Á, Araiz M, Astrain D, Branco Lopes E, Brito FP, et al. Computer simulations of silicide-tetrahedrite thermoelectric generators. Micromachines (Basel). 2022;**13**:1915. DOI: 10.3390/mi13111915

[79] Deng T, Qiu P, Yin T, Li Z, Yang J, Wei T, et al. High-throughput strategies in the discovery of thermoelectric materials. Advanced Materials. 2024. DOI: 10.1002/adma.202311278

[80] Tiryaki H, Yusuf A, Ballikaya S. Determination of electrical and thermal conductivities of n- and p-type thermoelectric materials by prediction iteration machine learning method. Energy. 2024;**292**:130597. DOI: 10.1016/j.energy.2024.130597

[81] Miao X, Zhang H, Ma F, Deng M, You E. Performance evaluation of AMTEC/TEG coupling system for nuclear power space stations in space exploration. Annals of Nuclear Energy. 2024;**207**:110742. DOI: 10.1016/j.anucene.2024.110742

[82] Palaporn D, Tanusilp S, Sun Y, Pinitsoontorn S, Kurosaki K. Thermoelectric materials for space explorations. Materials Advances. 2024;**5**:5351-5364. DOI: 10.1039/D4MA00309H

[83] Kim S, Lim S, Jeong MH, Kim W, Baik S, Suk JW. Flexible thermocouple using a thermoelectric graphene fiber with a seamless junction. Journal of Materials Science and Technology. 2024;**172**:15-22. DOI: 10.1016/j.jmst.2023.05.078

[84] LeBlanc S, Yee SK, Scullin ML, Dames C, Goodson KE. Material and manufacturing cost considerations for thermoelectrics. Renewable and Sustainable Energy Reviews. 2014;**32**:313-327. DOI: 10.1016/j.rser.2013.12.030

[85] Li J-W, Han Z, Yu J, Zhuang H-L, Hu H, Su B, et al. Wide-temperature-range thermoelectric n-type $Mg_3(Sb,Bi)_2$ with high average and peak zT values. Nature Communications. 2023;**14**:7428. DOI: 10.1038/s41467-023-43228-9

[86] Kim HY, Kim JE, Wee D. Assessment of greenhouse gas emissions and environmental impacts in the manufacturing process of thermoelectric coolers: A life-cycle impact perspective. Heliyon. 2025;**11**:e41527. DOI: 10.1016/j.heliyon.2024.e41527

[87] Wang T, Zhang C, Snoussi H, Zhang G. Machine learning approaches for thermoelectric materials research. Advanced Functional Materials. 2020;**30**. DOI: 10.1002/adfm.201906041

[88] Shi X-L, Wang L, Lyu W, Cao T, Chen W, Hu B, et al. Advancing flexible thermoelectrics for integrated electronics. Chemical Society Reviews. 2024;**53**:9254-9305. DOI: 10.1039/D4CS00361F

[89] Dai M, Zhang X, Wang QJ. 2D materials for photothermoelectric detectors: Mechanisms, materials, and devices. Advanced Functional Materials. 2024;**34**. DOI: 10.1002/adfm.202312872

[90] Aljafari B, Kalidasan B, Kareri T, Alqaed S, Bhutto YA, Pandey AK. Organic/carbon and organic/carbon-metal composite phase change material for thermoelectric generator: Experimental evaluation. Journal of Energy Storage. 2024;**78**:110082. DOI: 10.1016/j.est.2023.110082

[91] Kang Y-K, Lee S-J, Kim S, Nam Y, Jeong J-W. Performance analysis of a hybrid energy harvester incorporating a thermoelectric generator and phase-change material through annual experiments. Renewable Energy. 2025;**242**:122464. DOI: 10.1016/j.renene.2025.122464

Chapter 2

Thermoelectric Complete Response

Ronald Edgar Pirela La Cruz and
Sergio Rafael Velásquez Guzmán

Abstract

A new theory is introduced to describe the response of thermoelectric modules and materials in the presence and absence of both electrical and thermal power sources, assuming that the thermoelectric modules and materials are capacitors capable of storing electrical and thermal energy at the same time. The theory here described is designed by resolving the thermal components, and satisfies Newton's cooling law, Luttinger's thermal transport coefficients theory, and first and second-order electric circuit behavior. An especially useful feature is that a new equation for the thermoelectric figure of merit is given in terms of the ratio of two temperatures $Z\overline{T} = \Delta T'/\Delta T$. The successive analysis and mathematical operations yield equations for the thermal conductivity κ, electrical resistivity ρ, Seebeck coefficient α, and figure of merit $Z\overline{T}$; as well as, it includes the obtention of the thermal conductance K_o, electrical resistance R_m, thermal resistances related to thermal contacts $R_{C_{cold}}$ and $R_{C_{hot}}$, thermoelectric resistance R_{TE}, thermal capacitances related to thermal contacts C_C, thermoelectric capacitances C_{th} and C_{TE}, and among other less relevant specific variables. Additionally, through this theory, it is possible to determine the characteristic time constants τ_{TE}, τ_{th}, τ_o, τ_1, and relaxation time, as well as the characteristic angular frequencies ω_{TE}, ω_{th}, ω_o, and ω_1. Mathematical development and experiments considered non-ideal contacts and non-adiabatic conditions at room temperature, that is, $\overline{T} = 300K$.

Keywords: figure of merit, thermoelectric characterization, thermoelectric angular frequency, thermoelectric time constant, thermoelectric forced response, thermoelectric natural response, thermoelectric complete response

1. Introduction

Thermoelectric modules and thermoelectric materials (TEM) are solid-state thermodynamic engines that convert heat into electricity, a process known as Seebeck effect, or use electricity to pump heat from a cold side to a hot side, a process known as Peltier effect and Thomson effect. The term thermoelectric devices is used to refer without distinction to thermoelectric modules and thermoelectric materials (TEM), which usually consist of p and n type semiconductor material arrangements, connected thermally in parallel and electrically in series [1–4].

The thermoelectric performance of TEM is reduced to the determination of a single dimensionless quantity called figure of merit ($Z\overline{T}$), and a way of expressing it is presented in (1), where $\overline{T}$ is the average working temperature of the system, R_m is the electrical resistance, K_0 is the thermal conductance to the vanishing electric current, and α is the global Seebeck coefficient characterizing the thermoelectric coupling between the electric current and the heat flux through the TEM terminals [5]:

$$ZT = \frac{\alpha^2 \overline{T}}{R_m K_0} \tag{1}$$

In the late 1950s, a method was presented by Harman for testing the resistance in alternating current and determining the figure of merit of a thermoelectric material sample [6]. Also, the same methodology was used by Harman et al. to measure thermal conductivity by applying the Peltier effect [7]. In the early 1990s, a test methodology was developed by Buist, and this method is referred to as the transient test method, which is based on a similar concept but some fundamental differences compared with Harman's methods which gave rise to improvement in accuracy and reproducibility [8]. The fundamental similarity is that the techniques cited before are designed to solve the voltage components of a TEM and the fundamental difference is that Harman's method does this by measuring the resistive component, and transient test's method measures the Seebeck component. In 2022, a novel methodology was presented by Pirela and Velásquez, which was designed by determination of the thermal components, and it is intended to provide the ultimate solution for measuring the thermal conductance, thermal conductivity, and the figure of merit of TEM [9].

This chapter delves deeper into the study of TEM in the time domain and shows that both forced response and natural response correspond to fast perturbation and slow perturbation, respectively; as well as, related to small-signal models for frequency analysis and to existing DC models [10]. The theory proposed here follows a nomothetic order and it is consistent with the classical methods for studying first and second-order electrical circuits, providing theoretical foundations for the development of new characterization methods and enhancement of TEM [11, 12].

The chapter is structured as follows: in Section 2, the complete response of thermoelectric modules and materials is presented; then in Section 3, modeling and equations derivation; in Section 4, the simulations and experimental results, and then in Section 5 continues with thermoelectric characterization. Finally, the conclusions and bibliographical references.

2. Complete response of TEM

The unification of thermoelectric forced response (slow and fast perturbation) and thermoelectric natural response (absence of perturbation) theories allows to describe the complete response of thermoelectric modules and material, as well as consent to the study, characterization, and improvement of TEM [12–14].

2.1 Forced response of TEM

The thermoelectric circuit in **Figure 1** can be considered for both cases when a DC voltage source (V_s) is suddenly applied to the electrical terminals of TEM or correlatively when a temperature difference (ΔT) is suddenly applied across the thermal contacts of TEM. The electric current, thermal current, voltage, or temperature

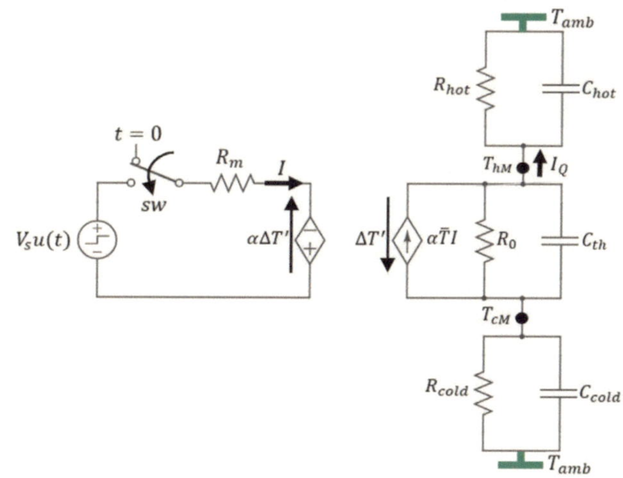

Figure 1.
Thermoelectric circuit $R_{hot}C_{hot}R_OC_{th}R_{cold}C_{cold}$ in cascade with external power source.

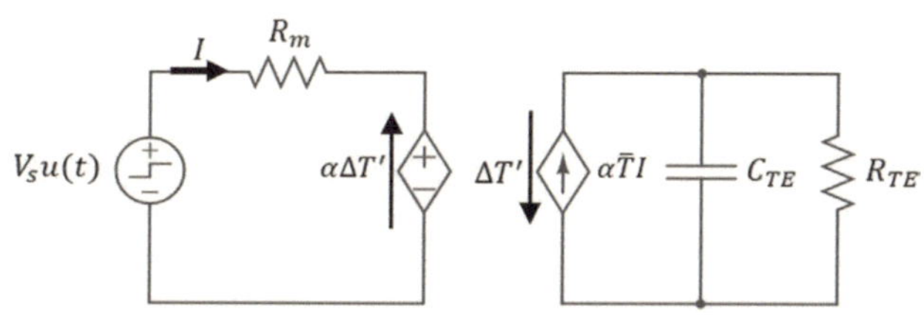

Figure 2.
Equivalent thermoelectric circuit $R_{TE}C_{TE}$.

sources can be modeled as a step function, and the response is known as a step response. The forced response of TEM represents its behavior when the excitation is a step function [13]. The thermoelectric circuit in **Figure 1** can be replaced by the equivalent thermoelectric circuit $R_{TE}C_{TE}$ in **Figure 2**; where $V_s u(t)$ is a step DC voltage source, $R_{TE} = 1/K_{TE}$ is the equivalent thermoelectric resistance, C_{TE} is the equivalent thermoelectric capacitance, α is the Seebeck coefficient, $\overline{T} = (T_{hot} + T_{cold})/2$ is the average temperature and the electric current through of TEM is $I = [V_s u(t) - V_\alpha]/R_m$, and $V_\alpha = \alpha \Delta T' = \alpha\,(T_{hM} - T_{cM})$. Therefore, C_{TE} and R_{TE} can be written as Eqs. (2) and (3), respectively [13]:

$$C_{TE} = C_{th} \| C_C = C_{th} + (C_{hot} \| C_{cold}) \tag{2}$$

$$R_{TE} = R_O \| R_C = R_O \| (R_{hot} + R_{cold}) \tag{3}$$

The temperature difference ΔT across the capacitor C_{TE} is selected as the response of the thermoelectric circuit to be determined. An initial temperature difference $\Delta T' = 0$ is assumed across C_{TE}; that is, inside of the TEM, and a temperature difference $\Delta T = 0$ across the thermal contact. Since the voltage of an electric capacitor cannot change instantaneously, then analogously, it is assumed that the temperature difference of a TEM cannot change abruptly. Therefore, these assumptions can be represented by Eqs. (4) and (5); where $\Delta T'(0^-)$ is the temperature difference across

C_{TE}, just before switching and $\Delta T'(0^+)$ is the temperature difference immediately after switching. For $t > 0$, the switch "*sw*" of the thermoelectric circuit in **Figure 1** is closed [13].

$$\Delta T(0^-) = \Delta T(0^+) = \Delta T \tag{4}$$

$$\Delta T'(0^-) = \Delta T'(0^+) = \Delta T'_0 \tag{5}$$

The Norton equivalent of the network connected to C_{TE} in **Figure 2** is obtained by applying Kirchhoff's Current Law (KCL), represented by Eq. (6) [11]. And dividing Eq. (6) by C_{TE}, it yields a first-order differential equation, Eq. (7), analogous to first-order electrical circuits. Solving Eq. (7), it shows that the temperature difference across the equivalent thermoelectric capacitor is defined as $\Delta T_{C_{TE}}(t)$, as shown in Eq. (8) [9, 13]:

$$C_{TE}\frac{d[\Delta T_{C_{TE}}(t)]}{dt} + \frac{\Delta T_{C_{TE}}(t)}{R_{TE}} = \alpha\overline{T}I \tag{6}$$

$$\frac{d[\Delta T_{C_{TE}}(t)]}{dt} + \frac{\Delta T_{C_{TE}}(t)}{R_{TE}C_{TE}} = \frac{\alpha\overline{T}I}{C_{TE}} \tag{7}$$

$$\Delta T_{C_{TE}}(t) = \alpha\overline{T}IR_{TE} + \left[\Delta T_{C_{TE}}(0) - \alpha\overline{T}IR_{TE}\right]e^{-t/(R_{TE}C_{TE})} \tag{8}$$

For $t > 0$, the time constant is $\tau_{TE} = R_{TE}C_{TE}$ and is determined considering that the amplitude increases by a factor "*e*" (63.2% of the amplitude it has) [9, 13]. Therefore, from the step response, it is possible to obtain the characteristic thermoelectric time constant of thermoelectric modules, and it corresponds to the inverse of the characteristic angular frequency of thermoelectric modules, that is, $\tau_{TE} = 1/\omega_{TE}$, which can be written as Eq. (10) [5]:

$$\tau_{TE} = R_{TE}C_{TE} = (C_{th} + C_C)\left(\frac{R_0R_C}{R_0 + R_C}\right) = \frac{C_{th} + C_C}{K_{th} + K_C} \tag{9}$$

$$\tau_{TE} = \frac{C_{th} + C_C}{K_0 + K_C} = \frac{1}{\omega_{TE}} \tag{10}$$

From **Figure 2**, it is appreciated that $\Delta T_{C_{TE}} = \Delta T' = \Delta T - \Delta T_{C_C}$. Then $\Delta T_{C_{TE}}$ represents the temperature difference behavior of TEM. However, when the power source is connected to the thermoelectric device abruptly, which corresponds to a quasi-instantaneous state change, that is, $t \to 0$, the impedance associated with the thermal contacts disappears. In fact, in this case, thermal contacts have no influence. This corresponds to the small-signal analysis when the frequency increases enough; at high frequency, the thermal capacitances create a bypass for the heat current and impose that $\Delta T_{C_{TE}} = \Delta T' = \Delta T$. For the AC and DC regimes, the resistance R_{TE} comes from the modification of the electromotive force $\Delta T'$ with the electric current. Thus, it yields Eq. (11) [5, 13]:

$$\Delta T(t) = \alpha\overline{T}IR_{TE} + \left[\Delta T'(0) - \alpha\overline{T}IR_{TE}\right]e^{-t/\tau_{TE}} \tag{11}$$

Analogously to the analysis of an *RC* type electrical circuit, Eq. (11) is a way of expressing the thermoelectric complete response to the sudden application of a DC voltage source at its electrical terminals, assuming that the TEM represents a thermoelectric capacitor initially discharged, then it must be noted that $\Delta T'(0) = 0$ in Eq. (11), so that the forced response can be obtained through the temperature difference generated by a TEM, from the Peltier effect, and is given by the following mathematical expressions Eqs. (12) and (13) [9, 11, 13]:

$$\Delta T(t) = \alpha \overline{T} I R_{TE}\left(1 - e^{-t/\tau_{TE}}\right) \tag{12}$$

$$\Delta T(t) = \alpha \overline{T} R_{TE}\left[\frac{V_s u(t) - V_\alpha(t)}{R_m}\right]\left(1 - e^{-t/\tau_{TE}}\right) \tag{13}$$

From Eqs. (12) and (13), it follows that $\tau_{TE} = R_{TE}C_{TE}$ is acquired considering that the amplitude of ΔT, T_{cold}, T_{hot} or V_α increases by a factor "e" (63.2% of the maximum amplitude that it has); where $V_\alpha = \alpha \Delta T$ (Seebeck voltage). And the time required for ΔT, T_{hot} or V_α to reach its final value is $5\tau_{TE}$.

If the measurement is performed on the cold side thermal contact, the time constant related to the thermal contacts τ_C is obtained and is the time required for T_{cold} to decrease by a factor of 1/e (36.8% of the initial value). Consequently, when the thermal contacts are absent, that is, the case where only thermoelectric material is present. Eq. (13) is reduced to Eq. (14) [13]:

$$\Delta T'(t) = \frac{\alpha \overline{T} R_0}{R_m}\left[V_s u(t) - V_\alpha(t)\right]\left(1 - e^{-t/\tau_{th}}\right) \tag{14}$$

Analogously to the analysis of electric circuits, the temperature difference $\Delta T(t)$ across the thermal contacts of the TEM has two components and there are two classical ways to decompose them into two parts. The first way is to divide it into "a forced response (independent source) and a natural response (stored energy)", and the second way is to divide it into "a steady-state response (permanent or stable part, which is the behavior of the thermoelectric circuit long after an external excitation is applied) and a transient response (temporary part, which will die out over time)" [9].

2.2 Natural response of TEM

To study the natural or transient response of TEM, it is necessary to refer to the model that represents a thermoelectric generator in **Figure 3**, which includes non-ideal thermal contacts [5, 9, 15]. For this model, two main scenarios can be considered; in the first scenario, it is assumed that for $t < 0$, the TEM has been connected through the thermal contacts to two thermal reservoirs at constant temperatures, one at hot temperature $T_h = T_{hot}$ and another at cold temperature $T_c = T_{cold}$, where $T_c < T_h$, and in the second scenario, it is assumed that for $t < 0$ a thermoelectric module has been connected to a direct current voltage source. The thermal contacts are characterized by finite thermal conductance, defined as $K_{hot} = 1/R_{hot}$ (hot side thermal contact) and $K_{cold} = 1/R_{cold}$ (cold side thermal contact), and $K_0 = 1/R_0$ (thermal conductance of the cold side thermal contact). Also, as a complement, the capacitances are defined as C_{th} (capacitance of the thermoelectric material), and C_{hot} and C_{cold} (capacitances of both hot and cold side thermal contacts, respectively) [14].

Consider the thermoelectric circuit shown in **Figure 3** as a cascade thermoelectric circuit, analogous to a cascade *RCRCRC* electric circuit, which belongs to a thermoelectric generator; where it is assumed that, $R_0 = 1/K_0$ is the thermal resistance of the

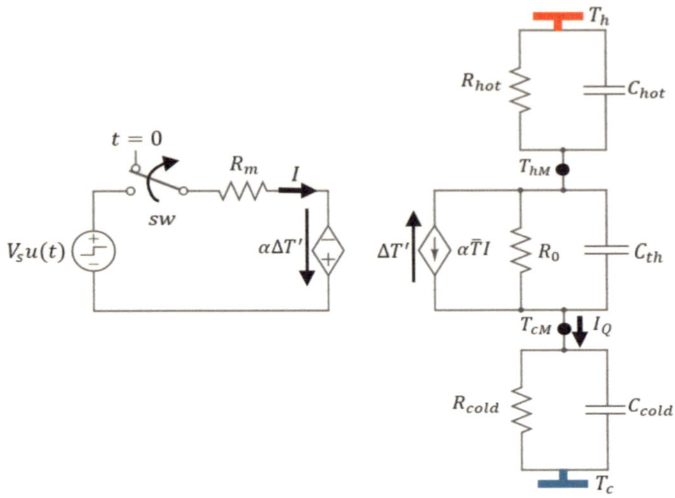

Figure 3.
Thermoelectric circuit $R_{hot}C_{hot}R_0C_{th}R_{cold}C_{cold}$ in cascade without external power sources.

thermoelectric material, and the thermal contacts are characterized by finite thermal resistance defined as $R_{hot} = 1/K_{hot}$ and $R_{cold} = 1/K_{cold}$. The thermoelectric circuit is excited by the energy initially stored in C_{th}, C_{hot}, and C_{cold}. Such energy is represented by the sum of the initial temperature of the hot side thermal capacitor $\Delta T_{C_{hot}}$, initial temperature of the cold side thermal capacitor $\Delta T_{C_{cold}}$ and the temperature of the thermoelectric capacitor $\Delta T_{C_{th}}$. At $t = 0$, the switch *sw* is open (sw interconnects the direct current voltage source $V_s u(t)$ with the thermoelectric circuit). Therefore, $T_{C_{th}}(0) = \frac{1}{C_{th}}\int_{-\infty}^{0} I_{Q_{C_{th}}} dt = \Delta T_{C_{th}} = \Delta T'$, $T_{C_{hot}}(0) = \frac{1}{C_{hot}}\int_{-\infty}^{0} I_{Q_{C_{hot}}} dt = \Delta T_{C_{hot}}$, and $T_{C_{cold}}(0) = \frac{1}{C_{cold}}\int_{-\infty}^{0} I_{Q_{C_{cold}}} dt = \Delta T_{C_{cold}}$; where the temperature difference provided by the thermal contacts is given by $\Delta T_C = T_{C_{hot}} - T_{C_{cold}}$, temperature difference provided by the thermoelectric material is $\Delta T_{C_{th}} = \Delta T' = T_{hM} - T_{cM}$, and the thermal currents through thermal contacts capacitors and thermoelectric capacitor are $I_{Q_{C_{hot}}}$, $I_{Q_{C_{cold}}}$, and $I_{Q_{C_{th}}}$, respectively [14].

According to Fourier's law, thermal currents through the thermoelectric resistor and thermal contact resistors are $I_{Q_{R_0}}(t) = \Delta T_{R_0}(t)/R_0$, $I_{Q_{R_{hot}}}(t) = \Delta T_{R_{hot}}(t)/R_{hot}$, and $I_{Q_{R_{cold}}}(t) = \Delta T_{R_{cold}}(t)/R_{cold}$, respectively. From equivalent cascade thermoelectric circuit $R_0C_{th}R_CC_C$ shown in **Figure 4**, the thermal current I_Q through the device is obtained, where $R_C = 1/K_C$ and $K_C = K_{hot}K_{cold}/(K_{hot} + K_{cold}) = R_{hot} + R_{cold}$. Therefore, I_Q can be represented as $I_Q(t) = I_{Q_{R_C}} = \Delta T_C(t)/R_C$. The thermal currents through each of the capacitors are given by $I_{Q_{C_{hot}}}(t) = C_{hot}d\left[T_{C_{hot}}(t)\right]/dt$, $I_{Q_{C_{cold}}}(t) = C_{cold}d\left[T_{C_{cold}}(t)\right]/dt$, $I_{Q_{C_{th}}}(t) = C_{th}d\left[\Delta T_{C_{th}}(t)\right]/dt$, and the thermal current through the equivalent capacitor of the thermal contacts $I_{Q_{C_C}}(t) = C_C d[\Delta T_{C_C}(t)]/dt$; where $C_C = C_{hot}C_{cold}/(C_{hot} + C_{cold})$.

Applying Kirchhoff's Voltage Law (KVL) along the outer mesh of the thermoelectric circuit $R_0C_{th}R_CC_C$, as shown in **Figure 4**, it allows us to obtain the temperature difference across of the device; that is, $\Delta T = \Delta T' + \Delta T_{C_C}$, and temperature difference across the equivalent thermal contacts capacitor $\Delta T_{C_C} = \Delta T - \Delta T'$. Remembering that, ΔT is the temperature difference between the two thermal reservoirs, and it is

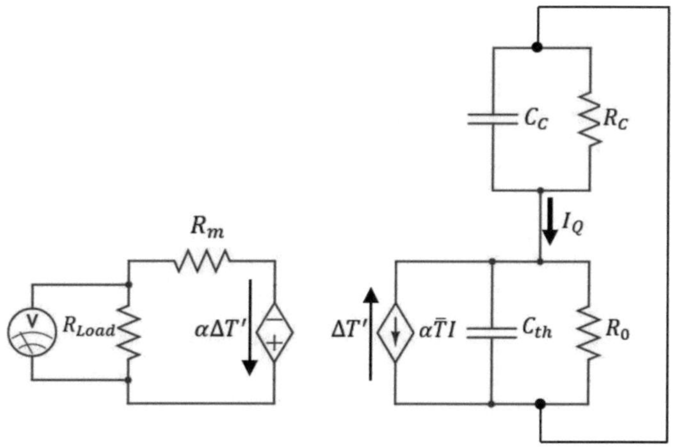

Figure 4.
Equivalent thermoelectric circuit $R_0C_{th}R_CC_C$.

defined by $\Delta T = T_{hot} - T_{cold} = T_h - T_c$. Therefore, ΔT seen by the thermoelectric generator inside the TEM satisfies $\Delta T' = T_{hM} - T_{cM} = \Delta T - R_CI_Q$. Since the temperature difference of capacitors cannot change abruptly, as occurs with electrical capacitors, then it is assumed that $\Delta T(0^-) = \Delta T(0^+) = \Delta T$.

Hence, it yields $R_C\left[I_Q(t) + I_{Q_{C_C}}(t)\right] + R_0\left[I_Q(t) + I_{Q_{C_{th}}}(t)\right] = \Delta T$. Para $t > 0$, sw is open as shown in **Figure 3** and all independent sources are off; thus, $\Delta T' + \Delta T_{C_C} = 0$ and $\Delta T_{C_C} = -\Delta T_{C_{th}}(t) = -\Delta T'$. Therefore, $R_C\left[I_Q(t) + I_{C_C}(t)\right] + R_0\left[I_Q(t) + I_{C_{th}}(t)\right] = 0$. Applying Kirchhoff's Current Law (KCL), other expressions can be obtained for $I_Q(t)$, which, when substituted in the expressions for the different temperature differences, it is possible to determine the general mathematical expression that models the natural response of the TEM. This mathematical expression is a second-order differential equation as a function of time, which is represented in Eq. (15) [14]:

$$R_CC_CR_0C_{th}\frac{d^2\left[\Delta T_{C_{th}}(t)\right]}{dt^2} + (R_CC_C + R_0C_{th})\frac{d\left[\Delta T_{C_{th}}(t)\right]}{dt} + \Delta T_{C_{th}}(t) = 0 \tag{15}$$

Solving the second-order differential equation requires two initial conditions, namely initial value of $\Delta T_{C_{th}}$ and its first derivative. With two initial conditions, Eq. (15) can be solved. Based on the theory of first-order electric circuits, it indicates that the solution is exponential [11]. Thus, $\Delta T_{C_{th}}$ can be written as Eq. (16), where A and s are constant to be determined. By replacing Eq. (16) in Eq. (15) and performing the necessary derivations, it gives Eq. (17):

$$\left|\Delta T_{C_{th}}\right| = \left|-\Delta T_{C_C}\right| = \left|Ae^{st}\right| \tag{16}$$

$$Ae^{st}\left[s^2 + \left(\frac{R_CC_C + R_0C_{th}}{R_CC_CR_0C_{th}}\right)s + \left(\frac{1}{R_CC_C + R_0C_{th}}\right)\right] = 0 \tag{17}$$

Since the solution under search is $\Delta T_{C_{th}} = Ae^{st}$, then only the expression in square brackets in Eq. (17) can be zero. Therefore, Eq. (18) represents a quadratic equation, and it is known as the characteristic equation of the differential Eq. (17), since its roots dictate the character of ΔT_{C_C} and $\Delta T_{C_{th}}$:

$$s^2 + \left(\frac{R_C C_C + R_0 C_{th}}{R_C C_C R_0 C_{th}}\right)s + \left(\frac{1}{R_C C_C + R_0 C_{th}}\right) = 0 \tag{18}$$

Hence, Eq. (18) represents a quadratic equation, and it is known as the characteristic equation of the differential equation Eq. (17). The compact and notable ways of expressing the roots of Eq. (18) are $s_1 = -\beta + \sqrt{\beta^2 - \omega_0{}^2}$ and $s_2 = -\beta - \sqrt{\beta^2 - \omega_0{}^2}$. The roots s_1 and s_2 are called natural frequencies, measured in Nepers per second (Np/s), because they are associated with the natural response of the TEM, where $j = \sqrt{-1}$, $\omega_d = \sqrt{(\omega_0{}^2 - \beta^2)}$ is called the damping frequency, ω_o is called the resonant frequency of TEM, or more strictly, the undamped natural frequency of TEM, and β is the frequency associated with thermal contacts. The characteristic thermoelectric angular frequencies emanate from the roots as $\beta = \omega_1 = (R_C C_C + R_0 C_{th})/2R_C C_C R_0 C_{th}$ and $\omega_0 = 1/\sqrt{2R_C C_C R_0 C_{th}}$. So, Eq. (18) can be written in terms of the characteristic angular frequency ω_1 and ω_o; that is, $s^2 + 2\omega_1 s + \omega_o^2 = 0$. When thermal contacts are equal, the damping ratio ζ is obtained through Eq. (19) [11, 13]:

$$\zeta = \frac{1}{2\omega_0}\left(\frac{\omega_0{}^2 + \omega_1{}^2}{\omega_1}\right) \tag{19}$$

The values s_1 and s_2 indicate that there are two possible solutions to ΔT_{C_C}, each one is expressed in the form of the assumed solution in Eq. (16); that is,

$$\Delta T_{Ccold}(t) = A_1 e^{s_1 t}, \Delta T_{Chot}(t) = A_2 e^{s_2 t} \tag{20}$$

Then, Eq. (15) is a linear equation. So, any linear combination of the two distinct solutions ΔT_{Ccold} and ΔT_{Chot} is also a solution of Eq. (21). A complete solution of Eq. (15) would require a linear combination of ΔT_{Ccold} and ΔT_{Chot}. Therefore, the natural response can be expressed in terms of thermal contacts by Eq. (21); where A_1 and A_2 are determined from the initial values of $\Delta T_{C_{th}}(0)$ and $d\Delta T_{C_{th}}(0)/dt$, or from the initial values of $\Delta T_{C_C}(0)$ and $d\Delta T_{C_C}(0)/dt$. This shows that the natural response of a TEM expressed in terms of temperature is an exponential drop of the initial temperature and obeys Newton's cooling law [16, 17]. It is also shown that the natural response of a thermoelectric module expressed in voltage is an exponential drop of the Seebeck voltage and complies with the theory of first and second-order electric circuits, considering that $V_\alpha = \alpha \Delta T$:

$$\Delta T_{C_C}(t) = A_1 e^{s_1 t} + A_2 e^{s_2 t} \tag{21}$$

The time constant associated with the thermal contacts is $\tau_C = R_C C_C = \tau_{C_{cold}} = \tau_1 = 1/\beta = 1/\omega_1$, and it is determined considering that the cold side thermal contact temperature is increased by a factor $1/e$ (36.8% of the amplitude it has). For the case where the cold side thermal contact is equal to the hot side thermal contact, τ_C is obtained experimentally from the temperature measurement on the cold side thermal contact. Thus, the capacitor related to the cold side thermal contact takes $5\tau_C$ to reach its final state [5, 9, 14].

The time constant associated with the capacitance of the thermoelectric material τ_{th} could be obtained at a slightly longer time than expected. Considering that the amplitude is reduced by a factor $1/e$, it can be obtained at a longer time with respect to the relationship $\tau_{th} = 1/\omega_{th}$, for which the case of an overdamped thermoelectric circuit is considered; when $\beta < \omega_0$. Therefore, the roots can be as $s_1 = -\beta + \sqrt{-(\omega_0{}^2 - \beta^2)} = -\beta + j\omega_d$ and $s_2 = -\beta - \sqrt{-(\omega_0{}^2 - \beta^2)} = -\beta - j\omega_d$; where both ω_0 and ω_d are natural frequencies because they contribute to determining the natural response of TEM; while ω_0 is often called the undamped natural frequency, ω_d is called the damped natural frequency or displacement frequency. The relaxation time associated with the capacitance of the thermoelectric material is obtained at $5\tau_{th}$. Therefore, the natural response with respect to the thermoelectric material is defined by Eq. (22) [5, 9, 14]:

$$\Delta T_{C_{th}}(t) = e^{-\beta_{th}t}(A_1 \cos \omega_d t + A_2 \mathrm{sen} \omega_d t) \tag{22}$$

With the presence of the sine and cosine functions it is trivial that the natural response for this case is exponentially damped and oscillatory. Such natural response has a time constant associated with the thermoelectric capacitance τ_{th} and a period $T = 2\pi/\omega_d$, and as the amplitude is reduced by a factor $1/e$, there are two time scales: T measures the time it takes to oscillate and τ_{th} the time it takes to dampen. The dimensionless quotient is represented by Eq. (23) [11]:

$$\frac{\tau_{th}}{T} = \frac{\omega_d}{2\pi\beta_{th}} \tag{23}$$

The time taken by ΔT to decay is given by the time constant $\tau_{th} = R_0 C_{th} = 1/\beta_{th}$. From Eq. (23) can be obtained an expression for the characteristic angular frequency related to the thermoelectric material between the thermal contacts, which is known as ω_{th} [6], and is given by Eq. (24) [5, 9, 14]:

$$\omega_{th} = 2\pi\beta_{th} = \frac{2\pi}{\tau_{th}} \tag{24}$$

3. Modeling and equations derivation

3.1 Thermoelectric coefficients and parameters

Using the forced response, it is possible to obtain K_0 using Eq. (10) as follows [5, 9, 13, 14]:

$$\tau_{TE} = \frac{C_{th} + C_C}{K_0 + K_C} = C_{TE} R_{TE} \tag{25}$$

$$\tau_{TE}(K_0 + K_C) = C_{th} + C_C \tag{26}$$

$$K_0 = \frac{C_{th} + C_C}{\tau_{TE}} - K_C = \frac{C_{TE}}{\tau_{TE}} - K_C \tag{27}$$

$$K_0 = \frac{1}{R_{TE}} - \frac{1}{R_C} \tag{28}$$

Also, from the forced response, it is possible to find R_{TE} using Eq. (12); considering that, at $t = 0$ the temperature difference $\Delta T_{C_{TE}}(0) = 0$ and that after $5\tau_{TE}$ the equivalent thermoelectric capacitor C_{TE} will be fully charged, obtaining Eq. (29). Eq. (29) can be expressed in a simpler form if it is taken into consideration that for a time greater than $t > 5\tau_{TE}$, $e^{-t/\tau_{TE}}$ tends to zero, then another expression for R_{TE} is shown in Eq. (30).

$$R_{TE} = \frac{\Delta T(t)}{\alpha \overline{T} I\left(1 - e^{-t/\tau_{TE}}\right)} \tag{29}$$

$$R_{TE} = \frac{\Delta T}{\alpha \overline{T} I} \tag{30}$$

Using the natural response, R_C can be obtained through Eqs. (16) and (21). The Eq. (21) shows that ΔT_{C_C} is the result of the temperature contribution of each thermal capacitor, and R_C is the result of the sum of the thermal resistances of the thermal contacts. Therefore, Eq. (21) is used considering that at $t = 0$, $A_1 = \alpha \overline{T} I_{TEG_{sc}} R_{C_{cold}}$ and $A_2 = \alpha \overline{T} I_{TEG_{sc}} R_{C_{hot}}$; where $I_{TEG_{sc}}$ is the short-circuit current of the TEM in thermal generator mode; that is, the load connected to the thermoelectric generator is zero Ohm. Thus, for $R_{Load} = 0\ \Omega$, $I_{TEG_{sc}} = V_\alpha / R_m$. And assuming that, $R_{C_{cold}} = R_{C_{hot}}$, $R_C = R_{C_{cold}} + R_{C_{hot}} = 2R_{C_{cold}}$, and $A_1 = A_2$. Therefore, R_C is obtained by Eq. (35); where R_m is given by Eq. (36). The minimum temperature value at the cold side thermal contact $T_{C_{cold}}$ is obtained at $t = 0$.

$$\Delta T_{C_C}(t) = \alpha \overline{T} I_{TEG_{sc}} R_{C_{cold}} e^{s_1 t} + \alpha \overline{T} I_{TEG_{sc}} R_{C_{hot}} e^{s_2 t} \tag{31}$$

$$T_{C_{cold}}(t) = \alpha \overline{T} I_{TEG_{sc}} \frac{R_C}{2} e^{s_1 t} \tag{32}$$

$$T_{C_{cold}}(t) = \alpha \overline{T} I_{TEG_{sc}} R_{C_{cold}} e^{s_1 t} \tag{33}$$

$$R_{C_{cold}} = \frac{T_{C_{cold}}(t)}{2\alpha \overline{T} I_{TEG_{sc}} e^{s_1 t}} \tag{34}$$

$$R_C = \frac{T_{C_{cold}}}{\alpha \overline{T}(V_\alpha / R_m)} \tag{35}$$

$$R_m = \frac{[V_s u(t) - V_\alpha]}{I} \tag{36}$$

Also, from the natural response, it is possible to obtain K_0 through Eq. (22). Considering that, at $t = 0$ the maximum value of the temperature differential $\Delta T_{C_{th}}(t) = \Delta T'$ is obtained, and $A_1 = A_2 = \alpha \overline{T} I_{TEG_{sc}} R_0$, $\operatorname{sen}\omega_d t = 0$ and $\cos \omega_d t = 1$; that is, a term in Eq. (22) does not contribute to the temperature difference $\Delta T_{C_{th}}(t)$. Then the relation for Eq. (22) is reduced to Eq. (37). Therefore, using the identity $R_0 = K_0^{-1}$ and $V_\alpha = \alpha \Delta T$, other expressions for K_0 are given by Eqs. (39) and (40).

$$\Delta T_{C_{th}}(t) = \Delta T' = \Delta T - T_{C_{cold}} = \alpha \overline{T} I_{TEG_{sc}} R_0 \tag{37}$$

$$R_0 = \frac{\Delta T_{C_{th}}(t)}{\alpha \overline{T} I_{TEG_{sc}}} = \frac{\Delta T - T_{C_{cold}}}{\alpha \overline{T}(V_\alpha / R_m)} = \frac{\Delta T - T_{C_{cold}}}{\alpha^2 \overline{T}(\Delta T / R_m)} \tag{38}$$

$$K_0 = \frac{\alpha^2 \overline{T}(\Delta T / R_m)}{\Delta T - T_{C_{cold}}} \tag{39}$$

$$K_0 = \frac{\alpha^2 \Delta T \overline{T}}{R_m \Delta T'} \tag{40}$$

Finally, the electric resistivity is obtained using Eq. (36) by multiplying R_m per A/L, obtaining Eq. (41). Also, the thermal conductivity of the thermoelectric material κ is obtained using Eq. (40) by multiplying K_0 per L/A, obtaining the Eq. (42), where L and A are the length and area of the specimen under test [9]:

$$\rho = \left[\frac{V_s u(t) - V_\alpha}{I}\right]\left(\frac{A}{L}\right) \tag{41}$$

$$\kappa = \left(\frac{\alpha^2 \overline{T} \Delta T}{R_m \Delta T'}\right)\left(\frac{L}{A}\right) \tag{42}$$

3.2 Figure of merit

Considering forced response, another mathematical expression for ZT is obtained, substituting Eq. (28) in Eq. (1). Now, ZT is given by Eq. (43) [9, 13]:

$$ZT = \frac{\alpha^2 \overline{T} R_{TE} R_C}{R_m (R_C - R_{TE})} \tag{43}$$

Using natural response, new mathematical expressions for ZT are obtained, substituting Eq. (39) in Eq. (1) and Eq. (40) in Eq. (1). Therefore, also ZT is given by Eq. (44) and Eq. (45) [9, 14, 18, 19]:

$$ZT = \frac{\Delta T - T_{C_{cold}}}{\Delta T} \tag{44}$$

$$ZT = \frac{\Delta T'}{\Delta T} \tag{45}$$

Eq. (44) is reciprocal to Buist's equation $ZT = V_{oa}/(V_{ia} - V_{oa})$ and Eq. (45) has the Harman's equation form $Z\overline{T} = V_\alpha/V_p$. However, equations for $Z\overline{T}$ discovered through the theory explained in this chapter are expressed as a temperature function instead of voltage function as the ones obtained by both Buist and Harman [6, 8].

For the case where the test configuration includes a Heat Sink (HS), such that the temperature of the hot side is set to the ambient temperature, then the temperature at the cold contact can be expressed as shown in Eq. (46) [13]:

$$T_{C_{cold}} = T_{amb} - \left[(T_{h_{No\ HS}} - T_{amb}) + T_{C_{HS}}\right] \tag{46}$$

4. Simulation and experimental results

To illustrate the complete response of thermoelectric modules and materials, a special thermoelectric module is considered as a sample [20], used by Lineykin and Ben-Yaakov [4, 18], Apertet and Ouerdane [5, 19]; as well as by Pirela and Velásquez [9, 13, 14, 21]. The parameters of the TEM at $\Delta T = 70\ K$ are presented in **Table 1**.

4.1 Simulation results

The simulations were performed using PSpice simulator (Orcad Family Release 9.2 Standalone).

$R_m (\Omega)$	K_0 $(W\,K^{-1})$	C_{th} $(J\,K^{-1})$	α $(V\,K^{-1})$	ZT
1.602	0.667	0.35	0.0532	0.795

For $\Delta T = 70\ K$, $\overline{T} = 300\ K$, $V_{max} = 15.9$ V, $I_{max} = 7.6\ A$, R_{ac} (295 K) = 1.5 Ω (Tolerance: +/– 10%), $Q_{max} = 75\ W$. Dimensions: 40 mm × 40 mm × 3.5 mm.

Table 1.
Kryotherm TB-127-1.4-1.2. Thermoelectric parameters.

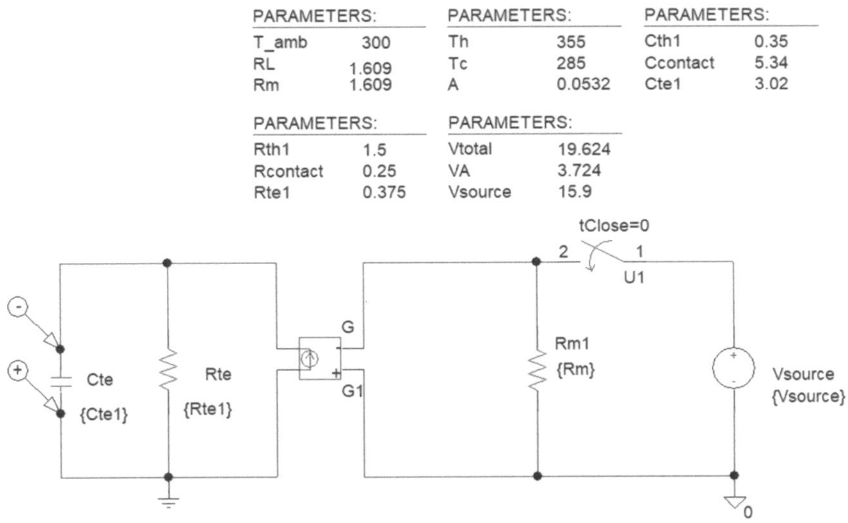

Figure 5.
Thermoelectric circuit for forced response simulation.

4.1.1 Simulation results (forced response)

Figure 5 shows the equivalent thermoelectric circuit for forced response simulation. **Figure 6** shows the temperature difference ΔT measured across the thermal contacts of TEM, and it is found from forced response simulation, where the time constant related to the thermoelectric module is $\tau_{TE} = 1.133\,s$. And it is obtained at 63.2% of the final value of ΔT. Also, τ_{TE} can be obtained by measuring temperature on the hot side thermal contact, temperature of the cold side thermal contact, or equally by measuring the Seebeck voltage generated by the TEM. Therefore, the angular frequency related to the thermoelectric module is $\omega_{TE} = 1/C_{TE}R_{TE} = 1/\tau_{TE} = 1/1.133\,s = 0.882\,rad/s$.

4.1.2 Simulation results (natural response)

Figure 7 shows the equivalent thermoelectric circuit for natural response simulation. **Figure 8** shows the temperature difference ΔT measured across the thermal contacts of TEM, and it is obtained through natural response simulation, where the time constant related to thermoelectric material is $\tau_{th} = 3.387\,s$. It is obtained at 36.8% of the initial value of ΔT. Also, τ_{th} can be obtained by measuring voltage on electrical terminals of the TEM, that is, from the Seebeck voltage, which is given by $V_\alpha = \alpha \Delta T$. And considering that the natural response of a thermoelectric capacitor presents an

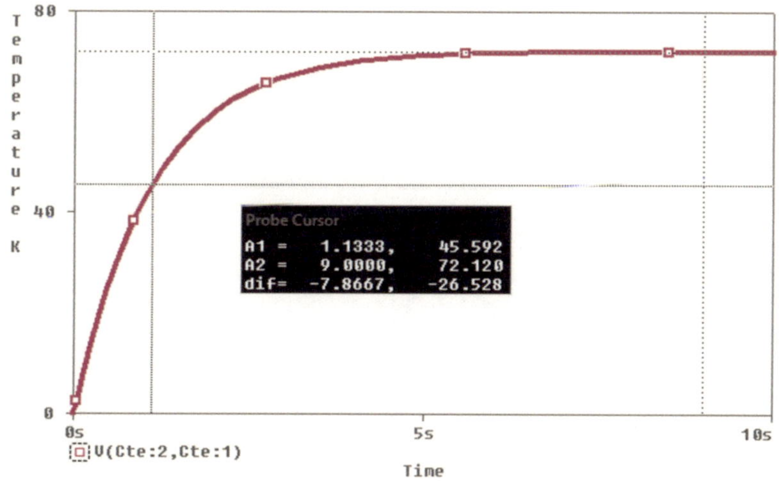

Figure 6.
Thermoelectric forced response of the TEM sample. The time constant $\tau_{TE} = 1.133$ s, *obtained from* ΔT *at 63.2% of its final value.*

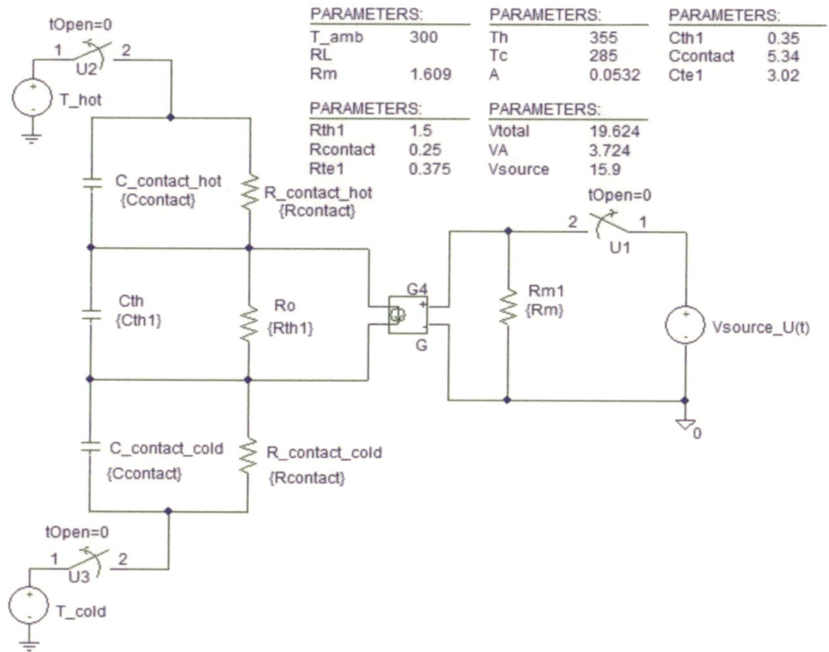

Figure 7.
Thermoelectric circuit for natural response simulation.

underdamped response, thus the angular frequency related to the thermoelectric material is $\omega_{th} = 2\pi/\tau_{th} = 2\pi/3.387\ s = 1.854\ rad/s$.

4.2 Experimental results

Suspended configuration has been used for the experimental demonstration, as shown in **Figure 9**. **Figure 10** shows the complete response of the thermoelectric

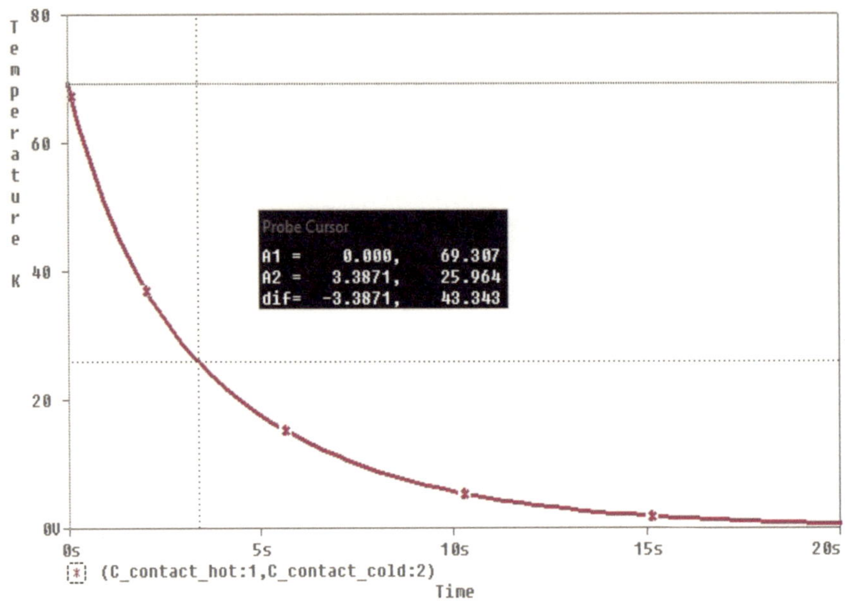

Figure 8.
Thermoelectric natural response of the TEM sample. The time constant $\tau_{th} = 3.387\ s$*, obtained from* ΔT *at 36.8% of its initial value.*

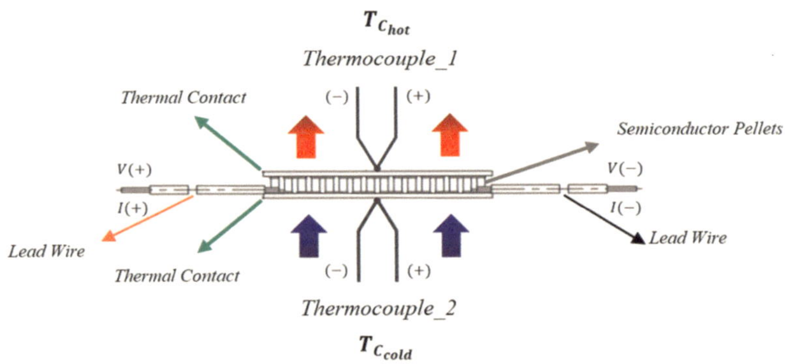

Figure 9.
TEM sample in suspended configuration and non-adiabatic conditions. Connection points for the characterization process.

module sample measuring ΔT, by the Peltier effect [2]. After the disconnection of the voltage source V_s, the natural response of the TEM is obtained, and it obeys Newton's cooling law [16, 20]. **Figure 11** shows the voltage measured on the electrical terminals of the sample, where the voltage measured during the forced response is given by V_s, and the voltage measured during the natural response is V_α, by the Seebeck effect [1].

4.2.1 Forced response test results

The thermoelectric time constant $\tau_{TE} = 1.130\ s$ is found experimentally, when V_s is active, measuring the temperature difference ΔT across the thermal contacts of the device and at 63.2% of its maximum, as shown in **Figure 10**. Therefore, the angular frequency is $\omega_{TE} = 1/\tau_{TE} = 0.884\ rad/s$.

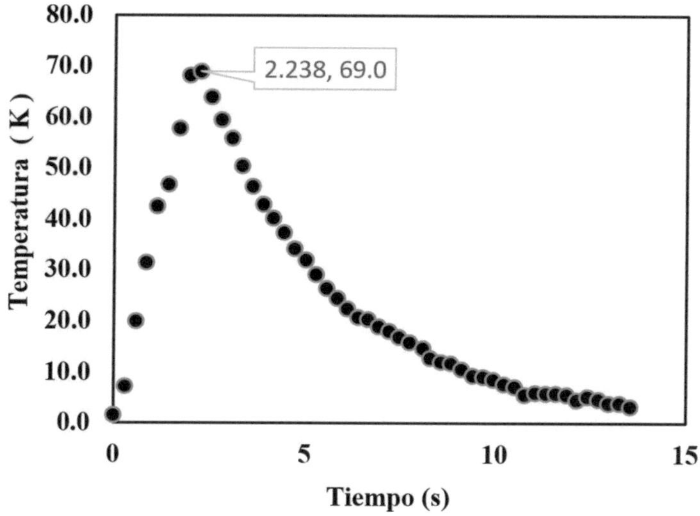

Figure 10.
Complete response of the sample through measurement of ΔT.

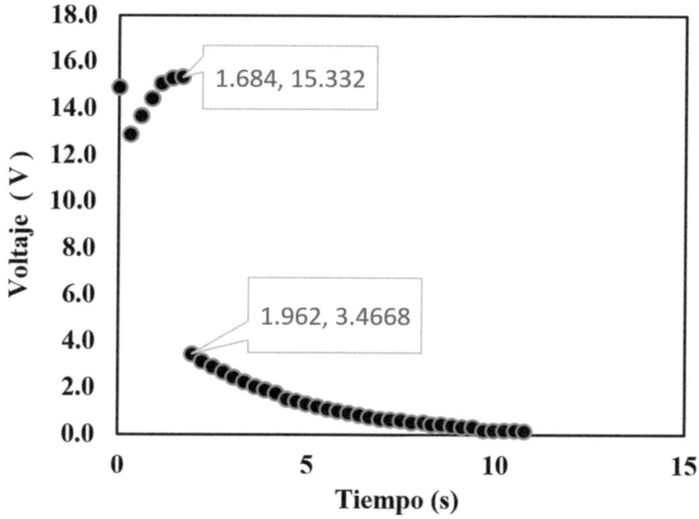

Figure 11.
Complete response of the sample through measurement of V_s *and* V_α.

4.2.2 Natural response test results

The time constant $\tau_{th} = 3.304\, s$ is found experimentally measuring the temperature difference ΔT, when V_s is inactive and is determined considering that the temperature difference at 36.8% of the maximum value of either ΔT or V_α, as shown in **Figures 10** and **11**, respectively. Considering that the natural response concerning the thermoelectric capacitor presents an underdamped response, then the angular frequency related to the thermoelectric material inside of the thermoelectric module sample is $\omega_{th} = 2\pi/\tau_{th} = 2\pi/3.304\, s = 1.901\, rad/s$.

The time constant $\tau_C = 1.370\, s$ is obtained experimentally through the measurement of the cold side temperature T_{cold} and is determined considering that the

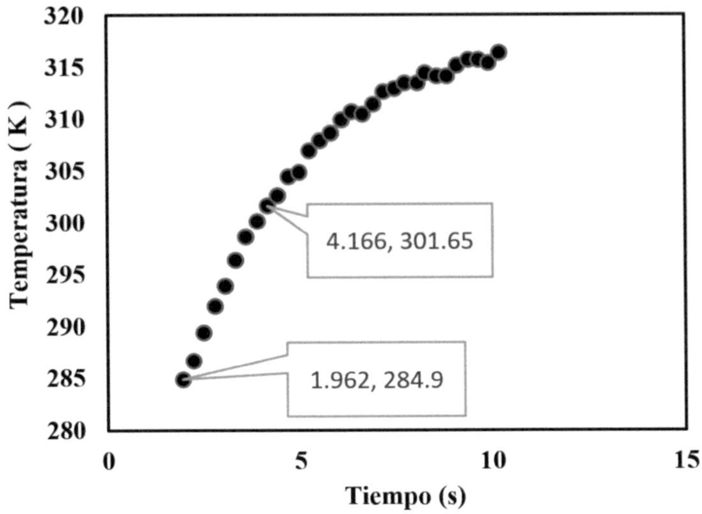

Figure 12.
Natural response of the sample through measurement of T_{cold}.

temperature increases by a factor 1/e (36.8% of the initial value), as shown in **Figure 12**. The angular frequency associated with thermal contacts is $\omega_1 = 1/\tau_C = 1/1.370\ s = 0.729\ rad/s$.

5. Thermoelectric characterization

5.1 Calculation of the thermoelectric coefficients

A complete thermoelectric characterization can be done using the equations found in Section 3 and either simulations or experimental results.

Therefore, combining both forced and natural responses, a complete characterization is made. Considering that at $t = 0$, both $T_{C_{cold}}$ and $T_{C_{hot}}$ must be equal to 300 K; that is, temperature offset errors must be corrected for each thermocouple before starting the test:

$$R_{TE} = \frac{\Delta T}{\alpha \overline{T} I} = \frac{352.75\ K - 283.5K}{(0.0508\ V.K^{-1})(300\ K)(11.518\ A)} = 0.394\ W^{-1}.K$$

$$R_C = \frac{T_{C_{cold}}}{\alpha \overline{T} I_{TEG_{sc}}} = \frac{300\ K - 283.5\ K}{(0.0508\ V.K^{-1})(300K)(2.061\ A)} = 0.525\ W^{-1}.K$$

$$K_0 = \frac{1}{R_{TE}} - \frac{1}{R_C} = \frac{1}{0.394\ \Omega} - \frac{1}{0.525\Omega} = 0.633\ W.K^{-1}$$

$$R_0 = \frac{1}{K_0} = \frac{1}{0.633\ W.K^{-1}} = 1.579\ W^{-1}.K$$

$$R_{C_{cold}} = R_{C_{hot}} = \frac{R_C}{2} = \frac{0.525\ W^{-1}.K}{2} = 0.262\ W^{-1}.K$$

$$K_{C_{cold}} = K_{C_{hot}} = \frac{1}{R_{C_{cold}}} = \frac{1}{R_{C_{hot}}} = \frac{1}{0.262\ W^{-1}.K} = 3.809\ W.K^{-1}$$

$$K_C = \frac{1}{R_C} = \frac{1}{0.525\ W^{-1}.K} = 1.904\ W.K^{-1}$$

$$C_{TE} = \frac{\tau_{TE}}{R_{TE}} = \frac{1.130\ s}{0.394\ W^{-1}.K} = 2.871\ J.K^{-1}$$

$$C_{C_{cold}} = C_{C_{hot}} = \frac{\tau_C}{R_C} = \frac{1.370\ \mathrm{s}}{0.525\ W^{-1}.K} = 2.609\ J.K^{-1}$$

$$C_C = 2C_{C_{cold}} = 2C_{C_{hot}} = 5.218\ J.K^{-1}$$

$$C_{th} = \frac{\tau_{th}}{2\pi R_0} = \frac{3.304\ s}{2\pi(1.579)\ W^{-1}.K} = 0.333\ J.K^{-1}$$

5.2 Calculation of the figure of merit

The figure of merit is obtained using either Eq. (42) and Eq. (1). Also, using other expressions for ZT:

$$ZT = \frac{\Delta T - T_{C_{cold}}}{\Delta T} = \frac{69.25\ K - 16.5\ K}{69.25\ K} = 0.762$$

$$ZT = \frac{\alpha^2 \overline{T}}{R_m K_0} = \frac{\left[\left(0.0508\ V.K^{-1}\right)^2\right](300\ K)}{(1.592\ \Omega)\left(0.633\ W.K^{-1}\right)} = 0.763$$

where R_m ca be obtained using Eq. (36). Also, through the mathematical expression $R_m = (V_{max}/I_{max})[(T_{hot} - \Delta T_{max})/T_{hot}]$ as follows [4, 22]:

$$R_m = (15.9\ V/7.6\ A)[(352.75\ K - 69,25\ K)/352.75\ K] = R_m = 1.592\ \Omega$$

The experimental results presented correspond to the characterization of a thermoelectric module, specifically, Kryotherm TB-127-1.4-1.2 [20], at $\Delta T = 69.25\ K$, in suspended configuration and without being thermally insulated; that is, it is in non-adiabatic conditions, where the environment or reservoir to which the module thermal contacts are exposed is at an ambient temperature of $\overline{T} = 300K$. For temperature measurements, two special K-type thermocouples were used with an error (Special Limits Error) of $+/- 1.1^{\circ}$C or $+/- 0.4\%$. The measurement process used is the methodology proposed by Pirela and Velásquez [9].

6. Conclusion

The unification of the forced response and natural response theories consent to describe the complete response of the thermoelectric modules and materials. The forced response expressed in terms of temperature difference is the exponential increase of the Peltier effect, and in terms of voltage, it is an exponential increase of the Seebeck effect and complies with the theory of first-order electrical circuits. The natural response expressed in temperature is an exponential drop in the temperature difference and satisfies Newton's cooling law, and in terms of voltage, it is an exponential Seebeck voltage drop and complies with second-order electrical circuits theory. Through the complete response, it is possible to perform direct measurement of the figure of merit with exactness and reproducibility. The subsequent computations yield values for the Seebeck coefficient, electrical resistance, and thermal conductance, as well as the famous thermoelectric transport coefficients α, ρ, and κ. Therefore, it is demonstrated that by combining forced response and natural response theory, there are enough fundamentals to develop new characterization methods and enhancement of thermoelectric modules and materials.

Acknowledgements

This science research work has been funded by the author (Ronald Edgar Pirela La Cruz). The author is an Engineering Science Doctor (EngScsD) from the *Universidad Nacional Experimental Politécnica "Antonio José de Sucre"* (UNEXPO), *Vicerrectorado Puerto Ordaz, Estado Bolívar, Venezuela*. Also, He is a professor of the Research and Postgraduate Department at UNEXPO. Currently, he is the Traction System Engineer in charge of fuel cell power pack development for trains powered by hydrogen, in *Alstom Ferroviaria S.p.A., Savigliano, Italy* (e-mail: repirelalc@estudiante.unexpo.edu.ve).

Author details

Ronald Edgar Pirela La Cruz* and Sergio Rafael Velásquez Guzmán
UNEXPO, Puerto Ordaz, Venezuela

*Address all correspondence to: repirelalc@estudiante.unexpo.edu.ve

References

[1] Seebeck TJ. Magnetic polarization of metals and minerals. Abhandlungender Deutschen Akademie der Wissenschafren zu Berlin;**265**:1822

[2] Peltier JC. Nouvelles experiences sur la caloricite des courans electrique. Annales de Chimie et de Physique. 1834; **LVI**:371-386

[3] Thomson W. 4. On a mechanical theory of thermo-electric currents. Proceedings of the Royal Society of Edinburgh. 1857;**3**:91-98[4] Lineykin S, Ben-Yaakov S. Modeling and analysis of thermoelectric modules. IEEE Transactions on Industry Applications. 2007;**43**(2):505-512

[5] Apertet Y, Ouerdane H. Small-signal model for frequency analysis of thermoelectric systems. Energy Conversion and Management. 2017;**149**: 564-569

[6] Harman TC. Special techniques for measurement of thermoelectric properties. Journal of Applied Physics. 1958;**29**(9):1373-1374

[7] Harman TC, Cahn JH, Logan MJ. Measurement of thermal conductivity by utilization of the Peltier effect. Journal of Applied Physics. 1959;**30**(9):1351-1359

[8] Buist RJ. A new method for testing thermoelectric materials and devices. In: 11th International Conference on Thermoelectrics; Vol. 1. Arlington, Texas. 1992

[9] Pirela RE, Velásquez SR. Novel methodology for characterization of thermoelectric modules and materials. Athenea Engineering Sciences. 2024; **5**(15):29-40

[10] Zlatic V, Monnie R. Modern Theory of Thermoelectricity. New York: Oxford University Press; 2014

[11] Alexander CK, Sadiku MN. Fundamentals of Electric Circuits. Vol. 4. New York: McGraw-Hill; 2009

[12] Luttinger JM. Theory of thermal transport coefficients. Physical Review. 1964;**135**(6A):A1505

[13] Pirela RE, Velásquez SR. Forced response of thermoelectric materials and devices. IEEE Latin America Transactions. 2022;**20**(8):2106-2113

[14] Pirela RE, Velásquez SR. Natural response of thermoelectric materials and devices. Athenea Engineering Sciences Journal. 2022;**3**(10):49-62

[15] Apertet Y, Ouerdane H, Glavatskaya O, Goupil C, Lecoeur P. Optimal working conditions for thermoelectric generators with realistic thermal coupling. EPL (Europhysics Letters). 2012;**97**(2):28001

[16] Anonymous. Scala graduum Caloris. Calorum descriptiones & signa. Philosophical Transactions. 1701; **270**(22):824-829

[17] Maruyama S, Moriya S. Newton's law of cooling: Follow up and exploration. International Journal of Heat and Mass Transfer. 2021;**164**:120544

[18] Sun Y, Liu Y, Li R, Li Y, Bai S. Strategies to improve the thermoelectric figure of merit in thermoelectric functional materials. Frontiers in Chemistry. 2022;**10**:865281

[19] Giri K, Wang YL, Chen TH, Chen CH. Challenges and strategies to optimize the figure of merit: Keeping eyes on thermoelectric metamaterials. Materials Science in Semiconductor Processing. 2022;**150**:106944

[20] Co K. Thermoelectric coolers for industrial applications: TB-127-1.0-1.2 and TB-127-1.4-1.2. In: Available in http://www.kryothermtec.com; Saint-Petersburg, Russia. 2023

[21] Ioffe AF. Physics of Semiconductors. London: Infosearch; 1960

[22] Lineykin S, Ben-Yaakov S. Analysis of thermoelectric coolers by a spice-compatible equivalent-circuit model. IEEE Power Electronics Letters. 2005; **3**(2):63-66

Chapter 3

Electron and Phonon Transport Mechanisms in Thermoelectric Materials

Hong Liu and Xinli Cheng

Abstract

Thermoelectric materials can directly convert thermal energy into electrical energy. Although the low efficiency of thermoelectric devices limits their use in the professional market, the development and improvement of condensed matter physics theory, as well as the emergence of various new material synthesis processes, have brought new research and development momentum to the utilization of thermoelectric materials. This chapter first starts from the basic theory of thermoelectricity, introduces the thermoelectric transport equation and the Boltzmann transport equation that controls phonon and electron transport, and then reviews the strategies to further improve the thermoelectric properties of materials in the past two decades. and evaluated the effectiveness and universality of existing strategies, pointing out their limitations. Subsequently, based on the superconducting Bardeen-Cooper-Schrieffer (BCS) theory and phonon drag effect, this chapter proposes a new physical paradigm different from the phonon-glass-electron-crystal method. The core of the new paradigm is to shift from decoupling the electron-phonon transport process to optimizing the electron-phonon coupling mechanism and provides a theoretical mechanism for improving the performance of thermoelectric materials by optimizing the electron-phonon coupling. Identified the material design concepts and challenges for implementing the new paradigm.

Keywords: thermoelectricity, Boltzmann transport equation, physical paradigm, electron-phonon coupling, phonon drag effect

1. Introduction

Thermoelectric technology, characterized by direct thermoelectric conversion, is increasingly recognized as a key component of emerging green energy technologies. In order to make thermoelectric equipment more economical and efficient, we must improve the energy conversion efficiency, which is evaluated by the material quality factor (ZT). $ZT = S^2 \sigma T/\kappa$ [1] depends on conductivity (σ), Seebeck coefficient (S), and thermal conductivity (κ, sum of phonon κ_l and electron κ_e components). ZT enhancement is typically achieved by increasing the power factor ($S^2\ \sigma$) and decreasing κ_l, which correspond to charge carrier transport and phonon propagation, respectively, and which is challenging as these requirements are often contradictory to each other in conventional materials [2].

High-performance thermoelectric materials rely on innovative thermoelectric transport mechanisms. Related strategies include the quantum confinement effect [3], resonant energy levels [4], band convergence [5], liquid ions [6], entropy engineering [7], anharmonicity [8], modulation doping [9], spin orbit coupling (SOC) [10], and Rashba effect [11], which enhance ZT in many thermoelectric material systems. However, the research progress on enhancing thermoelectric material ZT seems to have only partial and minor success. In order to further promote the development of thermoelectric materials, exploring the paradigm shift mechanism is crucial. This chapter will start from the basic theory of thermoelectricity and explore new thermoelectric transport mechanisms, highlighting the main physical parameters and related description methods that affect electron and phonon transport processes, providing new prospects for manipulating thermoelectric properties. Firstly, briefly introduce the basic principles of thermoelectric devices and materials (Section 2), and then summarize the current status of the optimization mechanism of thermoelectric transport in thermoelectric materials (Section 3). Then, we will discuss the thermoelectric transport mechanism in bulk materials, as well as the physical paradigm for synergistically reducing the thermal conductivity and improving the power factor of bulk materials (Section 4), and finally present the conclusions of this chapter (Section 5).

2. Thermoelectric transport

2.1 Transport equation in thermoelectric materials

The electrical and thermal transport processes in thermoelectric materials are not independent but interact with each other. Essentially, the charges in semiconductors and conductors carry away heat during their movement. The interrelated current density J_e and heat flux density J_q can be expressed as [12, 13]

$$J_e = L_{11}\left(-\frac{d\Phi}{dx}\right) + L_{12}\left(-\frac{dT}{dx}\right) \tag{1}$$

$$J_q = L_{21}\left(-\frac{d\Phi}{dx}\right) + L_{22}\left(-\frac{dT}{dx}\right) \tag{2}$$

Consider one-dimensional transport, where L_{11} is the conductivity and Φ is the potential, which is the electrochemical potential per unit charge. In an isothermal conductor, heat flux can be expressed as

$$J_q = \frac{L_{21}}{L_{11}} J_e = \Pi J_e \tag{3}$$

Proportional to the current, the proportionality constant is the Peltier coefficient Π. Although the Peltier coefficient is an inherent material property, as shown in **Figure 1**, due to the imbalance of Peltier heat between the inflow and outflow junctions, Peltier cooling and heating occur when two materials with different Peltier coefficients are connected together. Control volume analysis shows that the cooling or heating Q occurring at the junction is equal to

Figure 1.
Cooling or heating at the junctions of two materials occur because of the different Peltier coefficients of the two materials, which can be used as thermoelectric refrigerators or heat pumps.

$$Q = (\Pi_2 - \Pi_1)I \tag{4}$$

In Eq. (4), I is the current, and Π_1 and Π_2 represent the Peltier coefficients of the two materials. According to the definition of the Seebeck coefficient, taking $J_e = 0$ in Eq. (1), the Seebeck coefficient can be obtained as

$$S = \frac{-d\Phi/dx}{dT/dx} = -\frac{V}{\Delta T} = \frac{L_{12}}{L_{11}} \tag{5}$$

As shown in Eq. (5), the Seebeck voltage V between two points on a uniform material is independent of temperature distribution. Thermocouples measure temperature based on this material characteristic.

The Seebeck coefficient and the Peltier coefficient are not independent but interrelated and are expressed as $\Pi = TS$, this is also a generalized Onsager reciprocity relationship $L_{21} = TL_{12}$.

Furthermore, one can eliminate the electrochemical potential in Eq. (1) and express the heat current as

$$J_q = \Pi J_e - \kappa_e \frac{dT}{dx} \tag{6}$$

where the electronic contribution to thermal conductivity is

$$\kappa_e = L_{22} - \frac{L_{12}L_{21}}{L_{11}} \tag{7}$$

The transport coefficient can be further correlated with the electronic band structure of the material. Based on the Linearized Boltzmann Transport Equation (LBTE) formalism under the relaxation time approximation, the transport coefficient of isotropic materials with a single energy band can be expressed as [14].

$$L_{11} = -\frac{e^2}{3}\int v^2\tau \frac{\partial f^0}{\partial E} N(E)dE \tag{8}$$

$$L_{12} = \frac{e^2}{3T}\int v^2\tau\left(E - E_f\right)\frac{\partial f^0}{\partial E} N(E)dE \tag{9}$$

$$L_{22} = -\frac{1}{3T}\int v^2\tau\left(E - E_f\right)^2\frac{\partial f^0}{\partial E} N(E)dE \tag{10}$$

Among them, τ is the relaxation time, N (E) is the density of electronic states per unit volume per unit energy interval, f^0 is the Fermi Dirac distribution in equilibrium state, and E_f is the electrochemical potential based on doping modulation. According to Eqs. (5), (8), and (9), the Seebeck coefficient can be expressed as

$$S = -\frac{1}{eT}\frac{\int v^2\tau(E-E_f)\frac{\partial f^0}{\partial E}N(E)dE}{\int v^2\tau\frac{\partial f^0}{\partial E}N(E)dE} \tag{11}$$

According to Eqs. (1) and (2), considering a virtual process that does not generate temperature gradients, we can obtain:$\frac{J_q/T}{J_e} = S$, this means that the Seebeck coefficient is the entropy current carried by a unit current density. Moreover, temperature gradient (phonon non-equilibrium distribution) can generate non-uniform distribution of charge carriers and form potential gradient, and similarly, potential gradient (charge carrier non-equilibrium distribution) can lead to non-uniform distribution of phonons and form temperature gradient. Explanation of the evolutionary mechanisms for phonon non-equilibrium distribution and charge carrier non-equilibrium distribution in sections 2.2 and 2.3.

2.2 Boltzmann transport equation for phonon

The non-zero heat flux J can be generated by a temperature gradient of ∇T. Within the linear range, the component of thermal conductivity κ is determined by the following equation $J^\alpha = -\sum_\beta \kappa^{\alpha\beta}(\nabla\mathrm{T})^\beta$. $\boldsymbol{J}$ can be obtained through the following methods

$$\boldsymbol{J} = \sum_p \int f_\lambda \hbar\omega_\lambda \boldsymbol{v}_\lambda \frac{d\boldsymbol{q}}{(2\pi)^3} \tag{12}$$

Among them, λ is the phonon mode identifier, including phonon branching index p and wave vector q, ω_λ and $\boldsymbol{v}_\lambda$ are the angular frequency and group velocity of phonon mode λ, respectively, and f_λ is the corresponding phonon distribution function. In the state of thermal equilibrium without thermodynamic forces, the distribution of phonons is determined by the Bose-Einstein statistical distribution function $f_0(\omega_\lambda)$. When a temperature gradient of ∇T occurs, the distribution function f deviates from f_0, and the deviation can be obtained by iteratively solving the Boltzmann transport equation. The variation of the phonon distribution function comes from diffusion caused by ∇T and scattering caused by related mechanisms. When reaching steady state, the time rate of change of the distribution function is 0; The corresponding relationship can be expressed as [15]:

$$\frac{df_\lambda}{dt} = \frac{\partial f_\lambda}{\partial f}\Big|_{diffusion} + \frac{\partial f_\lambda}{\partial f}\Big|_{scattering} \tag{13}$$

where,

$$\frac{\partial f_\lambda}{\partial f}\Big|_{diffusion} = -\nabla T \bullet \boldsymbol{v}_\lambda \frac{\partial f_\lambda}{\partial T} \tag{14}$$

and $\frac{\partial f_\lambda}{\partial f}\Big|_{scattering}$ depending on the relevant scattering process, analyze and solve using perturbation theory. The phonon scattering in single crystal materials mainly comes from phonon-phonon scattering and impurity scattering such as isotopes. In addition, the size of actual block samples is always limited to the extent that boundary scattering dominates at very low temperatures.

In most cases, the norm of ∇T is small enough that f_λ can be expanded to the first-order term of ∇T, that is $f_\lambda = f_0(\omega_\lambda) + g_\lambda$, where g_λ depends on ∇T linearly. We can choose to express this dependence as $g_\lambda = -F_\lambda \nabla T\, df_0/dT$. When the scattering process is two phonons and three phonons, the linearized Boltzmann equation is [16]:

$$F_\lambda = \tau_\lambda^0\,(\boldsymbol{v}_\lambda + \Delta_\lambda) \tag{15}$$

In Eq. (15), τ_λ^0 is the relaxation time of the mode λ obtained from perturbation theory, which is commonly used in relaxation time approximation (RTA). Setting all Δ_λ to zero in Eq. (15) is equivalent to working in RTA. Therefore, the dimension of Δ_λ with velocity is a measure of the degree to which a population of specific phonon modes (and associated heat flux) deviates from RTA predictions. The calculation formula for Δ_λ and τ_λ^0 is

$$\Delta_\lambda = \frac{1}{N}\sum_{\lambda'\lambda''}^{+} \Gamma^{+}_{\lambda\lambda'\lambda''}(\xi_{\lambda\lambda''}\mathbf{F}_{\lambda''} - \xi_{\lambda\lambda'}\mathbf{F}_{\lambda'}) + \frac{1}{N}\sum_{\lambda'\lambda''}^{-}\frac{1}{2}\Gamma^{-}_{\lambda\lambda'\lambda''}(\xi_{\lambda\lambda''}\mathbf{F}_{\lambda''} - \xi_{\lambda\lambda'}\mathbf{F}_{\lambda'}) + \frac{1}{N}\sum_{\lambda'}\Gamma_{\lambda\lambda'}\xi_{\lambda\lambda'}\mathbf{F}_{\lambda'} \tag{16}$$

$$\frac{1}{\tau_\lambda^0} = \frac{1}{N}\left(\sum_{\lambda'\lambda''}^{+}\Gamma^{+}_{\lambda\lambda'\lambda''} + \sum_{\lambda'\lambda''}^{-}\frac{1}{2}\Gamma^{-}_{\lambda\lambda'\lambda''} + \sum_{\lambda'}\Gamma_{\lambda\lambda'}\right) \tag{17}$$

We discretize the Brillouin zone (BZ) into a regular grid of N=N1 × N2 × N3 q-points centered on Γ and use the abbreviation $\xi_{\lambda\lambda'} = \omega_{\lambda'}/\omega_\lambda$. In $\sum^{\pm}$ summation, the conservation of quasi-momentum requires $\mathbf{q}'' = \mathbf{q} \pm \mathbf{q}' + \mathbf{Q}$, which maps $\mathbf{q}''$ to the first Brillouin zone through translational reciprocal-lattice vector $\mathbf{Q}$. The symbols used here are slightly different from those used by some authors [17] in previous publications, where group velocity was not included in Δ_λ.

The quantity $\Gamma^{\pm}_{\lambda\lambda'\lambda''}$ in Eq. (17) is the scattering rate in the three-phonon process, which can be expressed as

$$\Gamma^{+}_{\lambda\lambda'\lambda''} = \frac{\hbar\pi}{4}\frac{f_0' - f_0''}{\omega_\lambda\omega_{\lambda'}\omega_{\lambda''}}\left|V^{+}_{\lambda\lambda'\lambda''}\right|^2\delta(\omega_\lambda + \omega_{\lambda'} - \omega_{\lambda''}) \tag{18}$$

$$\Gamma^{-}_{\lambda\lambda'\lambda''} = \frac{\hbar\pi}{4}\frac{f_0' + f_0'' + 1}{\omega_\lambda\omega_{\lambda'}\omega_{\lambda''}}\left|V^{-}_{\lambda\lambda'\lambda''}\right|^2\delta(\omega_\lambda - \omega_{\lambda'} - \omega_{\lambda''}) \tag{19}$$

To simplify the representation, f_0' refers to table $f_0(\omega_{\lambda'})$, and so on. $\Gamma^{+}_{\lambda\lambda'\lambda''}$ corresponds to the absorption process, where two phonons combine into one phonon, corresponding to energy $(\omega_\lambda + \omega_{\lambda'} = \omega_{\lambda''})$. The conservation of energy during absorption and emission is limited by the Dirac delta distribution in Eqs. (18) and (19). In order to calculate the value of $\Gamma^{\pm}_{\lambda\lambda'\lambda''}$, it is necessary to provide the scattering matrix element $V^{\pm}_{\lambda\lambda'\lambda''}$, which is given by [16].

$$V^{\pm}_{\lambda\lambda'\lambda''} = \sum_{i\in u.c}\sum_{j,k}\sum_{\alpha\beta\gamma}\Phi^{\alpha\beta\gamma}_{ijk}\frac{e^{\alpha}_{\lambda}(i)e^{\beta}_{p',\pm q'}(j)e^{\gamma}_{p'',-q''}(k)}{\sqrt{M_iM_jM_k}} \tag{20}$$

these in turn depend on the normalized eigenfunctions $e_{p,q}$ of the three phonons involved and on the anharmonic IFCs $\Phi_{ijk}^{\alpha\beta\gamma} = \frac{\partial^3 E}{\partial r_i^\alpha \partial r_j^\beta \partial_k^\gamma}$. In Eq. (20), i,j, and k represent atomic indices, while α, β, and γ represent Cartesian coordinates. The summation index i only traverses the atoms in the central unit cell, while j and k cover the entire system. r_i^α and M_i represent the alpha component of the displacement from the equilibrium position and the mass of the i-th atom, respectively. $e_\lambda^\alpha(i)$ is the alpha component of the mode λ eigenfunction at the i-th atom.

The above analysis process is based on particle models to calculate the diffusion and scattering mechanisms of phonons, without involving the wave characteristics of interface adaptation. The phonon transport mechanism in nanostructures is of great significance for the precise customization of thermal properties. By combining phonon particles and wave effects through different strategies, the study aims to achieve ultra-low or ultra-high thermal conductivity in various nanostructures. However, phonon particles and wave effects are coupled together, so their respective contributions to phonon transport have not yet been calculated clearly. It is generally believed that the wave effect is the main modulator that blocks phonon transport, and the particle effect is so weak that it can be ignored, but this may not be the case in reality.

In addition, the interaction between electrons and phonons can also affect the transport properties of phonons. The above analytical process only used the potential energy surface of electrons when calculating the interaction force constant using density functional perturbation theory (DFPT). The influence of electron-phonon interaction on phonon transport will be introduced in Section 4.3.2.

2.3 Boltzmann transport equation for electron

Assuming a solid with a temperature gradient, it can be inferred that the chemical potential at the hot end is lower than that at the cold end. Adding an electron requires less energy, and the electron naturally migrates toward the hot end. This provides another explanation for the thermoelectric phenomenon, where a thermal gradient can generate voltage without a net current. Because electrons are governed by two equilibrium forces: one is the so-called "thermal" force that pushes electrons from the cold end to the hot end, and the other is the opposite so-called "electric" force. Therefore, the thermoelectric phenomenon can be considered as a result of the temperature dependence of chemical potential, which is itself a result of the Fermi-Dirac distribution thermal broadening. **Figure 2** depicts the semiclassical picture of the thermoelectric effect.

The Boltzmann equation, a method historically used to calculate the transport coefficients of bulk materials in the Drude model [13, 15, 18].

The core concept in Boltzmann transport theory is the non-equilibrium distribution function $f_k(\boldsymbol{r})$, defined as the number of electrons per unit volume near the $(\boldsymbol{r}, \boldsymbol{k})$ phase space (in fact, the time, energy band, and spin are also specified, which have been omitted for simplicity). In the case of thermal equilibrium, where the temperature is uniform and there is no external field effect, the distribution function and equilibrium distribution function f^0 agrees.

When there is an external field, the distribution function will change due to the acceleration of electrons by the field (the wave vector distribution is no longer symmetrical); In addition, the distribution function may depend on location (non-uniformity in spatial distribution, such as temperature gradients, density gradients,

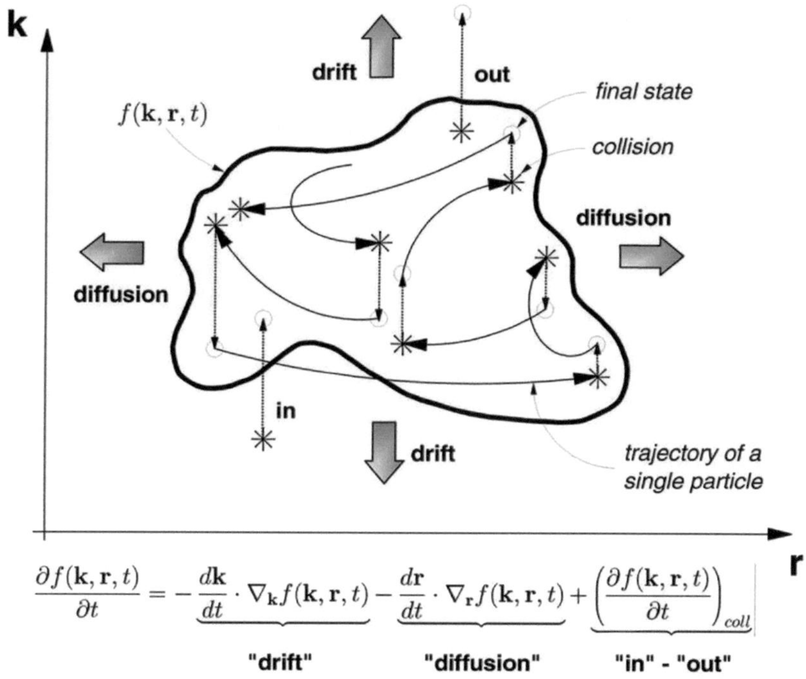

Figure 2.
The semi classical picture of thermoelectric effect.

etc.), which can occur when electrons have a tendency to diffuse; The third effect that causes changes in the distribution function is scattering—electrons interact with phonons, other electrons, impurities, defects, and surfaces. Therefore, the transport process can be attributed to the influence of external fields (electric fields, magnetic fields, etc.), diffusion, and collision scattering on the electron distribution function, leading to deviations from its relative Fermi-Dirac distribution, and the overall results of these three effects should cancel each other out.

Electron group velocity $\boldsymbol{v_k}$ as

$$\boldsymbol{v_k} = \frac{\partial \boldsymbol{\omega_k}}{\partial \boldsymbol{k}} = \frac{\mathbf{1}}{\hbar} \frac{\partial \boldsymbol{\varepsilon_k}}{\partial \boldsymbol{k}} \tag{21}$$

Current can be written as

$$J_e = \int \boldsymbol{v_k e f_k}(r) \boldsymbol{dk} \tag{22}$$

Energy flow can be written as

$$J_U = \int \boldsymbol{v_k \varepsilon f_k}(r) \boldsymbol{dk} \tag{23}$$

The heat flow (the amount of heat carried by electrons) is

$$J_q = J_U - \mu J_N = \int \boldsymbol{v_k}(\boldsymbol{\varepsilon} - \boldsymbol{\mu}) \boldsymbol{f_k}(r) \boldsymbol{dk} \tag{24}$$

Consider the near-equilibrium state (local equilibrium assumption and the time scale of interest Δt satisfy $\tau_{local} \ll \Delta t \ll \tau_{system}$). Based on relaxation time approximation:

$$\frac{df_k(\boldsymbol{r})}{dt}|_{scattering} = \frac{f_k(r) - f^0}{\tau} \tag{25}$$

The diffusion term caused by the temperature gradient is

$$\frac{df_k(\boldsymbol{r})}{dt}|_{Temp.grad} = \frac{d\boldsymbol{r}}{dt} \bullet \frac{\partial f_k(\boldsymbol{r})}{\partial \boldsymbol{r}} = \boldsymbol{v}_k \frac{\partial f_k(\boldsymbol{r})}{\partial \boldsymbol{r}} \approx \boldsymbol{v}_k \bullet \frac{\partial f^0}{\partial \boldsymbol{T}} \nabla T \tag{26}$$

The approximation in the above equation assumes that the influence of the temperature gradient on the distribution function is mainly due to the thermal broadening of the Fermi-Dirac distribution. Between two scattering events, electrons drift under the influence of thermal gradients.

Similarly, the influence of the external electric field on the distribution function can be expressed as

$$\frac{df_k(\boldsymbol{r})}{dt}|_{Elec.field} = \frac{d\boldsymbol{k}}{dt} \bullet \frac{\partial f_k(\boldsymbol{r})}{\partial \boldsymbol{k}} = \frac{e}{\hbar} \mathbf{E} \bullet \frac{\partial f_k(\boldsymbol{r})}{\partial \boldsymbol{k}} \approx e\boldsymbol{v}_k \bullet \frac{\partial f^0}{\partial \mathcal{E}_k} \mathbf{E} \tag{27}$$

Therefore, the linear Boltzmann equation can be written as

$$\boldsymbol{v}_k \bullet \frac{\partial f^0}{\partial \boldsymbol{T}} \nabla T + e\boldsymbol{v}_k \bullet \frac{\partial f^0}{\partial \mathcal{E}_k} \mathbf{E} + \frac{f_k(r) - f^0}{\tau} = 0 \tag{28}$$

The deviation of the local distribution function from the steady state comes from two disturbances, which are called drift terms. As shown in **Figure 3**, small changes in

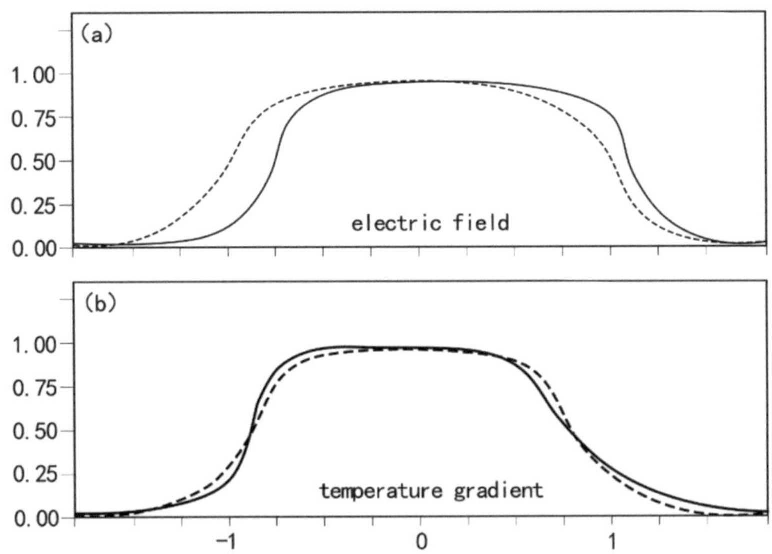

Figure 3.
The distribution function in the presence of an electric field and temperature gradient (the gray line represents the equilibrium distribution).

energy will shift the distribution function, while an increase in temperature will widen the distribution. Therefore, applying an electric field will cause electrons moving in one direction to have greater energy than electrons moving in the opposite direction. On the other hand, in the presence of a thermal gradient, the distribution on the cold side is steeper, and due to the difference in distribution width at both ends of the sample, there is a so-called "heat" that pushes electrons from the cold end to the hot end, which has a significant impact on the physical properties of the system.

According to Eqs. (22) and (28), the conductivity of the system without a thermal gradient is

$$J_e(\nabla T = 0) = \int \left(\boldsymbol{v_k} \boldsymbol{e} f^0 - e^2 \tau \boldsymbol{v}_k^2 \frac{\partial f^0}{\partial \boldsymbol{\mathcal{E}_k}} \mathbf{E} \right) \mathbf{dk} \tag{29}$$

There is no current in thermal equilibrium, so the first term on the right-hand side of the Eq. (29) is integrated to zero, so the conductivity is expressed as

$$\sigma = -\int e^2 \tau(k) \boldsymbol{v}_k^2 \frac{\partial f^0}{\partial \boldsymbol{\mathcal{E}_k}} dk \tag{30}$$

Similarly, in the absence of an electric field, the current density generated by the temperature gradient is

$$J_e(\mathbf{E} = 0) = -\int e \tau \boldsymbol{v}_k^2 \frac{\partial f^0}{\partial T} \nabla T \mathbf{dk} \tag{31}$$

The electrical conductivity generated by heat is derived as follows:

$$\sigma = \int e \tau(k) \boldsymbol{v}_k^2 \frac{\partial f^0}{\partial \boldsymbol{T}} dk \tag{32}$$

Eqs. (30) and (32) clearly demonstrate the k-dependence of relaxation time. How do these two equations tell us the relationship between conductivity generated by electric fields and heat? How are the two related to the Fermi-Dirac distribution? How are the two related to electron velocity and relaxation time? The latter two quantities vary depending on the material and have strong anisotropy.

As can be seen from the above, the conductivity is always positive, while the conductivity generated by heat can be positive or negative. Because $\frac{\partial f^0}{\partial \boldsymbol{\mathcal{E}_k}} = -\frac{1}{k_B T} \left(f^0\right)^2 \exp\left(\frac{\varepsilon - \mu}{k_B T}\right)$ is always negative, $\frac{\partial f^0}{\partial \boldsymbol{T}} = -\frac{\varepsilon - \mu}{T} \frac{\partial f^0}{\partial \boldsymbol{\mathcal{E}_k}}$ depends on $(\varepsilon - \mu)$.

Based on the above discussion, the carriers that play the main role in the two transport processes are different, as shown in **Figure 4**. In the case of charge transport (conductivity), electrons with energy exactly at the chemical potential contribute the most to conductivity (i.e., conductivity mainly depends on the contribution of electrons near the Fermi level); in the case of thermoelectric transport, these carriers do not contribute at all to the thermal conductivity, and the entire signal is determined by electrons with energy slightly higher or lower than the chemical potential, and the contributions of these two parts have opposite signs.

Similar to the above discussion and utilizing the properties of the Fermi-Dirac distribution function, the expression for heat flux can be derived as

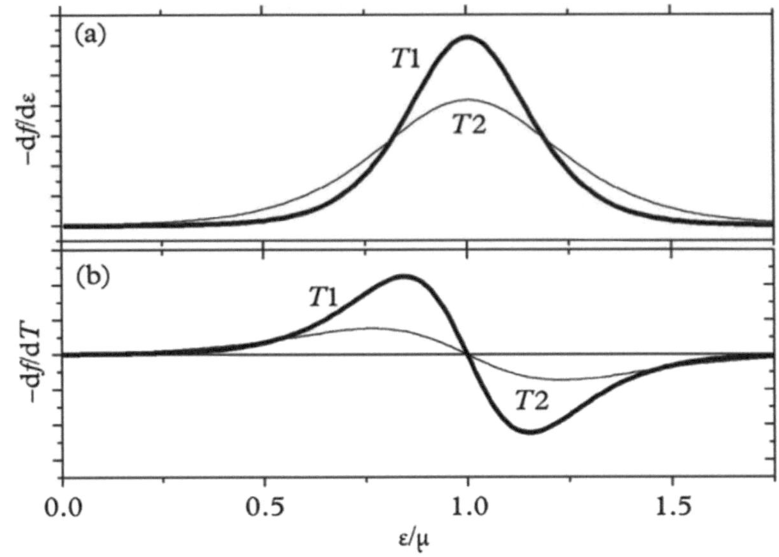

Figure 4.
Two types of derivatives of Fermi distribution, temperature T_1 is less than T_2.

$$J_q = \int \mathbf{v_k}(\boldsymbol{\varepsilon} - \boldsymbol{\mu})\boldsymbol{\tau}\mathbf{v_k} \bullet \left[\frac{(\boldsymbol{\varepsilon} - \boldsymbol{\mu})}{\boldsymbol{T}} \frac{\partial f^0}{\partial \mathcal{E}_k} \nabla T - e \frac{\partial f^0}{\partial \mathcal{E}_k} \mathbf{E} \right] \mathbf{dk} \tag{33}$$

According to the above, the transport coefficients corresponding to Eqs. (8), (9), and (10) can be derived.

The Boltzmann equation is a product of the classical particle Newtonian mechanics model and the quantum mechanics model of energy state transitions. If all coherence effects are ignored and simplified, the Boltzmann equation can be derived from the quantum transport model. The classical transport theory is based on the Boltzmann transport theory, and the basic assumptions of Boltzmann theory include: (i) Electrons and holes are both tiny particles; (ii) The particles are independent of each other, without coherence, and interact with each other through scattering; (iii) Particles can be described using Bloch theory; (iv) Scattering is a transient behavior without temporal and spatial persistence; (v) Only consider the scattering between two particles, without taking into account the combined effects between multiple particles. The Boltzmann transport equation is an equation used in the distribution function method, which is an equation satisfied by the non-equilibrium distribution function $f(k, r, t)$. Solving this equation can obtain $f(k, r, t)$ under different conditions, and then various transport parameters of electrons can be obtained.

The Boltzmann transport equation takes into account the velocity distribution of charge carriers and the directionality of scattering, making it relatively accurate. In the presence of external fields such as electric fields or temperature gradients, the Boltzmann equation can be established based on the drift changes caused by the distribution function due to external fields such as electric fields, magnetic fields, and temperature gradients, as well as the changes caused by scattering. Since the scattering term should be an integral of the scattering probability, the Boltaman equation is a differential integral equation. The solution of this equation is very complex, and approximation methods are usually used. One commonly used approximation method is the relaxation time approximation.

The Boltzmann equation is a high-dimensional equation with three-dimensional wave vector space (k), three-dimensional real space (r), and one-dimensional time (t), which is difficult to solve and is commonly simulated using Monte Carlo methods.

The difference in the description of heat-carrying phonons and conducting electrons by the Boltzmann transport equation is only reflected in the difference in statistical laws and the dynamic forces of distribution functions deviating from the equilibrium state. Both recover to equilibrium through scattering mechanisms, while phonons and electrons always have mutual influence through the correlation of material structure and environment. Therefore, co-modulating the transport process of both is the fundamental strategy to improve the performance of thermoelectric materials. The fourth section of this chapter will discuss in detail the relevant modulation paradigms.

3. The current status of the optimization mechanism of thermoelectric transport in thermoelectric materials

3.1 Strategy for enhancing Seebeck coefficient

3.1.1 Energy band engineering

Based on the quantum confinement effect of low-dimensional materials on band structure [19, 20], the band structure of materials can be modulated by alloying or chemically doping bulk materials (such as forming resonance energy levels and generating band convergence), thereby improving thermoelectric performance.

3.1.1.1 Resonance energy levels

Mahan and Sofo [21] believe that electronic systems with narrow energy distributions during transport can maximize thermoelectric performance. The so-called resonant energy level is called the "virtual bound state" [22–24], similar to the required delta-shaped energy level proposed by Mahan and Sofo [21]. Unlike doping in the bandgap to form electronic energy levels, resonant dopants will generate electronic energy levels located above the conduction band edge or below the valence band edge as shown in **Figure 5** [25], which will significantly alter the electronic density of states of the host material. The state of a resonant energy level is characterized by the energy of the resonant energy level (E_D) and the width of the resonant state (Γ).

The Bethe-Sommerfeld expansion for the Seebeck coefficient can be expressed as [25].

$$S = \frac{\pi^2}{3}\frac{k_B}{e}(k_BT)\left[\frac{1}{n(E)}\frac{dn(E)}{E} + \frac{1}{\mu(E)}\frac{d\mu(E)}{dE}\right] \tag{34}$$

Clearly, the Seebeck coefficient depends on the density of states and the carrier scattering mechanism. Resonance energy levels can provide additional energy states to the main band and increase the effective band mass (m_b), thereby increasing the density-of-states effective mass (m_d^*) [4], where $m_d^* = N_V^{2/3}m_b$ and N_V is the band degeneracy. Additionally, carrier scattering can generate a carrier filtering effect [26, 27].

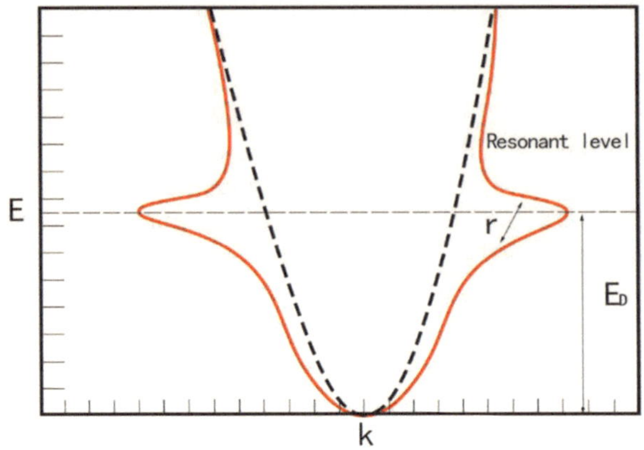

Figure 5.
Schematic view of the resonant level. Reproduced with permission from reference [25]. Copyright of The Royal Society of Chemistry 2012.

3.1.1.2 Energy band convergence

According to the relationship between the band degeneracy and density-of-states effective mass ($m_d^* = N_V^{2/3} m_b$), enhancement in density-of-states effective mass can also be realized by improving the band degeneracy. In principle, when there is no band extremum or the energy difference can be ignored in multiple bands, band degeneracy will increase. In addition, when multiple carrier pockets in the Brillouin zone degenerate, high band degeneracy can be obtained because they are symmetrically equivalent due to the symmetry of the crystal.

3.1.2 Carrier filtering effect

According to the Mott relation, the Seebeck coefficient can be expressed as [27].

$$S = \frac{-1}{e\sigma} \int \frac{\varepsilon - \mu}{T} \sigma(\varepsilon) \left[\frac{-\partial f^0(\varepsilon)}{\partial \varepsilon} \right] \tag{35}$$

In Eq. (35), ε is the energy of the carrier, μ is the chemical potential, $\sigma(\varepsilon)$ is the differential conductivity, and $f^0(\varepsilon)$ is the Fermi distribution function. Furthermore, as can be seen from the equation, the contribution signs of charge carriers with energy less than the chemical potential and charge carriers with energy greater than the chemical potential to the Seebeck coefficient are opposite. As shown in **Figure 6**, low-energy electrons in n-type SiGe alloys have a negative impact on the Seebeck coefficient (green area). Therefore, the absolute value of the Seebeck coefficient decreases due to this negative effect [27]. Here, if preferential scattering on low-energy carriers (i.e., carrier filtering effect) can be achieved, the Seebeck coefficient can be improved [28]. Due to the fact that the carrier filtering effect essentially manipulates the carrier scattering mechanism, the scattering term in Eq. (34) indicates the enhancement of the Seebeck coefficient caused by the carrier filtering effect.

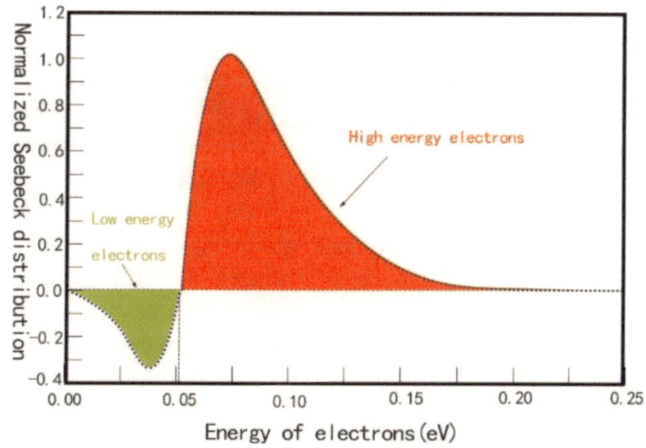

Figure 6.
Relationship between the energy of electrons and the normalized Seebeck distribution for heavily doped n-type SiGe alloy. Reproduced with permission from reference [28]. Copyright of The Royal Society of Chemistry 2009.

3.2 Mobility enhancement

In the case of elastic scattering, the carrier mobility of semiconductors with simple parabolic energy bands is [29].

$$\mu = \frac{e\langle\tau\rangle}{m^*} \tag{36}$$

where $\langle\tau\rangle$ is the average relaxation time and m^* is the effective mass. In simple terms, relaxation time can be regarded as a power function of carrier energy E, temperature T, and effective mass m^*, as shown below

$$\tau \propto E^r T^s m^{*t} \tag{37}$$

From Eqs. (36) and (37), it can be seen that the mobility depends on the band structure (i.e., m^*) and the carrier scattering mechanism (i.e., $\langle\tau\rangle$). Therefore, an effective strategy to improve carrier mobility is to regulate the effective mass and carrier scattering mechanism.

3.2.1 Effective mass engineering

Because the charge carriers are mainly scattered by acoustic phonons, Snyder pointed out that the nondegenerate mobility in multi-valley conduction can be expressed as [30].

$$\mu = \frac{2^{3/2}\pi^{1/2}\hbar^4 e C_1}{3m_I (m_b k_B T)^{3/2} \Xi^2} \tag{38}$$

where m_I is the inertial effective mass, C_1 is the composite elastic constant, and Ξ is the deformation potential coefficient. Hence, higher carrier mobility requires smaller band effective mass and inertial effective mass [30].

3.2.2 Manipulation of carrier scattering mechanism

When acoustic phonon waves propagate in a crystal, they cause compression and expansion of the local lattice. Therefore, it can cause disturbances in the band potential energy, leading to the scattering of charge carriers, known as acoustic phonon scattering [31]. Due to the constant existence of acoustic phonon scattering,

It can determine the upper limit of carrier mobility. Generally speaking, acoustic phonon scattering is one of the largest intrinsic scattering mechanisms in semiconductors, especially at high doping levels and high temperatures. According to Eq. (38), the deformation potential coefficient is a measure of the strength of electron-phonon interactions and a key parameter for evaluating thermoelectric performance. Low deformation potential coefficient of weak electron-phonon coupling leads to high carrier mobility [32–34]. However, the physical source of low deformation forces in materials is still unclear. The latest research indicates that the physical essence of low-strain materials is non-bonding orbitals that make significant contributions to electron transport [35].

3.2.3 Selecting doping sites

Based on the rigid band approximation, it is generally believed that dopants only change the Fermi level and carrier concentration. In the case of complete ionization of impurities, different dopants achieve the same effect. However, experiments have found that differences in dopants and doping sites may lead to variations in carrier mobility at high carrier concentrations.

The dopant does not alter the band structure to maintain the effectiveness of the rigid band approximation, which can be represented by the carrier concentration dependence of the Seebeck coefficient. Therefore, the difference in mobility should correspond to different measurements of scattering. Of course, there must be scattering of charge carriers by ionized impurities that generate charge carriers. However, when different dopants have the same valence, complete ionization, and random distribution, the ionization impurity scattering theory equivalently treats them. Therefore, different degrees of scattering should not be attributed to ionizing impurity scattering.

3.3 Carrier concentration optimization

The main thermoelectric parameters that determine the quality factor of thermoelectric materials are all related to carrier concentration and therefore heavily coupled with each other. It can be seen that the most effective strategy to improve thermoelectric performance is to optimize the carrier concentration. From a historical perspective, tuning of carrier concentration has been achieved through intentional introduction of exogenous impurities through chemical doping. In this way, the carrier concentration remains a constant value before bipolar conduction occurs, almost independent of temperature. However, the optimal carrier concentration (n_{opt}) is highly dependent on temperature, so thermoelectric performance is only optimized within a limited temperature range.

3.4 Phonon engineering

Given the relative independence of lattice thermal conductivity from other thermoelectric parameters, reducing lattice thermal conductivity is considered the most direct means of improving thermoelectric performance. The dynamic theoretical expression for the thermal conductivity of three-dimensional materials is [36].

$$k(\Lambda) = \frac{1}{N_k} \sum_{k\lambda}^{\Lambda_{k\lambda} < \Lambda} \frac{1}{3} v_{k\lambda} \Lambda_{k\lambda} C_{vk\lambda} \tag{39}$$

where v, Λ, C_v are the phonon group velocity, phonon mean free path, and specific heat, respectively; λ labels a phonon band index (or branch), k refers to a point in the first Brillouin zone (with a k-point mesh), and N_k is the number of points. The mean free path of phonons is determined by multiple scattering mechanisms. Assuming that different scattering mechanisms are independent of each other, the Matthiessen's rule can be used to estimate the phonon mean free path.

$$\Lambda^{-1} = \sum \Lambda_i^{-1} \tag{40}$$

The latest progress in simulating phonon transport from scratch has greatly deepened our understanding of the thermal conduction mechanism of thermoelectric materials [37]. At the same time, the relationship between lattice thermal conductivity and material parameters was clarified. The low lattice thermal conductivity is attributed to the low sound velocity, large cell volume, long bond length (i.e., weak bonding), and large Grüneisen parameters (i.e., strong anharmonic).

3.4.1 Phonon-phonon scattering

The phonon-phonon interaction mechanism is that the first phonon causes periodic elastic strain (through anharmonic interaction), which adjusts the elastic constant of the crystal in space and time. The second phonon perceives the modulated elastic constant and scatters it, producing a third phonon, known as the three-phonon process [38].

3.4.1.1 Umklapp scattering

When the total momentum (K) of the phononic gas remains constant (K1 + K2 = K3), the three-phonon collision process is a normal process (or N-process). The phonon energy during the N process is only redistributed from the incident phonon to the excited phonon, without generating any thermal resistance. **Figure** 7 **(a)** and **(c)** indicate the presence of phonon net flow between the phonon source and phonon sink during the N process.

There is another case of three-phonon collision, which requires an additional reciprocal lattice vector (**G**) to maintain total momentum conservation ($K_1 + K_2 = K_3 + G$). This collision process is called the Umklapp process (or U process). Due to the fact that the wave vector K describing the phonon state can only be located in the first Brillouin zone, any wave vector K generated outside the first Brillouin zone during a collision must be moved to the first Brillouin zone by adding G (**Figure** 7**(d)**). This leads to net phonon backscattering during the U-process, resulting in thermal resistance (**Figure** 7**(b)**). Based on the physical mechanism of the U-process, the lattice thermal conductivity of materials can be limited at high temperatures, with a temperature dependence of $\kappa_{lat} \sim T^{-1}$.

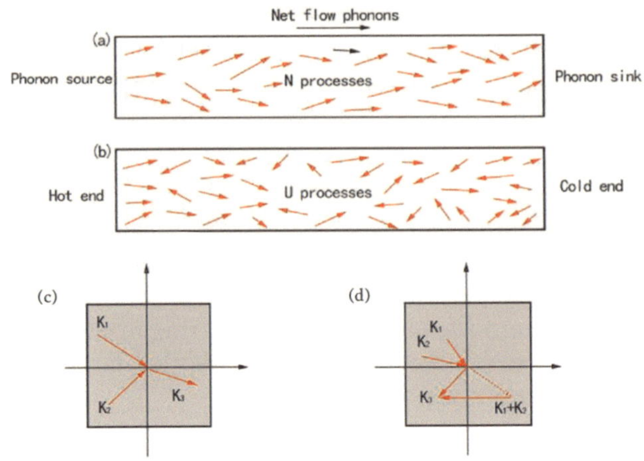

Figure 7.
N process (a) and U process (b) for the phonon gas. Phonon collision in a two-dimensional square lattice for (c) N process and (d) U process. Reproduced with permission from reference [38]. Copyright of 1976 by John Wiley & Sons, Inc.

3.4.1.2 Anharmonic effect

The theory of harmonic approximation may lead to some conclusions that are not consistent with reality: there is no thermal expansion (the equilibrium position of atoms does not depend on temperature), the force constant and elastic constant do not depend on temperature and pressure, the heat capacity at high temperatures is constant (Dulong Petit law), the isochoric heat capacity and isobaric heat capacity are equal ($C_V = C_p$), there is no interaction between phonons, the average free path and lifetime of phonons are infinite, and the thermal conductivity of harmonic crystals without impurities and defects is infinite; In the expansion of the potential energy of crystal atom interactions, terms of the third power and above are called non-harmonic terms. Many physical effects are caused by nonharmonic terms, such as vibration softening due to a decrease in the restoring force constant and scattering between phonons due to no longer independent vibration modes (phonons). Therefore, increasing the nonharmonic effect can be a beneficial strategy for limiting lattice thermal conductivity.

3.4.1.3 Acoustic and optical phonon coupling

Acoustic phonons: Low-frequency phonons formed by stimulated Brillouin scattering during the interaction between a light field and a medium. In acoustic mode, the crystal undergoes overall motion at a low frequency and can be regarded as a continuous medium elastic wave. Optical phonons: In crystals, atoms or ions vibrate around equilibrium positions, forming optical phonons. The frequency of optical mode is close to that of electromagnetic waves, making it easy to couple with electromagnetic fields. Acoustic phonons mainly contribute to lattice thermal conductivity due to their high group velocity. Based on the physical nature of the two types of phonons, significant thermal resistance can also be generated by mediating the coupling between optical phonons and acoustic phonons through relevant strategies.

3.4.1.4 Rattling modes

Phonon softening is an effective method to reduce lattice thermal conductivity. Designing lattice void structures filled with weakly bonded heavy atoms can significantly reduce lattice thermal conductivity, known as rattling modes. Slack and Tsoukala proposed the concept of phonon scattering for this type of rattling [39].

3.4.2 Phonon scattering by electrons

Acoustic phonon scattering is one of the important mechanisms of electron scattering. Due to the possible interaction between electrons and phonons, electron-phonon scattering is possible. Ziman was the first to study electron-phonon scattering [18]. Charge carriers are usually achieved through doping. As the impurity concentration increases, the wave functions of charge carriers overlap. When the overlap reaches a certain degree, it can be considered that the carriers are in a quasi-free state, forming an unfilled impurity band. The carriers in the band can collide with phonons and scatter phonons [40].

3.4.3 Scattering of phonons by microstructural defects

Figure 8 [41] shows that when the microstructure of solid materials contains point defects, dislocations, stacking faults, twin boundaries, grain boundaries, and nanostructures, it can distort the perfect crystal structure and hinder the propagation of phonons. The success of phonon engineering largely relies on shortening the phonon mean free path by introducing microstructural defects [41]. Based on different lattice environments and system temperatures, the average free path distribution of phonons ranges from a few angstroms to several hundred nanometers. The effective scattering of phonons of different wavelengths requires the design of defects with corresponding characteristic sizes [42–44]. Phonon scattering based on manipulating microstructural defects has been extensively studied and proven to be highly effective in minimizing lattice thermal conductivity and improving thermoelectric performance.

3.4.3.1 Phonon scattering by point defects

Point defects such as dopants, alloy atoms, isotopes, vacancies, and gaps can effectively scatter high-frequency phonons due to their smaller characteristic sizes. When the phonon wavelength is greater than the size of the point defect, the point defect will strongly scatter phonons. Without considering resonance or multiple scattering effects, the phonon relaxation time $\langle\tau\rangle$ is related to the phonon frequency ω [45, 46].

$$\frac{1}{\tau_{PD}} = \phi c^2 \omega^2 \qquad (41)$$

Where ϕ is the fractional concentration of point defects, and c is a contrast parameter such as mass or strain.

Atomic defects contribute to perturbation energy in three ways: (a) changes in kinetic energy due to differences in mass; (b) Potential energy generated by changes in the elastic constants of atomic bonds; (c) Potential energy caused by an elastic strain field that changes the elastic constants of atomic bonds [47, 48]. Callaway [49, 50], Klemens [51], and Abeles [48, 52] reported a theoretical study on the effect of point defects on lattice thermal conductivity.

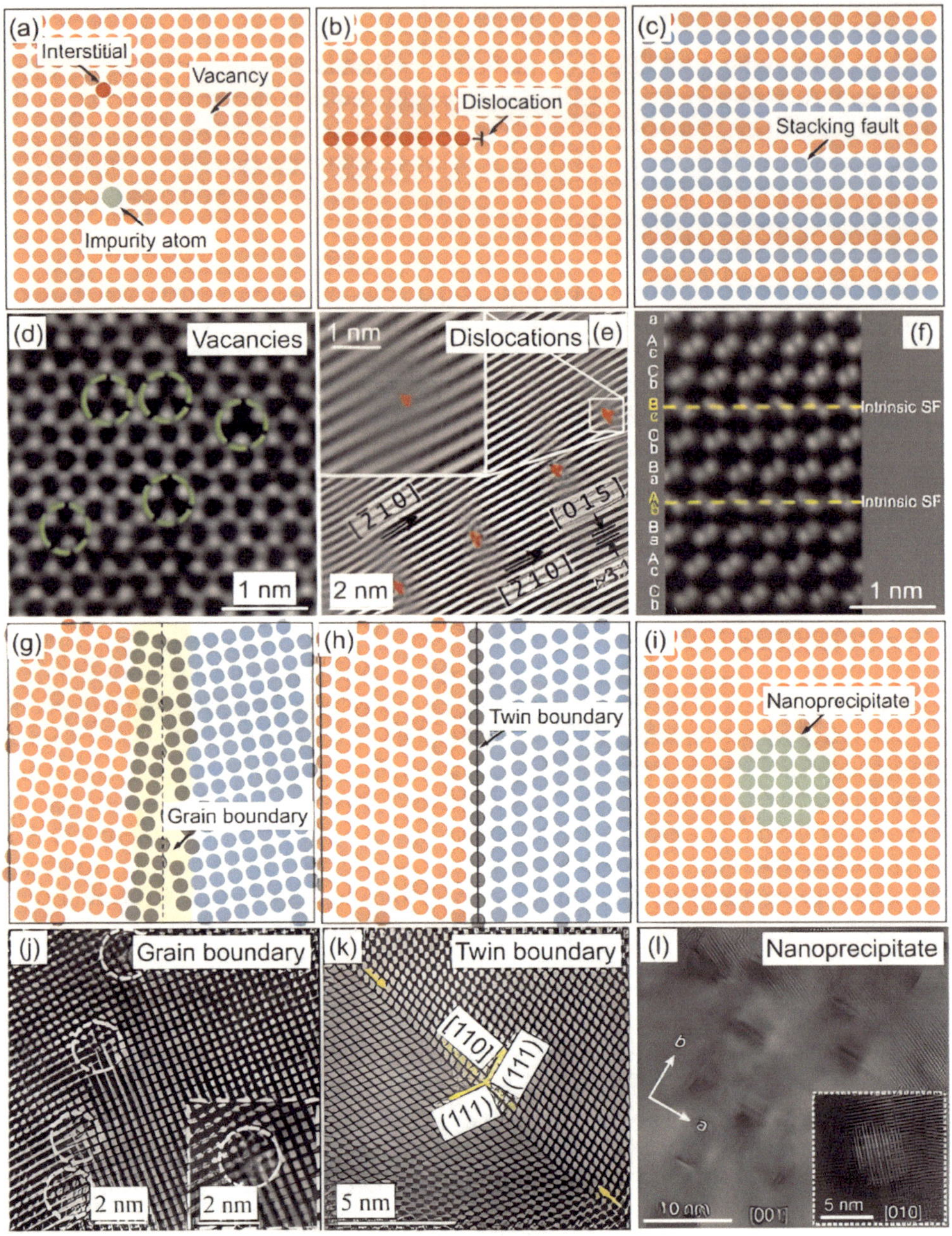

Figure 8.
Schematic views for (a) point defects, (b) dislocation, (c) stacking fault, (g) grain boundary, (h) twin boundary, and (i) nanoprecipitates. The rest of panels are the corresponding TEM images for the microstructural defect. Reproduced with permission from reference [41]. Copyright of 2018 Materials Research Society.

3.4.3.2 Phonon scattering by grain boundaries

For infinite lattices, the thermal resistance of insulating crystals rapidly decreases with decreasing temperature [53]. When the temperature of a given crystal with a finite size is below its corresponding critical temperature, its phonon free path is the size of the crystal. Once this situation occurs, the scattering of phonons by crystal boundaries will generate additional resistance [54]. The Casimir theory [54] provides the diffuse reflection of phonons from sample boundaries as the primary mechanism for achieving phonon equilibrium at very low temperatures. The relaxation time of boundary scattering can be expressed as

$$\frac{1}{\tau_{GB}} = \frac{v}{L} \tag{42}$$

where L is the average dimension of the sample.

3.4.3.3 Phonon scattering by dislocations

Dislocations are composed of dislocation lines and dislocation nuclei, and the crystal structure near the dislocation line is a finite strain, so phonons will be scattered when passing near the dislocation. The scattering degree of phonons by dislocation cores is much higher than that of phonons by dislocation stress fields. Dislocations can also move under the influence of phonons, and when dislocation oscillations occur, phonons will be re-emitted [55]. Klemens pointed out that both the type of dislocation and the dislocation line scatter phonons in the direction of the temperature gradient. The relaxation time of scattered phonons from dislocation lines perpendicular to the temperature gradient is [47].

$$\frac{1}{\tau_{D,line}} = C_1 \frac{b^2}{a^2} \omega \tag{43}$$

In Eq. (43), C_1 is a constant, b is the magnitude of the Burgers vector, and a is the lattice parameter. The relaxation time of phonon scattering by dislocation nuclei is expressed as

$$\frac{1}{\tau_{D,core}} = C_2 a^2 \omega^3 \left(\frac{\Delta M}{M} \right)^2 \tag{44}$$

where C_2 is a constant, M is the atomic mass of the unit cell, and ΔM is the mass difference.

3.4.3.4 Scattering of phonons by nanoparticles

In the synthesis process of bulk materials, nanoparticles and nanoprecipitate-embedded structures can be obtained through synergistic modulation of processes and components to reduce the thermal conductivity of the material. A. Shakouri provided the relaxation of phonon scattering by nanoparticles as [56].

$$\frac{1}{\tau_{NP}} = v \left(\sigma_s^{-1} + \sigma_1^{-1} \right)^{-1} \rho \tag{45}$$

$$\sigma_s = 2\pi R^2 \tag{46}$$

$$\sigma_1 = \pi R^2 \frac{4}{9} \left(\frac{\Delta D}{D} \right)^2 \left(\frac{\omega R}{v} \right)^4 \tag{47}$$

Where ρ and R are the density and radius of nanoparticles, respectively, and D and ΔD are the density of the matrix material and the density difference between nanoparticles and matrix material, respectively.

In summary, various modulations of material microstructures can be obtained by modulating atomic and molecular structures, creating nanostructures and multi-scale structures in TE materials to optimize electron and phonon transport. Various modulation methods can improve material properties, but the performance improvement is limited, and all methods also lack universality. In addition, all the material performance optimization strategies mentioned above are independently applied to phonon and electron transport. The actual situation is that the transport of electrons and phonons in materials is inseparable and permanently coupled. Therefore, exploring the optimized coupling of electrons and phonons to significantly improve the performance of thermoelectric materials may be an effective strategy and a good paradigm to break through the upper limit of the coefficient of merit of thermoelectric materials.

4. Physical paradigm for synergistically reducing the thermal conductivity and improving the power factor

According to the definition of the TE quality factor, high ZT values originate from material systems with high conductivity, a large Seebeck coefficient, and low thermal conductivity. In the past few decades, traditional methods for optimizing TE performance have relied on electron-phonon transport separation, electronic band engineering, and phonon scattering mechanism modulation. However, even at various optimization levels, only moderate ZT values can usually be obtained. This section explores the physical paradigm of synergistic electro/phonon transport based on the mechanism of electron-phonon interaction, starting from the phonon drag effect and superconducting BCS theory.

4.1 BCS theory overview

Experimental measurements of the electromagnetic properties of superconductors tell us that the charge of charge carriers in superconductors is twice that of electrons, implying that electrons always pair up. Secondly, superconductors can theoretically be effectively described by a "macroscopic wave function," which is a characteristic of boson systems at low temperatures. A composite particle composed of two fermions (spin half integer) has an integer spin, while a boson has an integer spin. It seems that a composite particle composed of two fermions is a boson. But in fact, this understanding is naive: the analysis under the second quantization system denies this point; however, the paired two fermions are somewhat similar to bosons, and superconductors do indeed exhibit Cooper pair condensation (BEC) at low temperatures, similar to boson condensation.

The measurement of the specific heat of superconductors at low temperatures tells us that superconductivity has an energy gap—to restore the superconducting state to normal through heating, a certain energy gap needs to be overcome, just like the existence of an energy gap in a semiconductor/insulator; the latter is non-conductive and does not match the conductivity of superconductors. Therefore, based on the above situation, a mechanism is needed to change the electronic system originally distributed to the Fermi surface at low temperatures so that some electrons pair up to form approximate bosons and form macroscopic condensation. At the same time, there should be an energy gap between the condensed paired electrons and the normal electron states to maintain this state and prevent it from being easily destroyed by external energy such as heating.

Bardeen, Cooper, and Schrieffer, three physicists, presented their superconducting theory: based on the interaction between electrons and phonons, they reasonably explained the above phenomenon.

Electron-phonon interaction can lead to effective attraction between electrons. Phonons indicate that the atomic nucleus has deviated from its equilibrium position, and the displacement can be represented by the phonon generation/annihilation operator:

$$\vec{u_i} = \sum_{\vec{q}} \left(\frac{\hbar}{2M\omega_{\vec{q}}} \right)^{1/2} \left(\hat{a}^{\dagger}_{\vec{q}} + \hat{a}_{-\vec{q}} \right) e^{i\vec{q}\vec{R_i}} \tag{48}$$

Where $\vec{q}$ represents the momentum of phonons in momentum space.

The change in potential energy generated by a small deviation of the atomic nucleus is

$$\delta V_{tot} = \sum_i \frac{\partial V\left(\vec{x_e} - \vec{R_i}\right)}{\partial R_i} \bullet \vec{u_i} \tag{49}$$

The Hamiltonian of electron-phonon interaction is

$$\hat{H}_{el-ph} = V_{tot} + \delta V_{tot} \tag{50}$$

In order to obtain an effective theory of the interaction between electrons, the path integral/partition function of the electron field and phonon field (coupling) can be considered, and the phonon field can be directly integrated to obtain the part where only electrons remain; it is also possible to simply consider the scattering of electrons and phonons—represented by Feynman diagrams (**Figure 9**). Just as exchanging photons between electrons creates an effective interaction potential—Coulomb

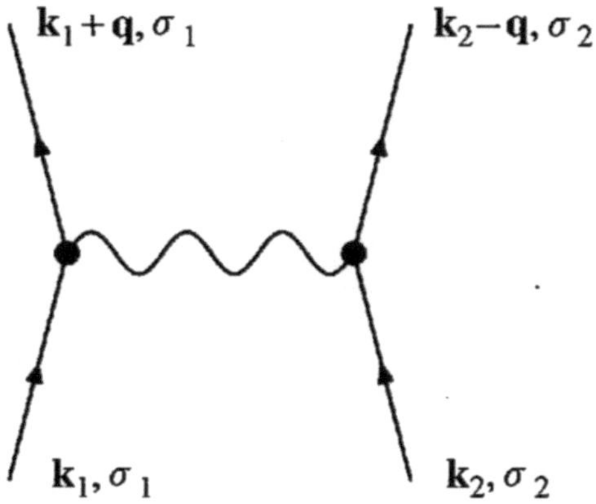

Figure 9.
Interaction of fermions via exchange of a boson.

potential—exchanging phonons between electrons also creates an effective potential energy.

Obtain the indirect interaction potential between electrons caused by phonons, which in momentum space is

$$V_{eff}\left(\vec{q},\vec{k}\right) = \left|g_{\vec{q}}\right|^2 \frac{\hbar\omega_{\vec{q}}}{\left(\epsilon_{\vec{k}} - \epsilon_{\vec{k}+\vec{q}}\right)^2 - \left(\hbar\omega_{\vec{q}}\right)^2} \tag{51}$$

When $\left|\epsilon_{\vec{k}} - \epsilon_{\vec{k}+\vec{q}}\right| < \omega_{\vec{q}}$, the sign of $V_{eff}\left(\vec{q},\vec{k}\right)$ is negative, which means an attractive force.

4.2 Phonon drag effect

Phonon drag effect refers to the fluctuation of crystal atomic density caused by the propagation of sound waves (longitudinal waves) in a crystal, forming additional potential field waves with the same period as the sound waves. In semiconductors, this potential field wave manifests as a small distorted periodic potential field; in piezoelectric semiconductors, it manifests as a large periodic piezoelectric potential field. Sound waves are essentially long acoustic waves of lattice vibration quantized into phonons, and the resulting potential field waves can be seen as the effect of phonons—phonon potential field waves.

When the average free path of an electron is less than the wavelength of the acoustic wave, the electron will be frequently scattered by phonons and captured by the valley of the phonon potential field wave. At this point, as the sound wave valley, electrons are dragged by the phonon potential field wave, resulting in the phonon drag effect, as shown in **Figure 10**. The phonon potential field in piezoelectric semiconductors has a larger amplitude, leading to a more significant traction effect. Ultrasonic waves can also generate a larger amplitude periodic potential field, thereby enhancing the drag effect on electrons.

The phonon drag effect causes an uneven distribution of electrons in semiconductors, where electrons are pulled forward by phonons, which inevitably leads to a greater concentration of electrons to one side, resulting in the generation of an electromotive

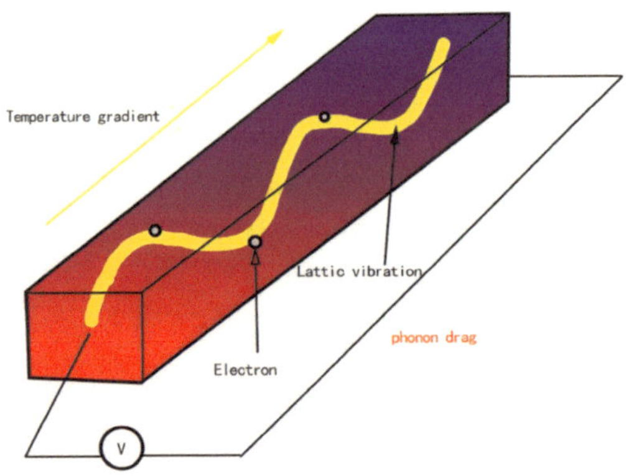

Figure 10.
Schematic diagram of phonon drag effect.

force. When an electric field acts on electrons, causing them to accelerate, the phonon drag effect should affect the propagation of sound waves. When the electron drift velocity is greater than the sound wave velocity, the electron pushes the phonon potential field wave, which enhances the sound wave energy and forms the basis of the ultrasonic amplifier. On the contrary, when the drift velocity of electrons is lower than that of sound waves, the energy of sound waves is transferred to electrons, resulting in a decrease in the amplitude of sound waves, that is, sound wave attenuation.

4.2.1 Non-diffusive thermoelectric transport caused by phonons (similar effects can also occur with magnons/spin waves)

Consider testing the Seebeck coefficient: apply a temperature gradient to the solid and measure the potential difference under an open circuit. The lattice waves generated by lattice vibrations have a certain frequency spectrum and propagate at a certain speed in the crystal. During the propagation process, thermal energy can be carried from the hot end to the cold end, resulting in a non-uniform distribution of phonons. This means that the scattering rate of charge carriers traveling along the temperature gradient is not equal to the scattering rate of charge carriers traveling in the opposite direction, resulting in a net current in the direction of the temperature gradient. However, the actual open circuit state cannot maintain charge flow, so the system has to establish a finite electric field to compensate for the excess scattering experienced by charge carriers traveling in one of the two directions. The additional electric field generated is the contribution of phonon drag to the Seebeck coefficient, in addition to the electron diffusion response.

Consider the Peltier test: measure the ratio of energy flow to current along the sample without any temperature gradient. There is a net current along the sample, which causes the scattering rates of electrons and phonons moving in each other's direction to be different from those in the opposite direction. As a result, the phonon distribution is no longer balanced, ultimately producing an additional energy flow in addition to the energy flow purely brought by electrons, which is the contribution of phonon drag to the Peltier coefficient.

4.2.2 Phonon drag theory in semiconductors (Herring theory)

For semiconductors, phonon drag is quantitatively described from the perspective of phonons. Herring equates the rate at which a phonon system gains crystal momentum from electrons with the rate at which it loses crystal momentum through U-processes and boundary scattering [57].

$$\left.\frac{d\mathbf{P}}{dt}\right|_e = f^{el.} ne\mathbf{E} \tag{52}$$

Here, P is the crystal momentum per unit volume, $f^{el.}$ is the proportion of electron momentum loss (with a value of 1 in the absence of impurity scattering), and $ne\mathbf{E}$ is the force per unit volume, where $\mathbf{E}$ is the electric field and n is the carrier concentration.

Assuming the relaxation time of electron-phonon scattering is τ_{ph-e}, we have

$$\left.\frac{d\mathbf{P}}{dt}\right|_e = \frac{\mathbf{P}}{\tau_{ph-e}} \tag{53}$$

The energy flux density j_q caused by crystal momentum is

$$\mathbf{J}_q = \mathbf{P}\overline{v}_s^2 \tag{54}$$

Where $\overline{v}_s$ is the average speed of sound, so

$$\mathbf{J}_q = \overline{v}_s^2 \tau_{ph-e} f^{el.} ne\mathbf{E} = \overline{v}_s^2 \tau_{ph-e} f^{el.} \frac{\mathbf{J}_e}{\mu} \tag{55}$$

Based on Eq. (6), the contribution of phonon drag to the Peltier coefficient is

$$\Pi_g = f^{el.} \frac{\overline{v}_s^2 \tau_{ph-e}}{\mu} \tag{56}$$

By using the Kelvin relationship, the thermoelectric potential generated by phonon drag can be obtained as

$$S_g = \frac{1}{T} f^{el.} \frac{\overline{v}_s^2 \tau_{ph-e}}{\mu} \tag{57}$$

The Herring scene discussed above has many simplifications. Electron-phonon coupling is viewed from the perspective of phonons, and their flow carries energy as a simple reaction to their collision with electrons.

4.3 A new paradigm for enhancing thermoelectric quality factors

4.3.1 Discussion on the mechanism of electron-phonon interaction for improving the performance of thermoelectric materials

In recent decades, the improvement of thermoelectric material performance in the scientific community has been based on the strategy of electron crystal-phonon glass. The specific implementation strategy is mostly to design various methods to decouple the correlation between electron and phonon transport. However, the actual performance improvement is limited. Although some material systems are based on modulation mechanisms that can greatly improve material performance, the stability of materials during thermoelectric operation has not been fully verified.

The most concerned thermoelectric semiconductor material in the industry at present is undoubtedly bismuth telluride polar semiconductor. Fröhlich interaction is an interaction between electrons and longitudinal optical phonons (LO phonons) in polar semiconductor materials. This interaction generates a macroscopic electric field, leading to long-range coupling between electrons and phonons, forming a quasiparticle called a polaron. The strength of Fröhlich interaction is measured by a dimensionless Frohlich coupling constant. In polarized semiconductors, due to the coupling between polarons and LO phonons, charge carriers transfer their initial kinetic energy to the lattice through a highly efficient relaxation mechanism.

According to Eqs. (1, 2), the thermal conductivity (open circuit), κ, under zero current conditions and the thermal conductivity (short circuit) under zero potential gradient can be obtained as follows:

$$\kappa|_{J_e=0} = \frac{1}{T^2} \frac{L_{11}L_{22} - L_{12}L_{21}}{L_{11}} \tag{58}$$

$$\kappa'|_{E=0} = \frac{1}{T^2} L_{22} \tag{59}$$

By using Eqs. (8–10) and defining thermal inducing electric conductivity as $\alpha = \frac{e}{T^2} L_{12}$, we can obtain

$$\kappa' = \kappa \left(1 + \frac{\alpha^2 T}{\sigma \kappa}\right) \tag{60}$$

Where $\frac{\alpha^2 T}{\sigma \kappa}$ is the thermoelectric figure of merit. Increasing thermal-induced electric conductivity can improve thermoelectric figure of merit (constructing an optimization strategy based on Eq. (32)). This also provides a new perspective for the design of high-performance thermoelectric materials.

Based on BCS theory and the phonon drag effect, we can design a multi-element thermoelectric structure material with uniformly distributed nanocrystals. However, one should always keep in mind that the presence of nanoinclusions might adversely affect the conducting carriers. To achieve this, we must face the following issues:

Firstly, the design of nanostructures is based on the preferential scattering of phonons carrying heat while maintaining the integrity of the charge carrier transport path. In order to match the band structure of carrier transport, nanoinclusions or nanoscale defects must have only a small band edge offset relative to the band edge of the matrix [58]. In n-type nanocomposites, the conduction band edges of the nanoinclusions and matrix should be closely matched, with an energy difference generally not exceeding a few $k_B T$, to minimize electron scattering at the interface potential barrier. Experimental and theoretical studies have shown that the interface and boundary morphology in nanostructured materials should meet the requirement of atomic coherence or alignment at very small angles to maintain high carrier mobility [59].

Secondly, based on Eq. (51) in BCS theory, phonon exchange can achieve the attractive potential energy of electrons, thereby optimizing carrier transport. The open questions are how to modulate the electron distribution near the Fermi surface and ensure that the energy change of electrons after exchanging phonons is less than the corresponding phonon energy.

Thirdly, design material structures based on the phonon drag effect to achieve efficient coupling between long wave acoustic phonons and electrons. The open questions are how to achieve effective transfer of electron-phonon momentum and large electron-phonon interaction relaxation time.

Fourthly, the calculation of interface phonon modes and the revelation of the interaction mechanism between interface phonons and electrons. Nano-embedded materials or bulk materials containing nano-inclusions inevitably contain a large number of grain boundaries, and interface phonon modes will inevitably affect electron transport and energy exchange. A deep understanding of the interaction mechanism between interface phonon modes and electrons is the basis for constructing high-performance thermoelectric materials.

Fifthly, constructing nanostructured phononic metamaterials and exploring the quantitative relationship between phonon fluctuations and particle effects on thermal conductivity in phononic metamaterials. Research on the modulation mechanism of thermal conductivity by local resonance hybridization based on phonon fluctuation, and obtain the physical basis for synergistic regulation of phonon fluctuation and particle transport.

4.3.2 Electron-phonon coupling mechanism and its strategy in optimizing the properties of thermoelectric materials

Nanostructured materials have achieved great success. The basic strategy is to decouple the correlation between electrons and phonons and introduce clusters or microstructures that generate multiple scattering to phonons to reduce material thermal conductivity, but they are gradually running out of power in further reducing thermal conductivity, and the lattice deformation potential caused by lattice vibration always interacts with electrons. It is becoming increasingly evident that any further benefits of ZT must now come from improvements in power factor. The current research on improving power factor aims to identify materials with good band structure characteristics, such as resonance states and low-dimensional "class" features in bulk materials, or band structure engineering, such as band convergence strategies. In fact, most promising TE materials have complex band structures with multiple energy valleys and bands that extend throughout the entire Brillouin zone (BZ). In summary, instead of forcibly decoupling the electron-phonon correlation, it is more advantageous to use a new paradigm of electron-phonon correlation to optimize the key properties of thermoelectric materials. Of course, it is computationally challenging to utilize the entire phonon and electron dispersions for full transport calculations; thus, it is common to employ reasonable approximations.

In periodic solids, the electron-phonon coupling effect is usually discussed in terms of the Hamiltonian [60].

$$\hat{H} = \hat{H}_e + \hat{H}_{ph} + \hat{H}_{e-ph} \tag{61}$$

Based on the Fock space, the generation annihilation operator can be established, and the Hamiltonian of fermions can be obtained as

$$\hat{H}_e = \sum_{nk} \varepsilon_{nk} \hat{c}^{\dagger}_{nk} \hat{c}_{nk} \tag{62}$$

describes noninteracting quasiparticles with crystalline momentum **k**, band index n, and dispersion $\varepsilon_{n\mathbf{k}}$, while

$$\hat{H}_{ph} = \sum_{q\nu} \omega_{q\nu} \left(\hat{a}^{\dagger}_{q\nu} \hat{a}_{q\nu} + \frac{1}{2} \right) \tag{63}$$

is the Hamiltonian associated with noninteracting phonons with wave vector q, mode index ν, and frequency $\omega_{q\nu}.\hat{c}^{\dagger}_{nk}$ and $\hat{c}_{nk}$ ($\hat{a}^{\dagger}_{q\nu}$ and $\hat{a}_{q\nu}$) are fermionic (bosonic) creation and annihilation operators. Finally,

$$\hat{H}_{e-ph} = \frac{1}{\sqrt{N_p}} \sum_{\substack{k,q \\ mnv}} g_{mnv}(k,q) \hat{c}^{\dagger}_{mk+q} \hat{c}_{nk} \left(\hat{a}_{q\nu} + \hat{a}^{\dagger}_{-q\nu} \right) \tag{64}$$

Described the first-order coupling of atomic displacement, where N_p is the number of unit cells in the Born von Karman supercell, and $g_{mnv}(k,q)$ is the element of the e-ph coupling matrix. In principle, Eq. (61) should include the so-called Debye Waller

(DW) Hamiltonian, which gives the second-order coupling of atomic displacement [60]. This term makes a significant contribution to the real part of self-energy (SE) and therefore must be included when calculating quasi-particle correction and temperature-dependent band structures. However, we will mainly focus on the imaginary part of e-ph SE, so the DW term will be ignored [60].

The relationship between the multibody Hamiltonian in Eq. (61) and density functional theory (DFT) was established by using the Kohn Sham (KS) band structure in Eq. (62) and the phonons in density functional perturbation theory (DFPT) in Eq. (63) [60]. Finally, the e-ph matrix element, $g_{mnv}(k,q)$, is computed using

$$g_{mnv}(k,q) = <\psi_{mk+q}|\Delta_{q\nu}V^{KS}|\psi_{nk}> \tag{65}$$

where ψ_{nk} is the KS Bloch state and $\Delta_{q\nu}V^{KS}$ is the first-order variation of the self-consistent KS potential,

$$\Delta_{q\nu}V^{KS} = \frac{1}{\sqrt{2\omega_{q\nu}}}\sum_{p\kappa\alpha}\frac{\partial V^{KS}}{\partial \Gamma_{\kappa\alpha}}\frac{e_{\kappa\alpha,\nu}(q)}{\sqrt{M_\kappa}}e^{i_q * R_p} \tag{66}$$

where $e_{\kappa\alpha,\nu}(q)$ is the α-th Cartesian component of the phonon eigenvector for the atom κ in the unit cell, M_κ its atomic mass, R_p are lattice vectors identifying the unit cell p,and τ_k the ion position in the unit cell. Following the convention of Ref. [60], V^{KS} consists of contributions from the Hartree part (V^H), the exchange-correlation (XC) potential (V^{xc}), and the bare pseudopotential term that, in turn, consists of the local (V^{loc}) and nonlocal (V^{nl}) parts. We can write Eq. (66) in terms of a lattice-periodic function $\Delta_{q\nu}V^{KS}$ defined as:

$$\Delta_{q\nu}V^{KS} = e^{i\mathbf{qr}}\Delta_{q\nu}v^{KS} \tag{67}$$

where,

$$\Delta_{q\nu}v^{KS} = \frac{1}{\sqrt{2\omega_{q\nu}}}\sum_{\kappa\alpha}\frac{e_{\kappa\alpha,\nu}(q)}{\sqrt{M_K}}\partial_{\kappa\alpha,q}v^{KS} \tag{68}$$

and,

$$\partial_{\kappa\alpha,q}v^{KS} = \sum_p e^{-i\mathbf{q}(\mathbf{r}-\mathbf{R_p})}\left.\frac{\partial V^{KS}}{\partial \Gamma_{\kappa\alpha}}\right|_{(\mathbf{r}-\mathbf{R_p})} \tag{69}$$

It is the first derivative of the KS potential (Kohn Sham single electron interaction potential), which can be obtained by self-consistently solving the Sternheimer equation system using DFPT under a given $(\kappa\alpha, q)$ perturbation [61].

According to the theoretical description of e-ph interactions mentioned above, e-ph interactions can modulate the transport characteristics of electrons. The electron lifetime due to the e-ph interactions is related to the inverse of the imaginary part of the e-ph Fan-Migdal SE (energy level correction under the interaction of electrons and phonons) [60]; the diagonal matrix elements of the SE in the KS basis set are given by

$$\sum_{nk}^{FM}(\omega,\varepsilon_F,T)=\sum_{m,\nu}\int_{BZ}\frac{dq}{\Omega_{BZ}}|g_{mn\nu}(k,q)|^2 \times\left[\frac{n_{q\nu}(T)+f_{mk+q}(\varepsilon_F,T)}{\omega-\varepsilon_{mk+q}+\omega_{q\nu}+i\eta}+\frac{n_{q\nu}(T)+1-f_{mk+q}(\varepsilon_F,T)}{\omega-\varepsilon_{mk+q}-\omega_{q\nu}+i\eta}\right] \tag{70}$$

where $f_{mk+q}(\varepsilon_F,T)$ and $n_{q\nu}(T)$ correspond to the Fermi-Dirac and Bose-Einstein occupation functions at temperature T and ε_F is the Fermi level. For simplicity, the temperature and Fermi level are directly considered as parameters. The integral in Eq. (70) is performed over the q points in the BZ of volume Ω_{BZ}, and η is a positive real infinitesimal.

In the $\eta\rightarrow 0^+$ limit, the imaginary part of the SE (Eq. (70)) is evaluated at the KS energy given by [60].

$$\lim_{\eta\rightarrow 0^+}\Im\left\{\sum_{nk}^{FM}(\varepsilon_{nk})\right\}=\pi\sum_{m,\nu}\int_{BZ}\frac{dq}{\Omega_{BZ}}|g_{mn\nu}(k,q)|^2 \times[(n_{q\nu}+f_{mk+q})\delta(\varepsilon_{nk}-\varepsilon_{mk+q}+\omega_{q\nu}) +(n_{q\nu}+1-f_{mk+q})\delta(\varepsilon_{nk}-\varepsilon_{mk+q}-\omega_{q\nu}] \tag{71}$$

and corresponds to the linewidth of the electron state *nk* due to the scattering with phonons. Finally, the electron lifetime τ_{nk} is inversely proportional to the linewidth of the SE evaluated at the KS energy [62]:

$$\frac{1}{\tau_{nk}}=2\lim_{\eta\rightarrow 0^+}\Im\left\{\sum_{nk}^{FM}(\varepsilon_{nk})\right\} \tag{72}$$

Based on the relaxation time approximation of the linear Boltzmann equation, the generalized transport coefficient can be expressed as:

$$L_{\alpha\beta}^{(m)}=-\sum_{n}\int\frac{dk}{\Omega_{BZ}}v_{nk,\alpha}v_{nk,\beta}\tau_{nk} \times(\varepsilon_{nk}-\varepsilon_F)^m\left.\frac{\partial f}{\partial\varepsilon}\right|_{\varepsilon_{nk}} \tag{73}$$

where $v_{nk,\alpha}$ is the $\alpha-th$ component of the matrix element v_{nk} of the electron velocity operator. Using the commutator of the Hamiltonian with the position operator, we can take v_{nk} as: [63].

$$v_{nk}=\frac{\partial\varepsilon_{nk}}{\partial k}=<\psi_{nk}|\hat{p}+i[V^{nl},r]|\psi_{nk}> \tag{74}$$

Among them, $\hat{p}$ is the momentum operator, as done in the DFPT calculation of the response to external electric fields [64]. The group velocity can be used to calculate the transport lifetime within the so-called momentum relaxation time approximation [65].

According to Eq. (70), the effective **k** and **q** points that determine the electron-phonon interaction can be obtained as shown in **Figure 11** (a) **k** and (b) **q** points for the computation of the SE in the case of a single parabolic conduction band at the Γ

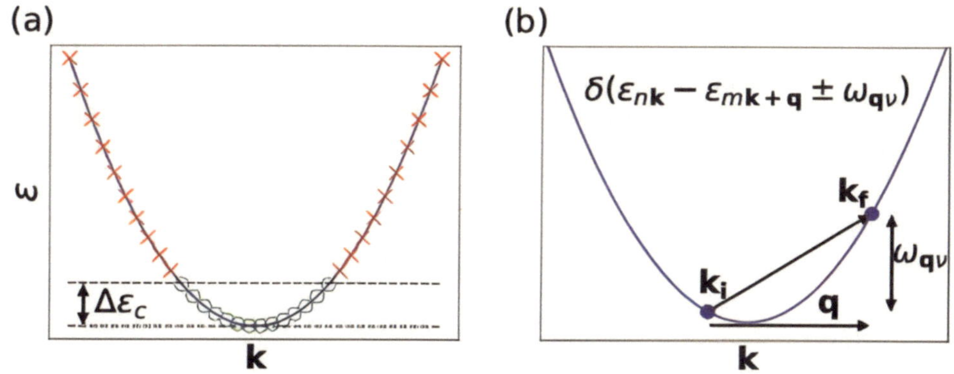

Figure 11.
(a) **k** *and (b)* **q** *points for the computation of the SE in the case of a single parabolic conduction band at the* Γ *point.*

point. (a) States represented by green dots (red crosses) are included (excluded) in (from) the transport computation according to the input variable $\Delta\varepsilon_c$. (b) For initial and final electron states $\mathbf{k_i}$ and $\mathbf{k_f}$, the wave vector $\mathbf{q} = \mathbf{k_f} - \mathbf{k_i}$ requires a phonon mode with frequency $\omega_{q\nu} = \varepsilon_{kf} - \varepsilon_{ki}$. If this condition is not fulfilled, the $\mathbf{q}$ point is ignored.

In the case of a determined composition, the thermal conductivity of a material is mainly determined by its thermal diffusivity and essentially by its bonding state. Based on the long-range coupling effect of electron-phonon, on the one hand, it can reduce short-range forces to soften phonon modes, and on the other hand, it can smooth out the lattice potential energy field that determines the electronic state and promote electron transport, simultaneously achieving energy filtering of electrons. Based on the hierarchical nanostructure design strategy, the electron-phonon coupling probability that is conducive to enhancing the thermoelectric effect is selected. Directed high-momentum electronic states are obtained by absorbing phonons, while small-momentum electronic states are filtered by emitting phonons. In addition, the phonon softening mechanism of the nanostructure is used to reduce the directional thermal conductivity of phonons and the adverse scattering of diffused electrons. Abnormal changes in the Debye-Waller factor can reflect the lattice softening phenomenon. In a normal electronic system, the electron-phonon interaction weakens the persistent current through the Debye-Waller factor, indicating that this factor can serve as an indirect indicator of the coupling strength between electrons and lattice vibrations. Finally, in a quantum well, an external electromagnetic field regulates the electron-phonon interaction, further affecting the polaron correction energy, which is also related to the correction of the Debye Waller factor.

5. Conclusion

This chapter starts from the basic theory of thermoelectricity and introduces the thermoelectric transport equation and the Boltzmann transport equation that controls phonon and electron transport. Subsequently, various innovative strategies for optimizing thermoelectric materials in recent decades were reviewed, and the related progress was evaluated. Finally, based on the superconducting BCS theory and phonon drag effect, this chapter proposes a new physical paradigm different from the phonon-glass-electron-crystal method. The core of the new paradigm is to shift from decoupling the electron-phonon transport process to optimizing the electron-phonon coupling mechanism and provides a theoretical mechanism for improving the performance of thermoelectric materials by optimizing the electron-phonon coupling. Identified the material design concepts and challenges for implementing the new paradigm.

Conflict of interest

The authors declare no conflict of interest.

Author details

Hong Liu* and Xinli Cheng
School of Physics and Technology, Suzhou University of Science and Technology, Suzhou, China

*Address all correspondence to: hliu@usts.edu.cn

References

[1] Snyder GJ, Toberer ES. Complex thermoelectric materials. Nature Materials. 2008;**7**:105-114

[2] Sootsman JR, Chung DY, Kanatzidis MG. New and old concepts in thermoelectric materials. Angewandte Chemie-International Edition. 2009;**48**: 8616-8639

[3] Zhu T, Liu Y, Fu C, Heremans JP, Snyder JG, Zhao X. Compromise and synergy in high-efficiency thermoelectric materials. Advanced Materials. 2017;**29**:1605884

[4] Heremans JP, Jovovic V, Toberer ES, Saramat A, Kurosaki K, Charoenphakdee A, et al. Enhancement of thermoelectric efficiency in PbTe by distortion of the electronic density of states. Science. 2008;**321**:554-557

[5] Pei Y, Shi X, LaLonde A, Wang H, Chen L, Snyder GJ. Convergence of electronic bands for high performance bulk thermoelectrics. Nature. 2011;**473**: 66-69

[6] Liu H, Shi X, Xu F, Zhang L, Zhang W, Chen L, et al. Copper ion liquid-like thermoelectrics. Nature Materials. 2012; **11**:422-425

[7] Liu R, Chen H, Zhao K, Qin Y, Jiang B, Zhang T, et al. Entropy as a gene-like performance indicator promoting thermoelectric materials. Advanced Materials. 2017;**29**:1702712

[8] Zhao LD, Lo SH, Zhang Y, Sun H, Tan G, Uher C, et al. Ultralow thermal conductivity and high thermoelectric figure of merit in SnSe crystals. Nature. 2014;**508**:373-377

[9] Pei YL, Wu HJ, Wu D, Zheng FS, He JQ. High thermoelectric performance realized in a BiCuSeO system by improving carrier mobility through 3D modulation doping. Journal of the American Chemical Society. 2014;**136**: 13902-13908

[10] Muchler L, Casper F, Yan B, Chadov S, Felser C. Topological insulators and thermoelectric materials. Physica Status Solidi RRL: Rapid Research Letters. 2013; **7**:91-100

[11] Uchida K, Takahashi S, Harii K, Ieda J, Koshibae W, Ando K, et al. Observation of the spin Seebeck effect. Nature. 2008;**455**:778-781

[12] Chen G. Nanoscale Energy Transport and Conversion: A Parallel Treatment of Electrons, Molecules, Phonons, and Photons. London: Oxford University Press; 2005

[13] Ashcroft NW, Mermin ND. Solid State Physics. Holt; 1976

[14] Nolas GS, Sharp J, Goldsmid HJ. Thermoelectrics: Basic Principles and New Materials Developments. Berlin: Springer; 2001

[15] Ziman JM. Electrons and Phonons: The Theory of Transport Phenomena in Solids. London: Clarendon Press; 1960

[16] Lindsay L, Broido DA. Three-phonon phase space and lattice thermal conductivity in semiconductors. Journal of Physics: Condensed Matter. 2008;**20**: 165209

[17] Li W, Lindsay L, Broido DA, Stewart DA, Mingo N. Thermal conductivity of bulk and nanowire $Mg_2Si_xSn_{1-x}$ alloys from first principles. Physical Review B. 2012;**86**:174307

[18] Ziman JM. The effect of free electrons on lattice conduction. Philosophical Magazine. 1957;**2**:292

[19] Dresselhaus MS, Chen G, Tang MY, YangR LH, Wang D, et al. New directions for low-dimensional thermoelectric materials[J]. Advanced Materials. 2007;**19**:1043-1053

[20] Mao J, Liu Z, Ren Z. Size effect in thermoelectric materials. npj Quantum Materials. 2016;**1**:16028

[21] Mahan GD, Sofo JO. The best thermoelectric. Proceedings of the National Academy of Sciences of the United States of America. 1996;**93**(15): 7436

[22] Friedel J. On some electrical and magnetic properties of metallic solid solutions. Canadian Journal of Physics. 1956;**34**(12A):1190-1211

[23] Anderson PW. Localized magnetic states in metals. Physics Review. 1961; **124**(1):41

[24] Blatt FJ, Schroeder PA, Foiles CL, Greig D. Thermoelectric Power of Metals. New York: Plenum Press; 1976

[25] Heremans JP, Wiendlocha B, Chamoire AM. Resonant levels in bulk thermoelectric semiconductors. Energy & Environmental Science. 2012;**5**:5510-5530

[26] Ravich YI. Selective carrier scattering in thermoelectric materials. In: Rowe DM, editor. CRC Handbook of Thermoelectrics. Boca Raton: CRC Press; 1995. p. 67

[27] Nishio Y, Hirano T. Improvement of the efficiency of thermoelectric energy conversion by utilizing potential barriers. Japanese Journal of Applied Physics. 1997;**36**(1a):170e4.1

[28] Minnich AJ, Dresselhaus MS, Ren ZF, Chen G. Bulk nanostructured thermoelectric materials: Current research and future prospects. Energy & Environmental Science. 2009;**2**:466-479. DOI: 10.1039/b822664b

[29] Ravich II. Semiconducting Lead Chalcogenides. New York: Springer; 2013

[30] Pei YZ, LaLonde AD, Wang H, Snyder GJ. Low effective mass leading to high thermoelectric performance. Energy & Environmental Science. 2012; **5**:7963

[31] Wang H, Pei Y, LaLonde AD, Snyder GJ. Material design considerations based on thermoelectric quality factor. In: Koumoto K, Mori T, editors. Thermoelectric Nanomaterials. Berlin: Springer; 2013. p. 3

[32] Wang H, Pei Y, LaLonde AD, Snyder GJ. Weak electron–phonon coupling contributing to high thermoelectric performance in n-type PbSe. Proceedings of the National Academy of Sciences of the United States of America. 2012;**109**:9705-9709. DOI: 10.1073/pnas.1111419109

[33] Liu X, Zhu T, Wang H, Hu L, Xie H, Jiang G, et al. Low electron scattering potentials in high performance $Mg_2Si_{0.45}Sn_{0.55}$ based thermoelectric solid solutions with band convergence. Advanced Energy Materials. 2013;**3**:1238

[34] Liu ZH, Wang YM, Mao J, Geng HY, Shuai J, Wang YX, et al. Lithium doping to enhance thermoelectric performance of MgAgSb with weak electron–phonon coupling. Advanced Energy Materials. 2016;**6**:1502269

[35] Zhou J, Zhu H, Liu T-H, Song Q, He R, Mao J, et al. Large thermoelectric power factor from crystal symmetry-protected non-bonding orbital in half-Heuslers. Nature Communications. 2018;**9**:1721

[36] Esfarjani K, Chen G, Stokes HT. Heat transport in silicon from first principle calculations. Physical Review B. 2011;**84**:085204

[37] Esfarjani K, Garg J, Chen G. Modeling heat conduction from first principles. Annual Review of Heat Transfer. 2014;**17**:9-47

[38] Kittel C. Introduction to Solid State Physics. 6th ed. New York: Wiley and Sons; 1986

[39] Slack GA, Tsoukala VG. Some properties of semiconducting $IrSb_3$. Journal of Applied Physics. 1994;**76**:1665

[40] Challis LJ, Cheeke JDN, Harness JB. The thermal conductivity of indium antimonide between 1.2 and 4.0°K. Philosophical Magazine. 1962;7:1941-1949. DOI: 10.1080/14786436 208213857

[41] Liu Z, Mao J, Liu T-H, Chen G, Ren ZF. Nano-microstructural control of phonon engineering for thermoelectric energy harvesting. MRS Bulletin. 2018; **43**:181

[42] Biswas K, He J, Blum ID, Wu C-I, Hogan TP, Seidman DN, et al. Correction: Corrigendum: High-performance bulk thermoelectrics with all-scale hierarchical architectures. Nature. 2012;**489**:414-418. DOI: 10.1038/nature11439

[43] He, Kanatzidis MG, Dravid VP. High performance bulk thermoelectrics via a panoscopic approach[J]. Materials Today. 2013;**16**:166-176. DOI: 10.1016/j. mattod. 2013.05.004

[44] Zhao LD, Dravid VP, Kanatzidis MG. The panoscopic approach to high performance thermoelectrics[J]. Energy & Environmental Science. 2014;7:251-268. DOI: 10.1039/c3ee43099e

[45] Klemens P. The thermal conductivity of dielectric solids at low temperatures (theoretical) [J]. Proceedings of the Physical Society, London, Section A. 1951;**208**(1092):108-133. DOI: 10.2307/98761

[46] Toberer ES, Baranowski LL, Dames C. Advances in thermal conductivity[J]. Annual Review of Materials Research. 2012;**42**:179-209

[47] Klemens P. The scattering of low-frequency lattice waves by static imperfections. Proceedings of the Physical Society, Section A. 1955;**A68**: 1113

[48] Abeles B. Lattice thermal conductivity of disordered semiconductor alloys at high temperatures[J]. Physics Review. 1963; **131**:1906

[49] Callaway J. Model for lattice thermal conductivity at low temperature[J]. Physics Review. 1959;**113**:1046

[50] Callaway J, von Baeyer HC. Effect of point imperfections on lattice thermal conductivity[J]. Physics Review. 1960; **120**:1149

[51] Klemens PG. Thermal resistance due to point defects at high temperatures[J]. Physics Review. 1960;**119**:507

[52] Abeles B, Beers D, Cody G, Dismukes J. Thermal conductivity of Ge-Si alloys at high temperatures. Physics Review. 1962;**125**:44

[53] Peierls R. On the kinetic theory of thermal conduction in crystals[J]. Annals of Physics. 1929;**3**:1055

[54] Casimir HBG. Note on the conduction of heat in crystals. Physica. 1938;**5**:495

[55] Nabarro F. The interaction of screw dislocations and sound waves. Proceedings of the Royal Society of London A. 1951;**209**:278

[56] Mingo N, Hauser D, Kobayashi NP, Plissonnier M, Shakouri A. "Nanoparticle-in-alloy" approach to efficient thermoelectrics: Silicides in SiGe. Nano Letters. 2009;**9**:711

[57] Conyers herring. Theory of the thermoelectric power of semiconductors. Physics Review. 1954;**96**:1163

[58] Zhao LD, Dravid VP, Kanatzidis MG. The panoscopic approach to high performance thermoelectrics. Energy & Environmental Science. 2014;**7**:251

[59] He Y, Lu P, Shi X, Xu F, Zhang T, Snyder GJ, et al. Ultrahigh thermoelectric performance in mosaic crystals. Advanced Materials. 2015;**27**: 3639

[60] Giustino F. Electron-phonon interactions from first principles. Reviews of Modern Physics. 2017;**89**: 015003

[61] Baroni S, de Gironcoli S, Dal Corso A, Giannozzi P. Phonons and related crystal properties from density-functional perturbation theory. Reviews of Modern Physics. 2001;**73**:515

[62] Ponce S, Margine ER, Giustino F. Towards predictive many-body calculations of phonon-limited carrier mobilities in semiconductors. Physical Review B. 2018;**97**:121201

[63] Del Sole R, Girlanda R. Optical properties semiconductors within independent-quasiparticle approximation [J]. Physical Review B. 1993;**48**:11789

[64] Gonze X, Lee C. Dynamical matrices, born effective charges, dielectric permittivity tensors, and interatomic force constants from density-functional perturbation theory. Physical Review B. 1997;**55**:10355

[65] Ma J, Nissimagoudar AS, Li W. First-principles study of electron and hole mobilities of Si and GaAs. Physical Review B. 2018;**97**:045201

Chapter 4

High-Perfection Bulk and Film Thermoelectrics Based on Indium-Doped Lead Telluride

Sergey Nemov, Mark Auslender, Rony Shneck and Zinovi Dashevsky

Abstract

Energy harvesting has been proposed as a promising method of supplying electric power to autonomous sensor network devices and mobile electrical instruments due to the possibility of obtaining electric power from unused environmental energy, such as heat, light, and vibrations. It was found that micro-generators for energy harvesting are more useful than standard electrical batteries. Lead telluride alloys are the first materials investigated and commercialized from micro- to macro-generators with electrical power up to a few hundred watts. However, their full potential for thermoelectrics has only recently been revealed to be greater than commonly believed. The maximal value of the thermoelectric figure of merit as a function of electron density is attained only for the specific location of the Fermi level E_F relative to the conduction band edge E_c. A systematic study of structural, micro-structural, and thermoelectric properties of bulk and film PbTe doped with indium is presented. Bulk samples were prepared utilizing a Spark Plasma Sintering. The films with high energy conversion efficiency were deposited on a polyimide substrate using a flash evaporation technique. A systematic study of the Seebeck coefficient, and electrical and thermal conductivity of $Pb_{1-x}In_xTe$ bulk and film samples over a wide temperature range has been performed.

Keywords: thermoelectric semiconductor materials, thermoelectric figures of merit (Z and ZT), lead telluride (PbTe), indium (In) and iodine (I) doping, spark plasma sintering (SPS), flash evaporation (FE)

1. Introduction

Among different energy converters (solar cells, wind generators, and hydropower), thermoelectric (TE) generators are currently promising alternative energy sources for the future [1]. Their advantage is a long lifetime, from 20 to 25 years [2]. Currently, the field of application of TE generators has expanded significantly. It was shown that TE micro-generators for energy harvesting were more useful than electrical batteries [3]. They have become realizable due to a decrease in power consumption of electronic components due to advances in the field of electric energy storage and

IntechOpen

their miniaturization [4–6]. In addition, they can generate power from low-temperature heat sources such as the human body [6–10]. Ioffe suggested a criterion for selecting an effective TE material, who prescribes it to possess electrical conductivity like a metal and thermal conductivity like an insulator. Ioffe deserves full credit for understanding that a "middle" class of materials; namely, semiconductors can provide the best TE performance as TE materials [11, 12]. The maximum efficiency of a TE generator (TEG) η_{max} can be calculated using the following expression [13]:

$$\eta_{max} = \frac{\Delta T}{T_h} \frac{\sqrt{1+(ZT)_{av}}-1}{\sqrt{1+(ZT)_{av}}+T_c/T_h}. \tag{1}$$

Here, T_h and T_c are the absolute temperatures of the hot and cold sides, respectively, and ΔT is the temperature (T) difference between these sides, *viz.*, $\Delta T = T_h - T_c$,

$$(ZT)_{av} = \frac{1}{\Delta T} \int_{T_c}^{T_h} ZT\,dT \tag{2}$$

is the averaged dimensionless TE figure of merit. In Eq. (2), the dimensional T-dependent figure of merit Z is given by

$$Z = \frac{S^2\sigma}{\kappa}, \tag{3}$$

where S is the Seebeck coefficient, and σ and κ are the electrical and thermal conductivities, respectively. Eq. (3) applies to the p- and n-type TE legs of any semiconductor TE generator, presenting the dependence of Z on T through the above carriers' transport characteristics of the legs.

Figure 1 shows the dependences of ZT on T for various TE materials over the range of $T = 300 - 900$ K [14–28]. For real applications in TEGs, along with a high TE efficiency, the material must also have a high mechanical strength [29–31]. Today, only two groups of TE materials are used at gas TEGs (operating $T_h = 850 - 900$ K and $T_c = 320 - 350$ K). The first group, targeted for use in a low-temperature range of $300 - 600$ K, includes solid solutions based on n- and p-type Bi_2Te_3; the second group proper for operating in a mid-range of $T = 600 - 900$ K involves those based on $A^{IV}B^{VI}$ semiconductors such as n-type PbTe and p-type GeTe [1, 32, 33].

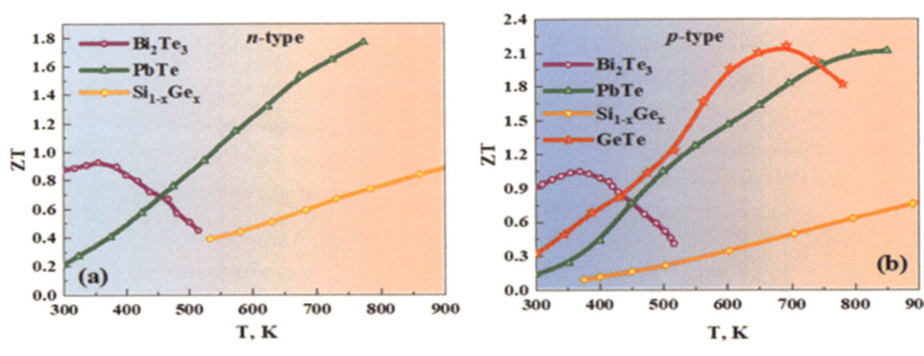

Figure 1.
The dimensionless figures of merit ZT as functions of temperature for different TE semiconductors over the temperature range of 300 – 900 K: *(a) n-type materials; (b) p-type materials.*

For optimal TE materials with a high value of carrier mobility and a small value of lattice thermal conductivity, the value of S determines the optimal value of Z, see Eq. (3). In the simplest case—for the quasi-elastic scattering of charge carriers, for example, on acoustic phonons and parabolic band structure, the Seebeck coefficient is calculated by Dashevsky and Skipidarov [12] with Eq. (A1), as shown in Appendix A. Let us remind that the parabolic conduction or valence band structure of a semiconductor is, in general, a multi-valley ellipsoidal one, characterized by the components m_1, m_2, m_3 of the effective-mass tensor along the principal axes of each ellipsoid, and the number of ellipsoids N equal 8 for PbTe.

The Seebeck coefficient S decreases with the growth of carriers' concentration, and S is close to zero for metals with this concentration being of $\sim 10^{22}$ cm^{-3}. At the same time, electrical conductivity σ increases with the growth of the charge carrier concentration. Therefore, the dependence of the product $S^2\sigma$ as a function of E_F has the bell-shaped form. The maximum value of $S^2\sigma$ is reached when E_F is close to E_c or E_v. In this case, the Seebeck coefficient for n-type material varies in a narrow range of $S \simeq -(200 \pm 20)\ \mu V/K$ at the electron concentration in the range of $n = 5. \times 10^{18} - 1. \times 10^{19}\text{cm}^{-3}$. In well-doped TE materials, the carriers' concentration, n for electrons or p for holes, is constant due to the dominant doped donor or acceptor impurity, respectively. As shown in Ref. [12], for a semiconductor with the above-noted parabolic band structure, the carriers' concentration is calculated using Eq. (A4) of Appendix A, which introduces an important notion of the density-of-states (DOS) effective mass. According to Eqs. (A1) and (A4), the chemical-potential location that enters the reduced potential expression μ^* defining Z moves to the forbidden band with increasing temperature, and, consequently, Z decreases.

This study uncovers the possibility of stabilizing the Fermi level in an optimal position over a wide temperature range, which proves practically independent of the concentration of doped impurity. This goal is achieved by doping indium (In) in lead telluride (PbTe). This semiconductor is widely known for its premier TE properties over the temperature range of $400 - 900$ K [34–36]. The behavior of In in IV-VI semiconductors was the subject of diverse scientific research [25, 37–42]. It was established that the In impurity in PbTe has a donor character; however, despite the high solubility of InTe in PbTe ($2 - 3$ mol%), the electron concentration n at $T =$ 300 K does not exceed the value of $(3 - 5) \times 10^{18}$ cm^{-3}, a level that corresponds to a dopant concentration of only $N_{In} \approx 0.03$ at.%.

The main features of the effect of the In doping of PbTe on its electro-physical properties were studied by V.I. Kaĭdanov [37], whose principal statements are as follows:

i. At low temperatures, the In impurity energy level E_{In} is close to EC. Besides, its location does not depend on N_{In} up to $N_{In} \sim 5$ at.%.

ii. Each In atom in the position of a Pb-lattice site contributes to one electron, and two allowed states to the impurity level. Therefore, the impurity level is half-filled.

iii. In atoms in the interstitial positions are neutral and do not influence the carrier concentration.

iv. The Fermi level E_F is almost wholly determined by E_{In}, leading to the so-called pinning of the Fermi level at low temperatures.

Due to item (iv) above, the doping of PbTe with In sets the electron density highly uniform over the sample of any structural form, single crystalline and polycrystalline, both bulk and thin-film ones, which is important for the TE applications. This is the uniqueness of such n-type doping compared to the doping by substituting Te with halogens. These donors remain ionized at all *T*'s and give rise to large electron-density fluctuations, unless In is doped.

To estimate the number of states in the In level, the experiments were made on additional (to In) doping (co-doping) with standard donor impurity at PbTe-iodine (I) [25, 37]. As long as the concentration of the I impurity, N_I is less than N_{In}, and the I co-doping effect is weak. For $N_I \geq N_{In}$, noticeable changes in the electron concentration emerge. **Figure 2**, anticipating further consideration of this phenomenon, shows schematically how the extrinsic physical conditions, such as N_{In} and $N_{In} + N_I$ (panel A), and T (panel B), would influence the position of the impurity level and its broadening into a narrow band, as well as these bands filling with electrons. At present, a few reviews, which are quoted in continuation of the article, have been devoted to PbTe-based TE materials with top *ZT*. However, most of them, though including PbTe as the main component, are rather intricate compounds, which are difficult to reproducibly fabricate for regular usage in industry. In contrast, In-doped PbTe, on which this review focuses, along with high TE efficiency, has the advantage of structural and chemical simplicity, and so the available well-reproducible synthesis routes.

Our scope encompasses single crystals and polycrystals. Section 2 overviews the currently available theoretical concepts on the physics of doping PbTe with In, as well as new first-principle band structure simulations of In-doped, and In and I co-doped PbTe. Sections 3 and 4 consider the fabrication and experimental studies of the thus

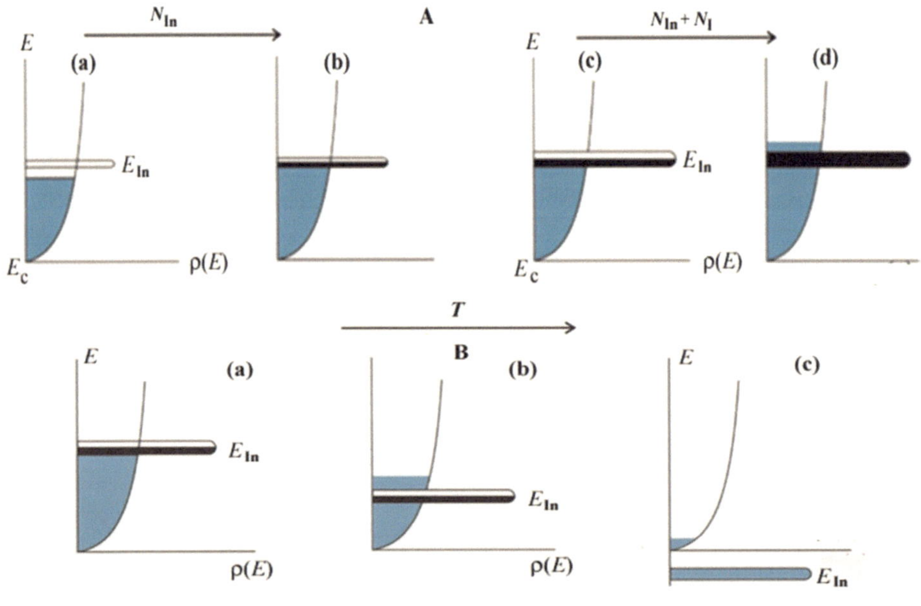

Figure 2.
Effects of increased doping at helium temperatures, and temperature on the band structure of PbTe. Abscissas—DOS, ordinates—energy. ***A:*** *(a)—low* N_{In}*,* E_F *is below* E_{In}*; (b)—higher* N_{In}*,* E_F *attains* E_{In} *and became pinned and half-filled; (c)—low* N_I*,* E_F *remains pinned,* E_{In} *half-filled and broadens; (d)—higher* N_I*,* E_F *overs the In impurity band that becomes filled.* ***B:*** *(a)—the same T as in* ***A****,* $N_{In} + N_I$ *at which* E_F *is pinned; (b)—higher* $T < 300$ *K,* E_F *moved above* E_{In}*; (c)* $T \geq 300$ *K,* E_c *moved above the In-impurity band.*

doped bulk and thin-film n-type PbTe, respectively, and present new datasets with tables and graphs. As concerns fabrication, subsection 3.1 recaps the Czochralski single-crystal growth method and uncovers a new idea of the functionally graded material (FGM) for the TE devices; subsections 3.2 and 4.1 describe the spark plasma sintering (SPS) and flash evaporation methods and their applications with a new in-house designed synthesis strategies for the manufacture of polycrystalline bulk and thin-film PbTe samples, respectively, doped and co-doped with In and I. Section 5 concludes the Chapter.

2. Theoretical modeling

2.1 Effect of the In impurity on the band structure of PbTe within semi-analytic approach

Deep and quasi-local resonant impurity states were discovered due to the pioneering study of transport phenomena in PbTe doped with In by Rosenberg and Wald [43] and Kaĭdanov [44]. As a result of extensive studies of lead, tin, and lead-tin chalcogenides, several new physical phenomena were discovered such as the above pinning (stabilization) of the Fermi level (chemical potential) by the impurity levels, resonant scattering of band electrons by the impurity centers, long-term relaxation of the non-equilibrium distribution of conduction electrons with characteristic times up to several hours at helium (He) temperatures, and superconductivity at this temperature range with a critical point unusually high for semiconductors [40].

Currently, no fully analytical models of quasi-local resonant impurity states exist that explain all features of the experimental data on the physical phenomena observed in In-doped and co-doped PbTe. A contemporary approach that meets this challenge is entirely numerical combined band structure—impurity simulations using such techniques as, for example, the Korringa–Kohn–Rostoker (KKR) and density-functional theory (DFT) jointly with coherent-potential approximation (CPA). Such an approach has recently been invoked [42] to analyze the formation of impurity levels and impurity bands associated with various impurities in PbTe. However, a simplified, at least semi-analytical modeling would be instructive and useful for understanding the physics of the phenomena in question. To this end, the clearest and most constructive model in our opinion, which adequately explains an entire set of available experimental data on transport phenomena at temperatures $T \geq 77$ K and allows one to predict the properties of materials upon changes in the doping composition, temperature, pressure, and additional doping, is due to Kaĭdanov et al. [37].

Localized states with energy levels located apart from the edges of the allowed bands are formed if the impurity potential is in short range. These deep impurity states are strongly localized, and the effective-mass approximation cannot be applied to describe such states. Rather, in these cases, the theory of Coster and Slater [45, 46] is customarily used. Specific to this theory is the approximation of the impurity potential by one localized within a one-unit cell, for which the quotient of the average perturbation energy introduced by the impurity over the perturbation volume is large enough. In wide-band-gap semiconductors, these impurity levels usually lie in the band gap, when their localized electronic, say s-type, state is well separated from the band of Bloch states. In narrow-band-gap semiconductors, such as in $A^{IV}B^{VI}$ at $T = 4.2$ K, and some of $A^{III}B^{V}$ ones, with the band gap E_g as small as $\sim 0.1 - 0.3$ eV, the deep impurity level can emerge in the allowed, either conduction or valence, band.

A theoretical proof of the possibility of such quasi-local states, today called *bound states in continuum*, is given in detail in Ref. [47]. At that rate, the energy of the impurity state ε_I coincides with that of the band (Bloch) states $\varepsilon\left(\vec{k}\right)$ on the equi-energetic surface, *viz.*, $\varepsilon\left(\vec{k}\right) = \varepsilon_I$, so the impurity states fall into resonance with the continuous band states of the same energy. For this reason, the states formed in this way are called resonant states, which are stationary, and their hybridized wave function is a linear combination of those of the impurity and band states. The result of solving the Schrödinger equation for the resonant state could be represented as a peak of the electronic DOS $\rho(\varepsilon)$ at the resonance energy level ε_I of a width Γ as shown in **Figure 3**. The necessary condition for observing these special states is that their energy smearing should be small compared to the distance of their energy level and the edge of the allowed band, that is, $\Gamma \ll \varepsilon_i$. Near the impurity atom, the dominant part of the hybridized-state wave function is spatially localized, which explains using the term "quasi-local."

Another view of the discussed problem suggests that the electron's wave function is only of the impurity nature. In this case, it is impossible to solve the Schrödinger equations exactly; therefore, the wave function is non-stationary. Thereby, the time τ_I of electron localization at an impurity center is finite and called the lifetime. In this picture, the electron passes into the allowed band of Bloch states without energy losses. A finite lifetime leads to an energy uncertainty $\Delta\varepsilon = \Gamma$ according to the celebrated uncertainty principle:

$$\Delta\varepsilon\tau_I = \Gamma\tau_I \approx \hbar, \tag{4}$$

where h and $\hbar = h/(2\pi)$ are the Planck and reduced Planck constants, respectively. The width of the DOS peak in **Figure 3** and the energy uncertainty are of the same order of magnitude. The energetic broadening of the resonant impurity level can be taken into account by introducing the complex energy:

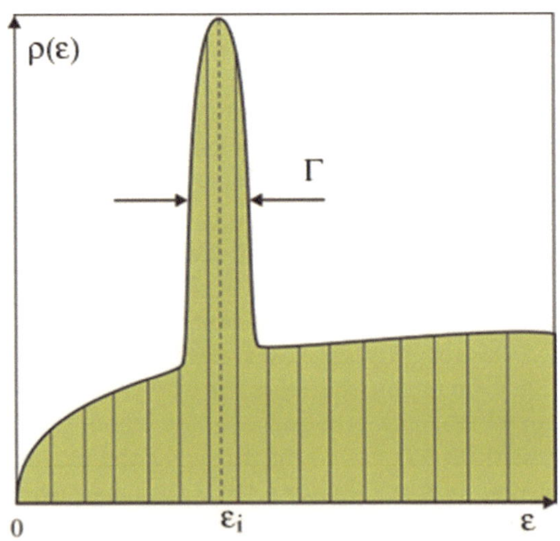

Figure 3.
Schematic of conduction band DOS $\rho(\varepsilon)$ in PbTe-based compounds with In dopant [42]; ε and ε_i are the electron and resonant impurity-level energies, respectively, counted from the band edge.

$$\varepsilon_{\mathrm{i}}^{*} = \varepsilon_{\mathrm{i}} - i\frac{\Gamma}{2}. \tag{5}$$

The presence of the imaginary part in the energy provides a temporal attenuation of the probability density of locating an electron on an impurity center, that is, the wave-function amplitude squared, according to the decay equation:

$$\left|\Psi_{\mathrm{i}}\left(\vec{r}, t\right)\right|^{2} \propto \exp(-\Gamma t/\hbar), \tag{6}$$

where $\Psi_{\mathrm{i}}\left(\vec{r}, t\right)$ is the corresponding wave function. Therefore, the decay time of $\left|\Psi_{\mathrm{i}}\left(\vec{r}, t\right)\right|^{2}$ equal $\hbar/\Gamma$ due to Eq. (6) coincides with the lifetime τ_{l} discussed above. Near ε_{i}, the density of impurity states, superimposed on the density of propagating carriers' states, is fairly approximated by the Lorentz function

$$\rho_{\mathrm{i}}(\varepsilon) \simeq \frac{N_{\mathrm{i}}}{\pi}\frac{\Gamma}{(\varepsilon - \varepsilon_{\mathrm{i}})^{2} + (\Gamma/2)^{2}}, \tag{7}$$

where N_{i} is the concentration of impurities forming the resonant states.

As noted above, the resonance level broadens due to the transitions of the electron from the impurity state to the band state. The reverse process leads to the resonant scattering of Bloch electrons. In a localized state, the electron has an initial quasi-momentum. Then, leaving an impurity state after a time equal to the lifetime τ_{l}, it acquires a new quasi-momentum, which differs significantly from the initial one. Hence, the resonant impurity-level broadening and resonant scattering are two sides of the same phenomenon, for which the scattering amplitude f due to one such impurity is given by the following expression [49].

$$f = -\frac{1}{k}\frac{\Gamma/2}{\varepsilon - \varepsilon_{\mathrm{i}} + i(\Gamma/2)}, \tag{8}$$

where k is the Bloch quasi-momentum of the scattered electron and other parameters are defined above.

The integral cross-section $\sigma_{\mathrm{sc}}(\varepsilon)$ of the scattering by one impurity center is related to the scattering amplitude by $\sigma_{\mathrm{sc}}(\varepsilon) = 4\pi|f|^{2}$, with which and Eq. (8), we obtain the Breit-Wigner-like formula:

$$\sigma_{\mathrm{sc}}(\varepsilon) = \frac{\pi}{k^{2}}\frac{\Gamma^{2}}{(\varepsilon - \varepsilon_{\mathrm{i}})^{2} + (\Gamma/2)^{2}}. \tag{9}$$

On the other hand, the inverse electron quasi-momentum relaxation time due to scattering by all resonant impurities is given by $\tau_{\mathrm{res}}^{-1}(\varepsilon) = N_{\mathrm{i}}\sigma_{\mathrm{sc}}(\varepsilon)v_{k}$, where v_{k} is the velocity of a band electron with the quasi-momentum k, see, for example, Ref. [48]. Combining this expression with Eqs. (7) and (9), one obtains the formula

$$\tau_{\mathrm{res}}^{-1}(\varepsilon) = \tau_{\mathrm{l}}^{-1}\frac{\rho_{\mathrm{i}}(\varepsilon)}{\rho(\varepsilon)}, \tag{10}$$

where $\rho(\varepsilon) = 2m^{*}k/h^{2}$ is the allowed-band DOS approximated by one with an effective mass m^{*}, which reveals that the scattering is resonant indeed.

Much experimental data concerning In-doped $A^{IV}B^{VI}$ semiconductors are summarized in Refs. [37, 49]. The clearest proof of the quasi-local resonant impurity states' emergence in PbTe is the stabilization or pinning of the chemical potential μ. Due to this effect, the electro- and thermo-physical properties as well as TE performance of the material are not so sensitive to fluctuations in the composition, the concentration of point defects, and the amount of introduced impurities [24] as they are in conventional heavily doped compensated semiconductors. It should be noted that the pinning effect is due to the partial filling of resonant impurity states with electrons.

Under the co-doping of PbTe with In and other electrically active impurities with a concentration N_i, such as donors, for example, I, or acceptors, for example, sodium (Na), there are no significant changes in the electron concentration as long as $N_i \leq N_{In}$. Moreover, the co-doping experiments allowed one to conclude that the band of resonant states contains two states per impurity level [37]. **Figure 4** shows a band diagram of the In-doped $Pb_{1-x}Sn_xTe$ alloys [40]. For the Sn-substitute amount in the range of $0.22 < x < 0.28$, the chemical potential μ is stabilized in the band gap of the alloy. The μ crosses the middle of the band gap at $x = 0.26$, and the n-to-p-type electrical conductivity conversion could be observed.

2.2 Basic principles. The In-impurity effects within the DFT approach

In this subsection, we overview the effects of the In-doping on the band structure and transport properties of PbTe in light of the DFT, which has already been well discussed in the literature [25]. Here, the temperature effects are not considered, so μ is called, as customarily, the Fermi energy and denoted by E_F.

The DFT simulation results of DOS for doped and co-doped PbTe are displayed in **Figure 5. Figure 5(a)** shows that In forms a density-of-states peak at the band gap of PbTe, as well as a hyper-deep state at about −5 eV below the conduction band (CB) edge [25] adopted *ad hoc* as a point from which all energies are counted. It is the 5*s* orbitals of In that form these peaks, see **Figure 5(b)**. The in-gap quasi-local narrow impurity band is formed just below the CB edge and pins E_F, as seen in **Figure 5(c)**.

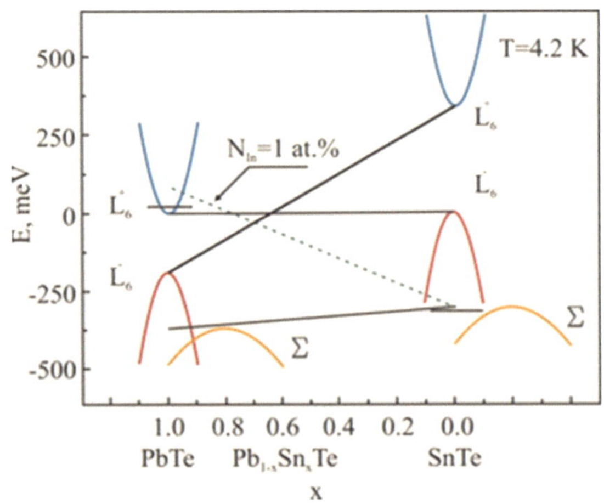

Figure 4.
The position of the In-impurity level vs. the $Pb_{1-x}Sn_xTe$ solid solutions composition. L and Σ are the Brillouin zone points of band structure extrema. L locates the lowest conduction-band bottom and highest valence-band top, and Σ a lower valence-band top. $L_6^{\pm}$ are irreducible representations of $Fm\bar{3}m$ space group.

This band is half-filled since E_F lies in the middle of it. It appears that In, doped into PbTe, affects the band structure differently from how it does when doped into SnTe [50].

The difference turns out to be more drastic if we compare, for example, the In and Tl dopants in PbTe [51–53]. Among them, the latter, another III-group element, forms a resonant energy level that submerges with the valence band. **Figure 5(d)** presents the effect of In and I co-doping on the PbTe band structure. Here, one observes that the same amounts of I and In, doped to PbTe, result in pushing E_F in between the In impurity level broadened into impurity band (IB) and CB edge due to the further occupation of IB with additional electrons supplied by the I dopants. Furthermore, the distance between the In-level derived IB and CB edges is much smaller for the co-doped system than for the In-doped PbTe. A smaller gap between IB and CB will lead to a lower T at which IB contributes to the transport properties of n-type PbTe.

The In-doping influence on electronic dispersion relations was shown using the calculated Bloch spectral density functions (BSF) in Ref. [25]. They proved to be useful in describing the electronic band structure of the system that contains impurities that do not follow a rigid band behavior [52–55]. **Figure 6** shows the BSF of I-doped and co-doped PbTe. In the case of only-I-doped PbTe, E_F is moving deep into the conduction band edge and a slight smearing near the band edge can be observed, see in the left sub-figure of **Figure 6** [25]. In the right sub-figure of **Figure 6**, it is seen

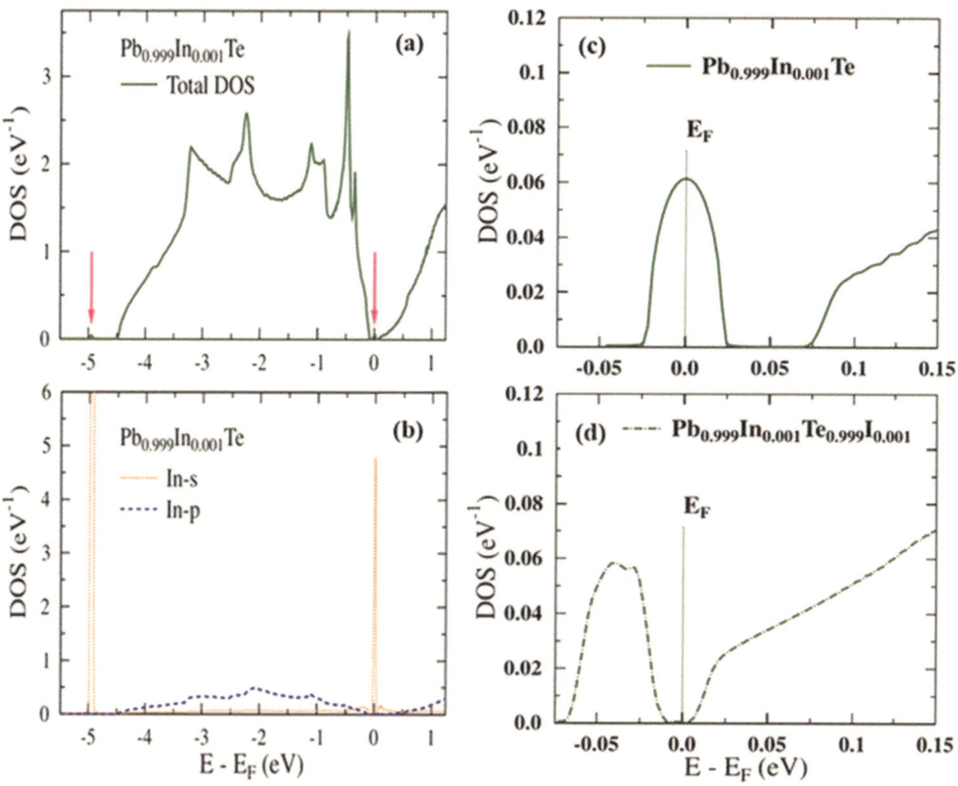

Figure 5.
Electronic DOS of doped PbTe calculated with DFT. Total DOS of $Pb_{0.999}In_{0.001}Te$, arrows indicate the hyper-deep state (−5 eV) and in-gap state (0 eV)—(a); partial s- and p-orbitals derived DOS of an In atom in PbTe—(b); DOS near the CB edge of $Pb_{0.999}In_{0.001}Te$—(c); and $Pb_{0.999}In_{0.001}Te_{0.999}I_{0.001}$—(d) [25].

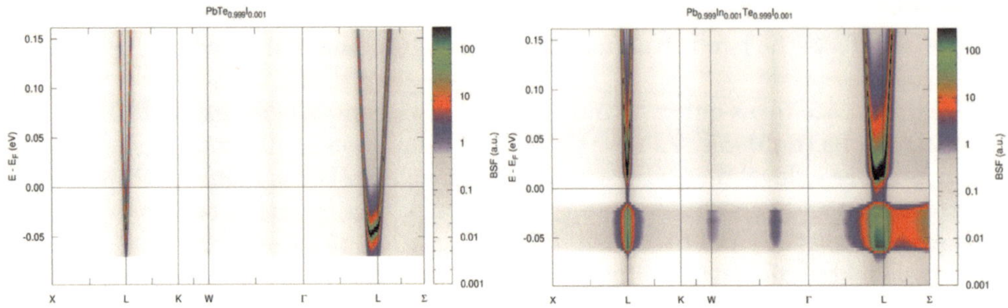

Figure 6.
Two-dimensional projections of the BSFs color maps of $PbTe_{0.999}I_{0.001}$ (left) and $Pb_{0.999}In_{0.001}Te_{0.999}I_{0.001}$ (right) calculated within DFT [25]. Abscissa and ordinate show directions between high symmetry points of Brillouin zone and distance between the current energy and Fermi level, respectively.

that In forms an additional quasi-local level with a high spectral weight at the L point and between the L- and Σ-points in the Brillouin zone.

The results presented above, especially as concerns the In and I doping and co-doping effects, agree well with the previous DOS calculations, see Ref. [24] and references therein, and the semi-analytical approaches discussed in the previous subsection. Though the reported results do not include the electron-transport properties of such doped n-PbTe, computed in the DFT framework [24], they support the experimental findings indirectly, due to compliance with the conclusions of subsection 2.1. The DFT simulation parameters used and some other features are detailed in Appendix B.

3. Bulk TE materials based on n-type doped PbTe

3.1 Single crystals prepared by Czochralski method

The single crystals of PbTe were grown using the Czochralski method that allows unidirectional solidification from the melt [56]. For growth, the synthesized PbTe was placed in a crucible, which was heated by a furnace to a temperature higher than the melting point of the PbTe compound ($T_m \approx 1200$ K). A PbTe seed crystal with an $\langle 100 \rangle$ orientation was used to pull the crystal out of the melt. Liquid encapsulation was used to prevent the melt's decomposition during crystal growth. A molten boron trioxide (B_2O_3) layer about 1 cm thick was used for liquid encapsulation. Molten B_2O_3 is inert to PbTe at the solidification temperature. The crystals were grown from a PbTe melt with an excess Te (up to 1 at. %) that resulted in the formation of Pb vacancies V_{Pb}^{2+} acting as acceptors in PbTe. The T gradient at the crystallization front and crystal pull rate ranges were $20 - 25$ K/cm and $5 - 10$ mm/h, respectively. To enhance the equalization of the heat flux through the lateral surface of the crystal, the crucible and the crystal were rotated in opposite directions with an angular velocity of $1\ s^{-1}$. Sufficiently rapid cooling partially quenches the non-stoichiometry defects, largely inhibiting their contribution to p-type electrical conduction.

A fabricated 10 cm-long PbTe single-crystal ingot with a diameter of 4 cm was demonstrated in Ref. [56]. The $\langle 100 \rangle$ crystal orientation was determined by the Laue method. The back reflection of an oriented PbTe crystal unequivocally revealed the fourfold symmetry and the absence of other reflections. The X-ray diffraction (XRD) was used to test the phase content of the fabricated crystals. The XRD pattern of an

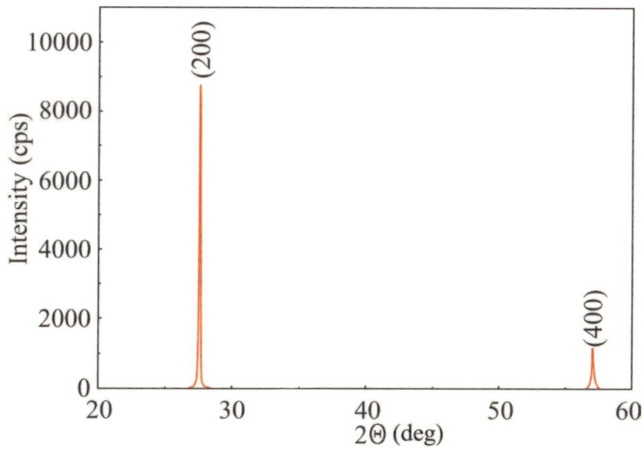

Figure 7.
The XRD pattern of a ⟨100⟩-oriented PbTe crystal [54].

⟨100⟩-oriented PbTe crystal is shown in **Figure 7**. Only one (200) peak and its multiple, (400) peak of much lesser amplitude, are seen in the XRD angular spectrum, which indicates that the specimens are strictly single phase. Moreover, the Rietveld refinement yields the FWHM value of 20 arcsec for the main XRD peak in **Figure 7**, which evidences high crystallographic quality.

To proceed to the physical properties study, PbTe wafers of various sizes were cut from the ingot and polished using mechanical and electrochemical methods. More wafer preparation details are found in Ref. [57]. The dislocation patterns were revealed by etching the well-polished surfaces of the wafers. Using the etch pits analysis [34], which determined the dislocation densities in the range of $10^6 - 10^7\,\text{cm}^{-2}$. The as-grown crystalline samples showed expected conductivity of p-type with the holes, and hence vacancies, density of $\sim 10^{18}\,\text{cm}^{-3}$. As was discovered [43, 44], doping PbTe with In leads at first to self-compensation and next to conductivity conversion from p- to n-type.

Thermally diffused doping with In was used by exposing the sample to In from a gaseous source of In_4Te_3 [58]. The adopted concentration of the In dopant is in the range of $0.1 - 4$ at. % and can be varied by variation of the exposure time and annealing temperature T_a from 900 to 1000 K. The surface concentration C_s of In, determined by energy-dispersive X-ray spectroscopy (EDXS), as a function of T_a, is presented in **Figure 8** [58].

The Seebeck coefficient S was measured in the direction parallel to the diffusive penetration of In into a p-type PbTe. The S values for all crystals, in which In had diffused, are large negative at the front of the crystal, $x = 0$, where x is the in-depth distance from the sample surface in the diffusion direction. Similar measurements were carried out in the Na-doped p-type PbTe crystals, which had no excess of Te, and hence, no structural vacancies but, due to the presence of Na at the Pb-lattice sites, had some initial concentration of holes, see in **Figure 9**.

Let $C(x,t)$ is the concentration of In (N_{In}) diffused at a distance x for a time t. This concentration is given by the appropriate solution of the diffusion equation

$$C(x,t) = C(0)\,\text{erfc}\left(\frac{x}{2\sqrt{Dt}}\right), \tag{11}$$

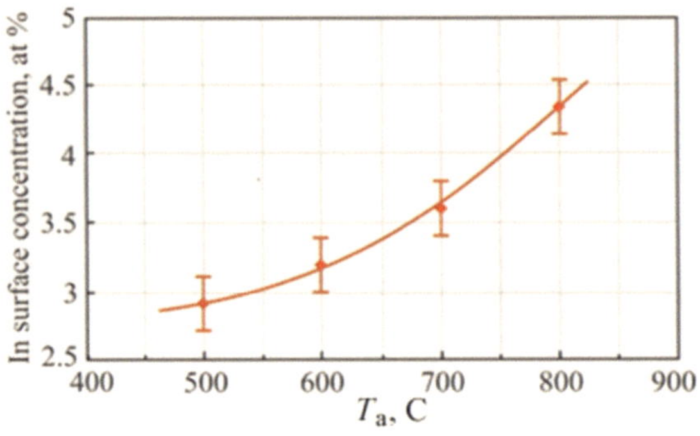

Figure 8.
The dependence of C_s on T_a, as determined by EDXS [58].

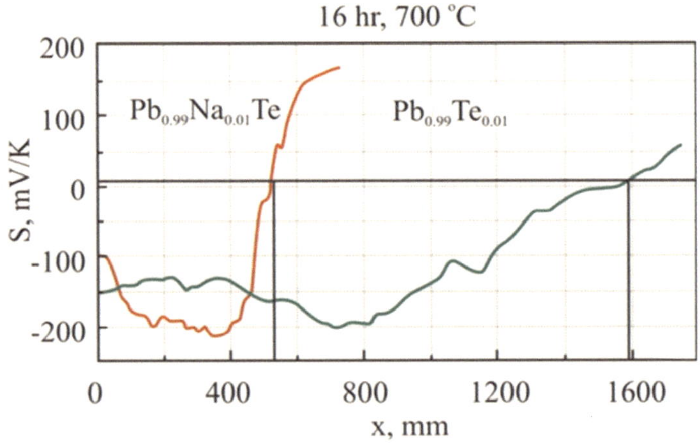

Figure 9.
Spatial profile of the Seebeck coefficient in crystalline $Pb_{0.99}Te_{1.01}$ (right) and PbTe doped with 1 at.% of Na (left), after diffusion of In for 16 hours at $T = 700°C$ [58].

where the initial value is $C(0) = C_s$, erfc is the complementary error function, and D is the In diffusion coefficient in PbTe. As seen in **Figure 9**, there exists a location x_0, at which S zeroes, and changes the sign above it. At this point, the contribution of electrons to S due to the presence of In at Pb-lattice sites counterbalances the TE effect of initial holes present due to the Te excess. With $S = 0$ in Eq. (A1) and using Eq. (A4), we can extract μ^* at x_0 and calculate corresponding n, respectively. With these data, we can estimate N_{In} at the point in question, that is, $C(x_0, t)$. In turn, knowing the latter, C_s, and t, an approximate value of D can be obtained by fitting $C(x_0, t)$ to Eq. (11). The value of D for $Pb_{0.99}Te_{1.01}$ at $T = 700°C$ is as high as $D(700°C) \approx 4.5 \times 10^{-9}$ cm^2/s, whereas for Na-doped PbTe it is an order-of-magnitude smaller. Further measurements, performed at the different temperatures, allowed revealing the activation-like dependence of D vs. T for both diffusion processes, as shown in **Figure 10**, while estimating a similar activation energy of $E_a \approx 1.5$ eV.

The two end surfaces of n-type samples that emerged after doping the PbTe crystals, the graded one *via* the In diffusion and homogeneous doping by I, were polished and positioned between two flat surfaces. The lower surface was kept at the

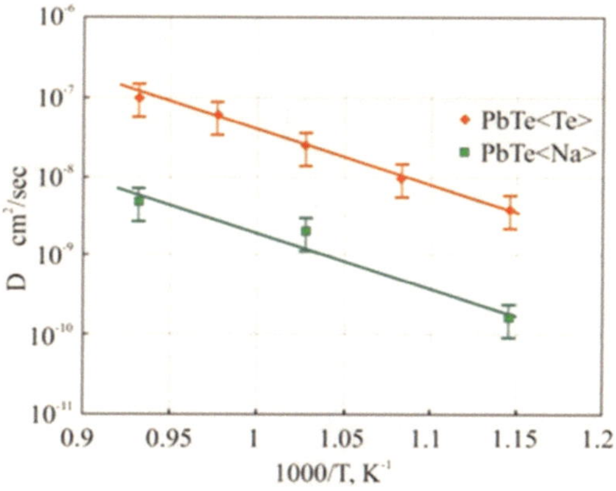

Figure 10.
Diffusivity of In atoms as a function of the reciprocal temperature in the $Pb_{0.99}Te_{1.01}$ and Na-doped PbTe crystals [58].

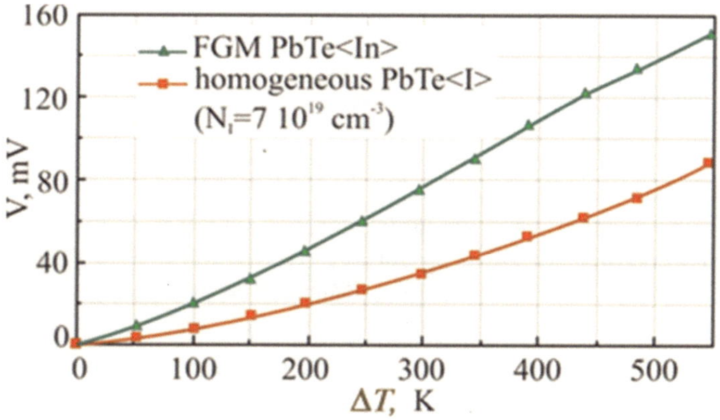

Figure 11.
The open-circuit voltage as a function of temperature difference in doped n-type PbTe crystals described in the text. Functionally-graded material (FGM) crystal—green; homogeneous crystal—red [58].

constant temperature $T_c = 320$ K, and the temperature of the heated upper surface T_h was gradually increased up to 870 K. The open-circuit voltages $V(\Delta T)$ measured at the varying temperature difference $\Delta T = T_h - T_c$ for the above samples are presented in **Figure 11**. It is worth noting that the $V(\Delta T)$ curves for samples of the two types are rather different. For the graded In-doping, $V(\Delta T)$—green line in **Figure 11**—is close to linear, and so the Seebeck coefficient $S = -d\Delta V(\Delta T)/d(\Delta T)$ extracted from this line is close to a constant value, equal $S \approx -250\,\mu\text{V/K}$ over a T-range of $320 - 850$ K. On the other hand, for the homogeneous I-doping, $V(\Delta T)$—red line in **Figure 11**—is notably concave so that $dV(\Delta T)/d(\Delta T)$ increases rather than saturates like for the green curve.

3.2 Bulk PbTe doped by In and I, prepared by SPS technology

3.2.1 Synthesis, sample preparation, and characterization

The experimental procedure of preparation and measurements of bulk polycrystalline PbTe samples doped by different impurities was presented in Ref. [24].

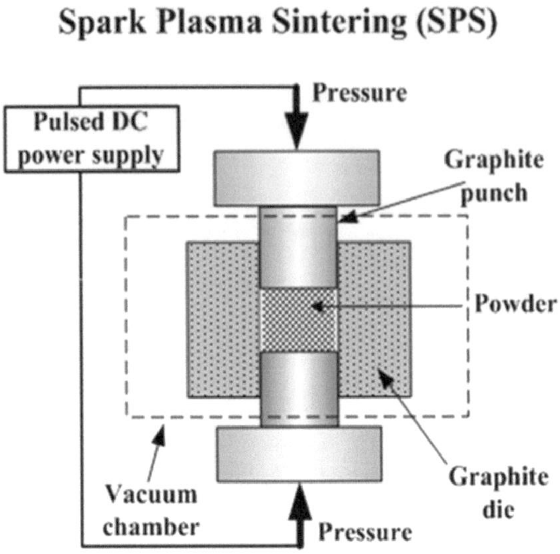

Figure 12.
Schematic view of Spark Plasma Sintering (SPS) apparatus.

The materials' synthesis was performed in quartz ampules evacuated to a pressure of 10^{-4} mbar. The ampules were thoroughly purified, which included washing in HNO_3: H_2SO_4 mixture and frequent cleaning with distilled water and isopropanol. Polycrystalline samples of $Pb_{1-x}In_xTe_{1-y}I_y$ were synthesized by melting the high-purity constitutive elements at a temperature of $T_s \simeq 1000^\circ C$ in a rocking furnace, whose inclination can be varied in the range of $\pm 30^\circ$ with a period of 15 sec to force mixing. The ampules were taken out and quenched in water at a temperature of $\sim 900^\circ C$. The obtained ingots were crushed into fine powders by ball milling and then densified by the SPS method, known also as pulsed electric current sintering (PECS), at $T = 500^\circ C$ for 30 min in a graphite mold under stress of 60 mPa in an argon (Ar) atmosphere. A schematic view of the SPS apparatus is presented in **Figure 12**. Disk-shaped samples of a high, over 98% theoretical, density with a height of 15 mm and cross-sectional diameter of 10 mm were thus output. Then, from each of them, 2 mm thick disks were cut and polished for the thermal diffusivity measurements, whereas the rest was used for the electrical and TE property measurements.

The phases and crystal structure were identified with the Bruker D8 Advance X-ray diffractometer using a Cu K_α radiation source ($\lambda = 1.5406$ Å, $\Delta\Theta = 9$ arcsec). The lattice parameter extraction and other crystallographic calculations, such as Rietveld refinement, were performed with the software GSAS-II. The structural analysis of the samples was performed using scanning electron microscopy (SEM) with a high-resolution SEM (HRSEM) apparatus JSM-6460LV (JEOL Ltd.). The instrument is equipped with EDXS facility for chemical analysis.

S and σ were measured by the SBA 458 Nemesis apparatus [24]. Measurements were made in Ar in the temperature range from 25 to 500°C. Thermal diffusivity α was measured with the Netzsch LFA 457 equipment [24]. The accuracy of the S and σ measurements was 6%, whereas that of R_H and the thermal conductivity κ measurement were estimated to be within 8%.

3.2.2 Results and discussion

3.2.2.1 Powder X-ray diffraction and microstructure analyses

Figure 13 represents the powder XRD (PXRD) patterns and scatter plot of lattice parameters of the $Pb_{1-x}In_xTe_{1-y}I_y$ specimens after synthesis vs. the In and I contents, in sub-figures **(a)** and **(b)**, respectively; also, in sub-figures **(c)** and **(d)**, the figure shows a top-view SEM image and EDXS elemental map, respectively, for the polished surface of a representative specimen [25].

In contrast to the two-peak XRD scan for single crystals in **Figure** 7, each PXRD pattern seen in **Figure 13(a)** shows several sharp peak characteristic of various crystallographic orientations of a rock-salt FCC single crystal, which indicates the single-phase, textured polycrystalline structure of the samples. A few independent peaks show up in order of descending amplitudes, namely (200), (220), (222), and (420), whereas the others are their crystallographic equivalents. Since the first peak hugely dominates over the others, we conclude these samples have a strong (200) texture; that is, most of their crystallites have the associated crystallographic orientation. Rietveld refinement of the XRD spectra was applied to the XRD spectra from fresh ingots and after thermoelectric measurements on cut samples. The FWHM, thus assessed of, for example, the dominant (200)-peaked XRD pattern, proved only

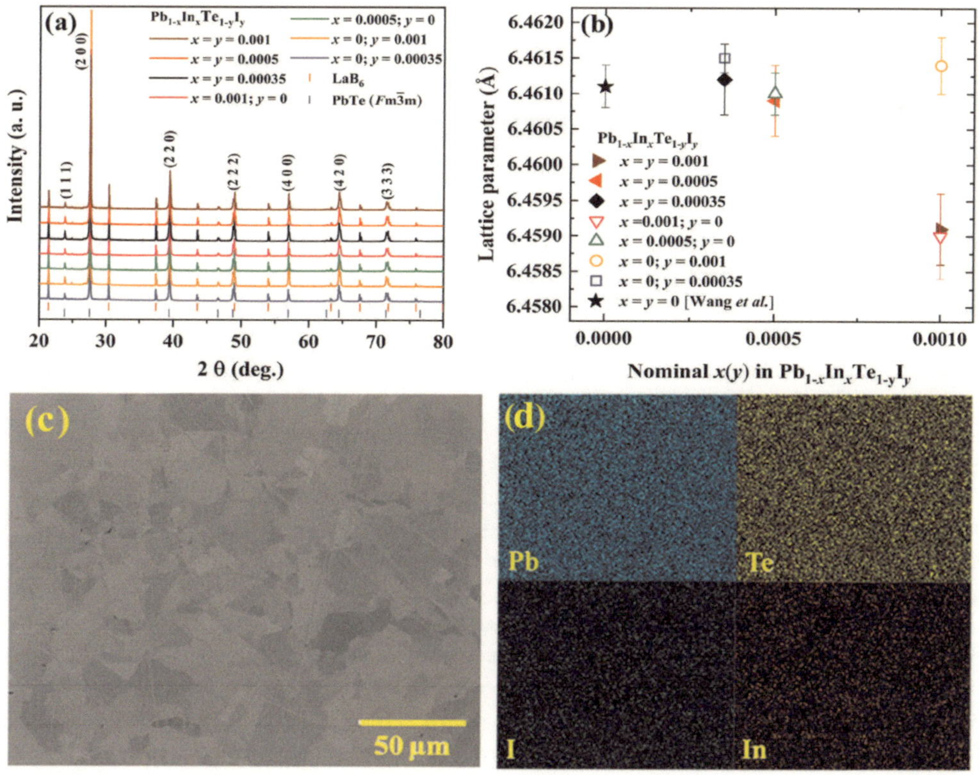

Figure 13.
The PXRD patterns (a) and lattice parameters scatter (b) for the $Pb_{1-x}In_xTe_{1-y}I_y$ compounds obtained with SPS. The lattice parameter for stoichiometric PbTe was taken from Ref. [59] for comparison and marked by a black star. SEM image (c) and EDXS elemental maps (d) for the $Pb_{0.999}In_{0.001}Te_{0.999}I_{0.001}$ specimen.

slightly larger than obtained for the single-crystal counterpart, which supports good crystallinity of the grains.

The lattice parameters for all investigated samples were determined with high accuracy due to applying LaB_6 as an internal standard, see **Figure 13(b)**. The lattice parameter of I-doped PbTe samples decreases only slightly with increasing y due to very similar ionic radii of I and Te, equal 2.06 Å and 2.07 Å, respectively. With increasing x, or x and y concurrently, the lattice parameters of In-doped PbTe or co-doped samples, in respect, decrease more significantly because of a smaller ionic radius of 0.94 Å for In^{3+} compared to that of 1.33 Å for Pb^{2+}. The estimated lattice parameters prove consistent with the lattice parameter value $a = 6.4611(3)$ Å for the stoichiometric PbTe crystal [59]. The microstructure analyses made using the SEM images of the type shown in **Figure 13(c)** discover varying degrees of darkness at the samples' surface due to different crystallographic orientations of the grains and show either rare well-cured grain boundaries or no distinct ones. No pores, voids exits, or cracks were found on the surface of all specimens. Studying the fracture surface of the pellets after SPS indicates that the grain (crystallite) sizes vary in the range from 20 to 60 μm, for In-doped, and to 200 μm for In and I co-doped samples. **Figure 13(d)** shows the distribution of the constitutive elements in the characteristic co-doped sample examined *via* the EDXS analysis. No demarcated regions containing detached elements Pb, Te, In, or I were detected, which supports the conclusion of the single-phase nature of the studied samples.

3.2.2.2 Electric, TE, and heat transport properties

Experimental investigation of the electric, TE, and heat transport properties, that is, σ, R_H, S, and κ, respectively, of $Pb_{1-x}In_xTe$ samples with $x = 0;\ \ 0.0005;\ \ 0.001;\ \ 0.01$ over the T-range of $300 - 800$ K was carried out and reported [24]. In Ref. [25], these properties for In and I co-doped PbTe, *viz.*, $Pb_{1-x}In_xTe_{1-y}I_y$, specimens with $x = y = 0.00035;\ \ 0.0005;\ \ 0.001$ were measured. For all samples, the Hall concentration n_H and Hall mobility were determined from the Hall effect and conductivity measurements using the well-known expressions:

$$n_H = (eR_H)^{-1}, \quad \mu_H = R_H\sigma \tag{12}$$

Due to accuracies of R_H and σ reported above, n_H and μ_H assessed using Eq. (12) will be accurate to within 6 and 14%, respectively.

For investigated samples, **Table 1** presents the transport properties measured at room temperature along with the corresponding values of n_H and μ_H, estimated due to Eq. (12). Note that the values of S in **Table 1** for the only I-doped PbTe specimens agree with the previously reported results [23, 60].

The I-doping efficiency practically equals unity; that is, it amounts to nearly one electron per I atom. For the In-doped PbTe specimens, the situation is much more intricate, with increasing N_{In} the electron concentration just slightly increases, from 4.8×10^{18} to 5.1×10^{18} cm^{-3}, which confirms that In is not a regular donor [37]. The following trend was observed when co-doping PbTe with both In and I. The electron concentration in 0.1 at. % I-doped PbTe specimens is 1.4×10^{19} cm^{-3}. When the PbTe specimen is co-doped with 0.1 at. % of In, the electron concentration drops to $\sim 8 \times 10^{18}$ cm^{-3}; thus, upon the In-doping many of band electrons become trapped at the quasi-local In levels. The concentration of electronic states at the quasi-local levels

$Pb_{1-x}In_xTe_{1-y}I_y$	S (μV/K)	σ (S/cm)	κ (W/mK)	n_H (10^{18} cm^{-3})	μ_H (cm^2/V·s)
$x = 0, y = 0.00035$	-175 ± 10.5	1020 ± 61	2.2 ± 0.18	5.4 ± 0.3	1180 ± 165
$x = 0, y = 0.001$	-92 ± 5.5	2590 ± 155	2.9 ± 0.23	14 ± 0.8	1150 ± 161
$x = 0.0005, y = 0$	-200 ± 12	600 ± 36	2.2 ± 0.18	4.8 ± 0.3	780 ± 109
$x = 0.001, y = 0$	-191 ± 11.5	660 ± 40	2.2 ± 0.18	5.1 ± 0.3	810 ± 113
$x = y = 0.00035$	-168 ± 10	800 ± 48	2.4 ± 0.19	6.6 ± 0.4	760 ± 106
$x = y = 0.0005$	-156 ± 9	1290 ± 77	2.0 ± 0.16	7.5 ± 0.45	1070 ± 150
$x = y = 0.001$	-155 ± 9	1150 ± 69	2.5 ± 0.20	7.6 ± 0.46	940 ± 132

Table 1.
The Seebeck coefficient, transport properties, Hall concentration, and mobility at $T = 300$ K of $Pb_{1-x}In_xTe_{1-y}I_y$ specimens fabricated by SPS [25].

is much smaller than that at the CB; therefore, the latter embeds the majority of doped electrons.

It can qualitatively be deduced that for 0.1 at. % I and 0.1 at. % In co-doped PbTe, the distance between the quasi-local level and the CB bottom is ~ 25 meV. In this case, the Hall coefficient R_H within a model of two bands, one of which is the CB and the other an impurity band (IB) emerging from broadening the In levels, is the weighted average value of the Hall coefficients of these bands [61]:

$$R_H = R_{H,c}\left(\frac{\sigma_c}{\sigma}\right)^2 + R_{H,i}\left(\frac{\sigma_i}{\sigma}\right)^2, \sigma = \sigma_c + \sigma_i. \tag{13}$$

Here, $R_{H,c}$ and σ_c are the Hall coefficient and conductivity of electrons in the CB, whereas $R_{H,i}$ and σ_i are their IB counterparts. In our case, $\sigma_i \ll \sigma_c$, so $(\sigma_c/\sigma)^2 \approx 1$ and $(\sigma_i/\sigma)^2 \ll 1$. Thus, the first term in Eq. (13) dominates the right-hand side, that is, $R_H \approx R_{H,c}$, unless $R_{H,i}/R_{H,c} \gg (\sigma/\sigma_i)^2$, which seems highly improbable.

Figure 14 presents the coefficient S vs. electron concentration, known as the Pisarenko plot, at $T = 300$ K for various samples specified in the figure. The experimental data are plotted together with the results of calculations for the single parabolic band (SPB) model and experimental data for the I-doped PbTe specimens from Refs. [23, 60]. The SPB model calculations are done for the electron effective mass of $m^* = 0.27\, m_0$, where m_0 is the free-electron mass, which is close to the DOS effective mass of ~ $0.24\, m_0$ reported for the stoichiometric PbTe [34] and assuming that the electron-momentum relaxation time $\tau(E)$ as a function of the electron energy E follows the simplified power law

$$\tau(E) \propto E^{\lambda - 0.5} \tag{14}$$

where λ is the so-called scattering parameter.

Usually, $\lambda = 0$ represents acoustical-phonon scattering, and $\lambda = 0.5$ evidently results in an energy-independent relaxation time $\tau(E) = \tau_0$, typical of electron scattering by neutral impurities. For n-type PbTe, the choices of these two λ's are the most sensible, though acoustical-phonon scattering is considered to be dominant [34, 62]. In some theoretical studies [34, 63], longitudinal optical (LO)-phonon scattering was

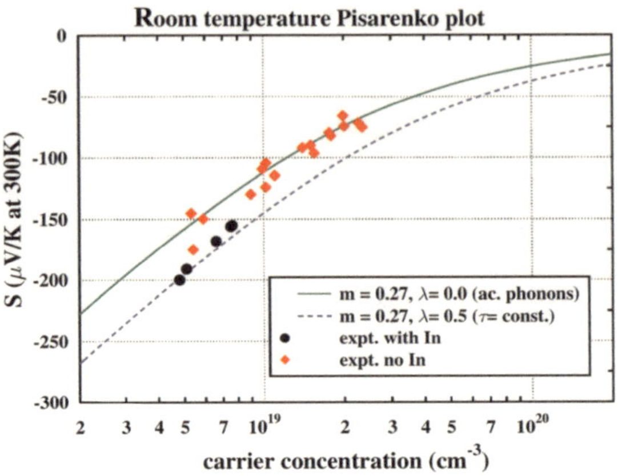

Figure 14.
The Seebeck coefficient as a function of the electron concentration for the $Pb_{1-x}In_xTe_{1-y}I_y$ specimens. The lines indicate theoretical predictions by the SPB model with $\mathbf{m} = m^*/m_0$ and two λ's as shown in the legend, and the symbols present selected experimental data [25].

found to be a dominating mechanism at low electron concentrations $n_H \leq 10^{18}\mathrm{cm}^{-3}$. Two above scattering parameters were used to analyze S within the SPB approximation, and consistency was found between its results and available experimental data for the single- and co-doped PbTe samples [25]. This confirms that practically no effective mass change occurs upon the In-doping, in contrast, for example, to the case of Tl-doped PbTe [53]. Therefore, solved In does not essentially modify CB in PbTe, yet forming the quasi-local level broadened when merging into the Bloch-states continuum at low T. As T rises the distance between the level and the CB bottom shortens, the latter proves above the former at $T \geq 300$ K, as sketched in **Figure 2B**.

Figure 15 shows the temperature dependences of S and σ in the sub-figures (a) and (b), respectively, for the n-type $Pb_{1-x}In_xTe_{1-y}I_x$ samples of seven compositions shown in the figures' legends. As was described above, the In atoms in PbTe form quasi-local impurity states [37], see also in **Figure 2**. In addition to the In-doping, the co-doping with I donates extra electrons, which can occupy empty states, in both the CB and IB. The three $S(T)$ curves in **Figure 15 (a)** for $x = y$ confirm this suggestion. These samples' $S(T)$ dependences are not typical of narrow band-gap semiconductors with bipolar conduction, which were also observed for only In- and I-doped PbTe samples. Hence, for co-doped PbTe samples, the intrinsic conduction range is effectively avoided due to a source of electrons provided by fully occupied quasi-local In-impurity states. On the other hand, **Figure 15 (b)** shows that for samples doped only with I and only with In and having close values of $S(300\ \mathrm{K})$ coefficient, so similar carrier concentration, the values of $\sigma(300\ \mathrm{K})$ are also close. Therefore, we can suggest that the values of electron mobility μ_H for these samples are also similar [23, 60].

The heat transport is determined by the total thermal conductivity κ. For a semiconductor at the range of one type of carriers (in this case, electrons), it consists of two components: the lattice thermal-conductivity κ_{lat} and electronic one κ_{el} due to the phonons and band electrons transport, respectively [34]:

$$\kappa = \kappa_{lat} + \kappa_{el}. \tag{15}$$

For PbTe at temperature $T = 300$ K, $\kappa_{lat} \approx 2$ W/mK [34], which is relatively high in comparison with low-temperature TE materials based on Bi_2Te_3 [12]. For doped PbTe

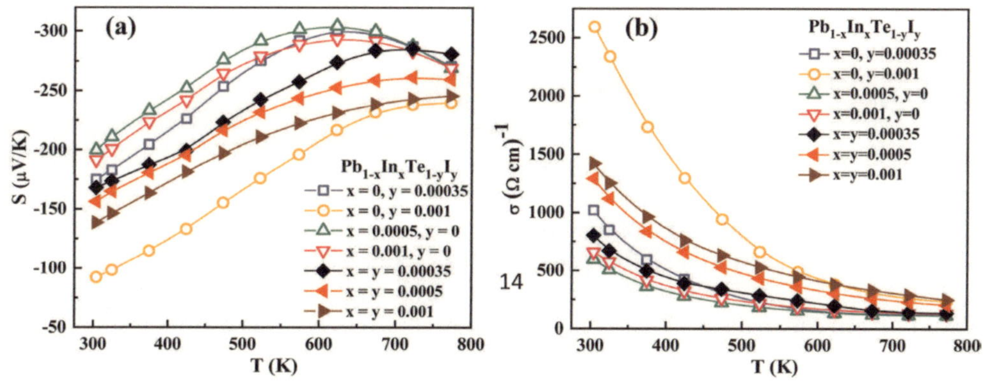

Figure 15.
Temperature dependences of the Seebeck coefficient (a) and electrical conductivity (b) for the $Pb_{1-x}In_xTe_{1-y}I_y$ specimens with the dopant fractions shown in the corresponding legends [25].

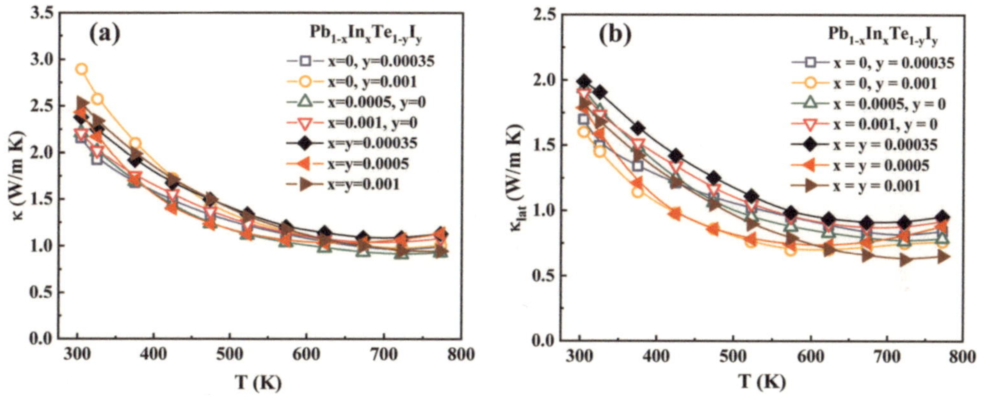

Figure 16.
The thermal conductivities as a function of temperature for the $Pb_{1-x}In_xTe_{1-y}I_y$ specimens: (a)—the measured total $\kappa(T)$; (b)—the lattice contribution $\kappa_{lat}(T)$ extracted as described in the text. The In and I fractions are shown in the sub-figures legends [25].

with electrical conductivity $\sigma \approx 10^3\ \Omega^{-1}\mathrm{cm}^{-1}$, the electronic thermal-conductivity at the same T equals $\kappa_{el} \approx 0.4\,\mathrm{W/mK}$, which is five times smaller than κ_{lat}. For this case, lattice contribution κ_{lat} practically dominates the TE figure of merit as follows:

$$Z = \frac{S^2\sigma}{\kappa_{lat}}. \tag{16}$$

Figure 16 presents the measured temperature dependence $\kappa(T)$ and extracted one $\kappa_{lat}(T)$ in sub-figures (a) and (b), respectively, for the $Pb_{1-x}In_xTe_{1-y}I_y$ specimens with the dopants fractions shown in the legends to the corresponding sub-figures. It is worth noting that the last five samples noted in **Figure 16(a)** coincide with those in **Figure 16(b)**, and that the extracted values of $\kappa_{lat}(300\ \mathrm{K})$ for all inquired samples are similar to those for stoichiometric PbTe equal $\sim 2\mathrm{W/mK}$ [34]. The $\kappa(T)$ curves shown in **Figure 16(a)** decrease over most of the investigated temperature range of $300-780\ \mathrm{K}$ indicating a quasi-metallic behavior. However, at $T > 650$ K they demonstrate either saturation or weak bending up. A slight effect of the intrinsic carriers is observed in samples with lower carrier concentrations. The lowest value of κ observed at $T = 780$ K equals 0.95 W/mK, typical for I-doped PbTe samples [25]. Theoretically, the lattice thermal conductivity κ_{lat} is defined by [34]

$$\kappa_{lat} = \frac{1}{3} c_V v, \tag{17}$$

where c_V is the specific heat at constant volume, and v is the propagation velocity of elementary wave excitations of lattice vibrations—phonons.

The values of $\kappa_{lat}(T)$ shown in **Figure 16(b)** were extracted by subtracting the ones of $\kappa_{el}(T)$ from the measured data on $\kappa(T)$ shown in **Figure 16(a)** as follows

$$\kappa_{lat}(T) = \kappa(T) - \kappa_{el}(T). \tag{18}$$

In turn, the values of $\kappa_{el}(T)$ were calculated using the celebrated Wiedemann-Franz law:

$$\kappa_{el}(T) = L\sigma(T)T, \tag{19}$$

where L is the Lorenz number. According to Refs. [5, 34], L can be calculated with Eq. (A2) presented in Appendix A for a general λ. Due to Eq. (A2), L depends on the physical parameters of the semiconductor under consideration *via* the reduced chemical potential μ^*, defined in Appendix A, so L is *not* a constant unless μ^* is such. Given the parameter λ that emerges in Eq. (14), the dependence $\mu^*(T)$ can be extracted by fitting the experimental data on $S(T)$ to Eq. (A1).

3.2.2.3 TE efficiency

Figure 17 presents the power factor (PF) $S^2\sigma$ and dimensionless figure of merit ZT as functions of temperature, in sub-figures (a) and (b), respectively, for $Pb_{1-x}In_xTe_{1-y}I_y$ specimens with the same compositions for both sub-figures as shown in their legends. As seen in **Figure 17(a)**, the co-doped specimen with $x = y = 0.0005$ presents the maximal PF of $S^2\sigma \approx 32$ $Wcm^{-1}K^{-2}$ at $T = 300$ K, and shows its decrease to about 16 $Wcm^{-1}K^{-2}$ at $T = 780$ K [25]. The enhancement in PF for the three co-doped

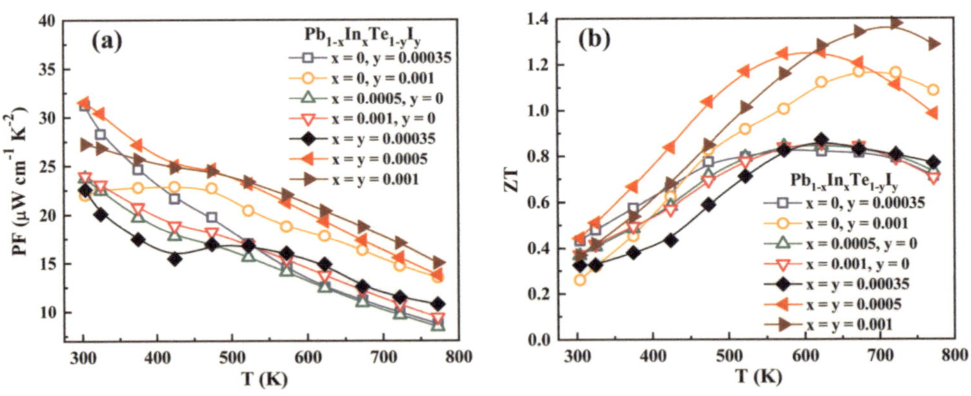

Figure 17.
PF—(a) and ZT—(b) as functions of temperature for the $Pb_{1-x}In_xTe_{1-y}I_y$ specimens. The same solved In and I fractions are shown in the sub-figures legends [25].

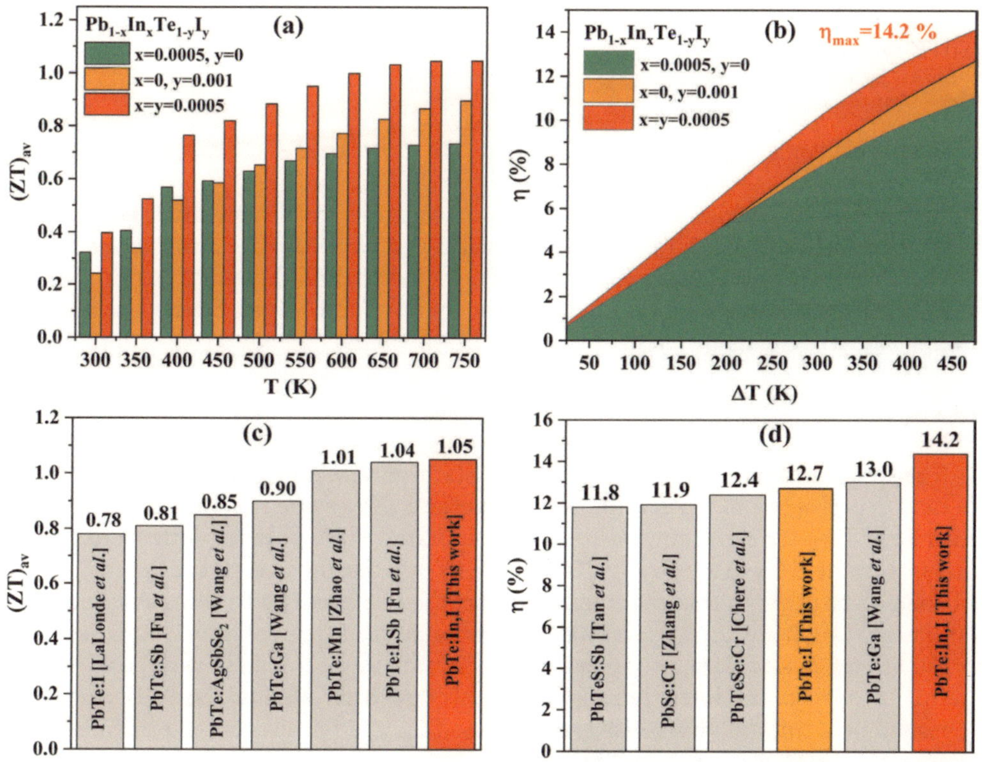

Figure 18.
The chart-graph of $(ZT)_{av}$ vs. T—(a) and graph of η vs. ΔT at fixed $T_c = 300$ K—(b) for $Pb_{1-x}In_xTe_{1-y}I_y$. The comparison charts of $(ZT)_{av}$—(c) and η—and (d) at $\Delta T = 480$ K for several high TE-performance n-type PbTe-based compounds.

specimens with $x = y$ as compared to only In- or I-doped specimens is caused by a high value of S with a simultaneous increase in σ over the entire temperature range of $300 - 780$ K. **Figure 17(b)** shows that the maximum value of ZT reached is $(ZT)_{max} \approx 1.25$ at $T = 725$ K for sample with $x = y = 0.001$, which is the high value reported for n-type PbTe [25]. However, due to Eq. (1), to achieve the maximal value of TE efficiency η for energy applications, a high value of ZT over the operation temperature range is necessary. The estimation of the integral average $(ZT)_{av}$, which emerges in Eq. (1), was done due to Eq. (2) by numerically calculating the areas under the ZT vs. T curves shown in **Figure 17(b)** and dividing the results by ΔT.

The results of the above-outlined $(ZT)_{av}$ calculations for the specimens in question over the T range from 300 to 780 K are presented in **Figure 18(a)**. As seen, the high value of $(ZT)_{av}$ up to ~ 1.05 for eventual $\Delta T = 480$ K is achieved with the co-doped sample, which at present is the highest one reported for single-composition n-leg TE element [25], as shown in **Figure 18(c)**. The η for the TE element consisting of n- and p-type legs has been estimated and compared with the literature data as shown in **Figure 18(b)** and **(d)**. The calculation was carried out, assuming the same TE properties of the p-legs attached to each n-type leg. It was shown that the p-type leg made from a $Ge_{1-x}Pb_xTe$ specimen has the TE properties over $300 - 780$ K temperature range close to those of n-type $Pb_{1-x}In_xTe_{1-y}I_y$ leg [64]. Also, η was calculated as a function of varying temperature difference ΔT for the specimens doped only with I or

In, and co-doped with In and I. High TE efficiency up to $\eta \approx 14.2\%$ for the above $\Delta T = 480$ K was achieved for the specimens co-doped with In and I [25, 65–70].

4. In-doped polycrystalline PbTe films on polyimide

In this section, we describe using the flash evaporation (FE) method, which is a specialized technology within the wider category of physical vapor deposition techniques. For the preparation of Bi_2Te_3-based specimens, and doped PbTe thin films on various substrates, using FE method was put forward and forged by the Dashevsky group [70].

4.1 Experimental arrangements

The schematic view of the setup for the film preparation by the above-noted method is presented in **Figure 19**. The FE method allows one to repeat in the synthesized PbTe film the same composition of evaporated PbTe powder with small dopant amounts in the $0.1 - 3$ at.% range. These are In or I, for the n-type specimens, and Na for the p-type ones. In this chapter, we report the In-doped PbTe films evaporated on a thinnest, $12 \pm 2\,\mu$m thick, polyimide substrate. More thick polyimide substrates up to 100 μm thickness were also investigated to study and optimize the deposition and thermal processing regimes thoroughly. Such a sandwich, that is, the PbTe film deposited on a thin polyimide film, can be used for the fabrication of a flexible-film TE module (FFTEM) in micro-TEGs [6–8].

We also practiced the PbTe deposition on both sides of the polyimide layer to boost the TE efficiency. The benefits of the polyimide material are low thermal conductivity of $\sim 3.5 \times 10^{-3}\,\mathrm{Wcm^{-1}K^{-1}}$ and flexibility [71]. Also, polyimide is a fully amorphous polymer, which does not require lattice matching, while excellently complying with PbTe regarding the thermal expansion.

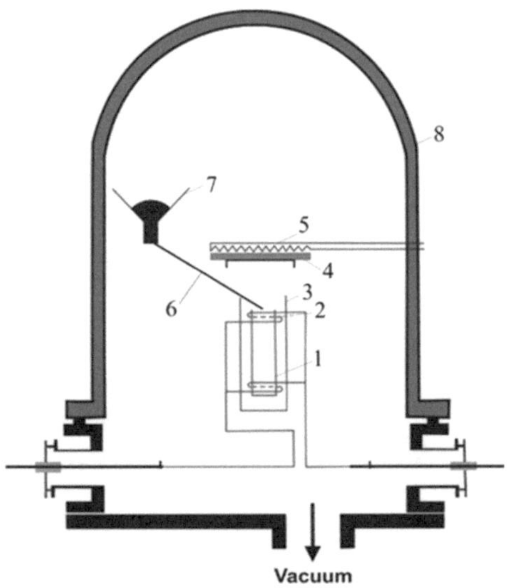

Figure 19.
The schematic view of the flash evaporation setup. 1—quartz crucible, 2—crucible heater, 3—heat shield, 4—substrate, 5—substrate heater, 6—guide, 7—powder vessel, 8—vacuum chamber [70].

The substrate temperatures T_s held during the film preparation ranged from 100 K to 573 K, which started with cooling the substrate by liquid nitrogen and gradually proceeded to heating it above room temperature. The film deposition rate upon FE, controlled by a piezo-sensor, was maintained at ~ 1.5 nm/s. This rapid evaporation/deposition process ensures a high film uniformity. The film's thickness was assessed by the predefined deposition time, while double-checking by direct measurements using a micro-interferometer MII-4 with an accuracy of 0.1 μm. For the chosen polyimide substrate, the thickness scalability for evaporated films sustains up to 4 μm and then stops. For thicker polyimide, the deposition can achieve thicknesses up to 6 μm at most.

A batch of films with In contents of $x = 0.01, 0.02, 0.03$, and the thicknesses of 1,2,3,4 μm each, were obtained. A part of the batch was removed from the evaporation chamber, and the remained samples were annealed at $T_a = 673 - 723$ K for 0.5 h in a pure Ar atmosphere at pressure $P = 0.9$ atm.

The method of XRD structural analysis, developed specifically for the PbTe films [72], was employed. The scanning electron microscopy (SEM) images were taken with a Quanta 200 environmental high-resolution scanning electron microscope (HRSEM) equipped with an energy-dispersive X-ray spectroscopy (EDXS) facility. An AFM apparatus, Autoprobe CP (Park Systems Inc.), was employed to study the morphology and topography of the films' surfaces.

For the investigation of the TE and transport properties, that is, S, σ, and R_H, of the thin films under consideration over a wide temperature range of $100 - 500$ K, a unique measurement setup has been developed and used [72].

The measurement of R_H was carried out in permanent magnetic fields up to 2T in two directions of the electrical and magnetic fields, and the final results are averaged over the recorded ones for these measurements. These measurements were performed with an 8% accuracy. The measurement accuracy of T was $0.1 - 0.2$ K and of the magnetic field $\pm 3\%$. The uncertainty of the S and σ measurements was $\pm 6\%$ for each.

Heat transfer through the films was measured using the method of dynamic lattices [73]. The accuracy of thermal diffusivity α measurements was $\pm 10\%$. Schematics of the appropriate measurement setup can be found in Ref. [72]. With the α data, the total thermal conductivity κ is calculated according to the well-known equation:

$$\kappa = \alpha \rho c_P. \tag{20}$$

Here ρ is the single-crystal density, and c_P is the heat capacity at constant pressure, which was calculated with the Dulong-Petit (intermediate- and high-temperature) limit. Hence, the calculated κ accuracy can safely be taken as ± 10 %.

4.2 Results and discussion

4.2.1 Structural properties

Figure 20 shows the XRD spectra of the as-grown films with $x = 0.01$, fabricated at different T_s shown in the figure caption. Like PbTe single crystal, the films tend to grow mostly in the $\langle 001 \rangle$ or $\langle 111 \rangle$ directions and equivalent to them, while the former is yet preferable to the latter due to lower surface energy. That the films at all T_s have a strong (200) texture with small admixture of the equivalent (220) one that notably decreases with increasing T_s, as is seen from the curves' evolution in **Figure 20**. Upon that increase, the additional XRD peaks corresponding to other growth planes almost completely disappear, whereas the main one sustains its several orders of magnitude

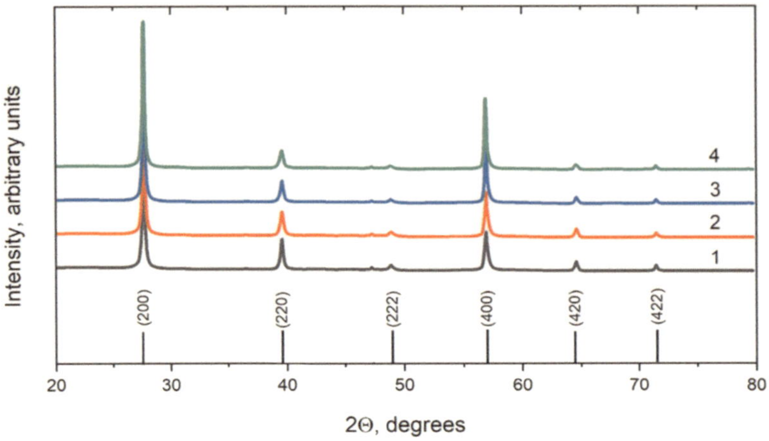

Figure 20.
The XRD spectra of an as-grown 1 μm thick $Pb_{0.99}In_{0.99}Te$ film vs. the substrate temperature T_S. The numbers near the right ordinate marking the curves relate to the following T_S: 1—373 K; 2—423 K; 3—473 K; 4—523 K. The main peaks are Rietveld-refined.

dominance. These data evidence a good crystallinity of the grains, which improves by preheating the substrate. Also, no XRD peaks corresponding to the foreign phases were found, which indicate that the films are single phase.

Figure 21 presents the results of structural and surface morphology examination for a $Pb_{0.99}In_{0.01}Te$ film. Its sub-figures (a) and (b) show a HRSEM close top-view

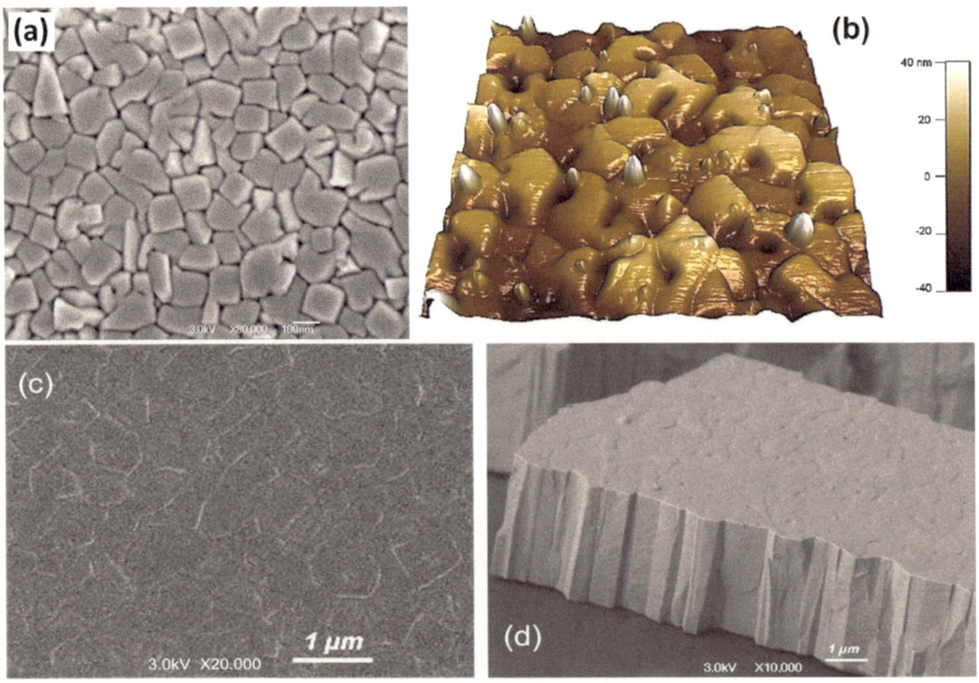

Figure 21.
Top close-view HRSEM image—(a) and AFM 3D image—(b) of an as-deposited 1μm thick $Pb_{0.99}In_{0.01}Te$ film. Post-annealing top-view HRSEM image of the same film—(c) and cross-sectional/3D view HRSEM image of a 4μm thick film with the same composition—(d).

image and AFM 3D image of the as-grown film with a thickness of 1 ± 0.1 μm, respectively, whereas the sub-figures (c) and (d) show the top-view and cross-sectional HRSEM images of two annealed samples of the same composition and thicknesses 1 ± 0.1 and 4 ± 0.1 μm, respectively, as specified in the figure caption. **Figure 21(a)** uncovers the grain structure of the as-deposited film with the grain-boundary (GB) exits. The film's surface shows no cracks or pores, while the GBs' exit fingerprint lines on the surface occupy a negligible area. The averaged lateral grain size in this film is 180 ± 30 nm and scales about proportionally with the film thickness up to the limit of film growth scalability noted above. AFM scan in **Figure 21(b)** shows a detailed film-surface morphology over 5×5 μm area and allows for estimating the RMS surface roughness of ~ 25 nm. **Figure 21(c)** and **(d)** evidences that the annealing facilitates closer packing of the crystallites in the film, which smooths the GBs fingerprints on the surface. On the other hand, **Figure 21(a)** and **(d)** shows that the films have a columnar grain structure altogether, which tightens after annealing.

4.2.2 Transport and TE properties

Figure 22 demonstrates the T dependences of R_H, S, and σ in the range of temperatures from 10 to 400 K for the prepared $Pb_{1-x}In_xTe$ films with $x = 1$, 2, and

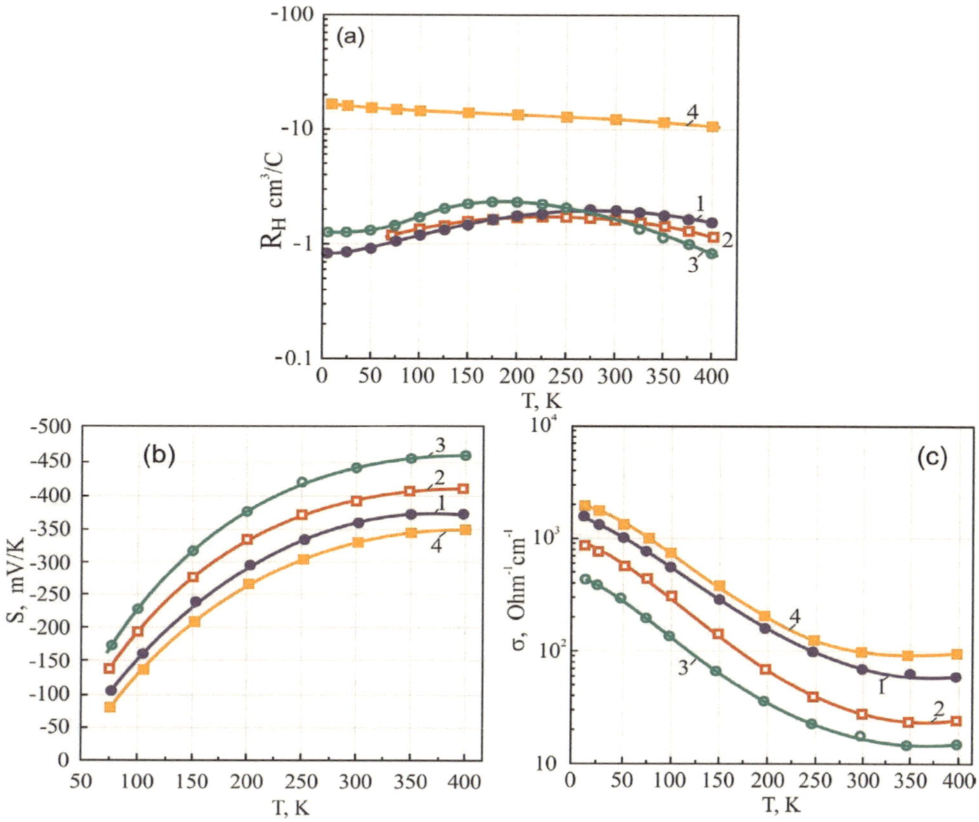

Figure 22.
The Hall coefficient—(a); Seebeck coefficient—(b); and electrical conductivity—(c) as functions of temperature for the annealed $Pb_{1-x}In_xTe$ films: $x = 0.01$—1; $x = 0.02$—2; $x = 0.03$—3; and $PbTe_{0.999}I_{0.001}$ film—4; the films' thickness is 2 μm.

$Pb_{1-x}In_xTe_{1-y}I_y$	S (μV/K)	σ (Ω^{-1}cm^{-1})	κ (W/mK)	κ_{lat} (W/mK)	$Z \times 10^3$ (K^{-1})	Reference
$x = 0.01, y = 0^a$	-350 ± 21	80 ± 5	1.5 ± 0.15	1.4	0.90	This work
$x = 0.02, y = 0^a$	-400 ± 24	40 ± 2	1.4 ± 0.1	1.4	0.80	This work
$x = 0.03, y = 0^a$	-440 ± 26	20 ± 1	1.3 ± 0.1	1.3	0.75	This work
$x = 0, y = 0.0001^a$	-330 ± 20	100 ± 6	1.8 ± 0.2	1.7	0.60	[72]
$x = 0.001, y = 0^b$	-205 ± 12	600 ± 36	2.2 ± 0.2	1.9	1.10	[24]
$x = y = 0.001^b$	-155 ± 9	1100 ± 66	2.5 ± 0.25	2.0	1.05	[25]

[a] *Polycrystalline-film samples.*
[b] *Polycrystalline-bulk samples.*

Table 2.
The Seebeck coefficient, electrical and thermal conductivities, and TE figure of merit of the n-type $Pb_{1-x}In_xTe_{1-y}I_y$ polycrystalline-film and polycrystalline-bulk (SPS) specimens at $T = 300$ K.

3 at. %. The figure also includes the reference curves for a $PbTe_{0.999}I_{0.001}$ film of the same thickness. For all the films, **Figure 22(a)** shows $R_H < 0$ over the entire temperature range, which confirms the electronic (n-type) conductivity. The electron Hall concentration n_H for the prepared $Pb_{1-x}In_xTe$ films was determined from measured R_H using Eq. (12). For all the films up to heavily doped $Pb_{0.97}In_{0.03}Te$ one, the n_H does not exceeded 5×10^{18} cm^{-3}. As seen in **Figure 22(a)**, the non-monotonic temperature dependence of R_H is observed in the In-doped PbTe films, which differs from that for the film doped with the I impurity that is a regular donor in PbTe [72].

Assuming that experimental n_H may substitute n in Eq. (A4), and adopting the SPB model with the electron effective mass $m_n^* \approx 0.27\ m_0$ suitable for bulk PbTe crystal, this equation can be solved for μ^*, and so the Fermi-level energy E_F be extracted. It turns out that the thus estimated E_F lies $0.06 - 0.08$ eV above E_c at $T = 80$ K. As the temperature increases, E_F rapidly decreases, crossing the conduction band bottom in the $T = 250 - 350$ K range.

The Seebeck coefficient S of the studied n-type films is also negative over the $80 - 400$ K temperature range, see **Figure 22(b)**, which is typical of semiconductors with the n-type conductivity. The value of S for $Pb_{0.97}In_{0.03}Te$ film achieves the value on a level of $S \approx -450\,\mu$V/K at 300 K. **Figure 22(c)** presents the temperature dependence electrical conductivity $\sigma(T)$. However, the behavior of $\sigma(T)$ differs from the temperature dependence of the Seebeck coefficient $S(T)$. An increase in the In dopant content in PbTe leads to a strong decrease in σ over the entire range of operation temperatures. Such an unusual behavior can be explained by a strong decrease in the electron drift mobility, represented by experimental μ_H, given by Eq. (12), with the growth of In concentration N_{In}.

Table 2 summarizes the TE, electric, and thermal properties, such as Seebeck coefficient S, electrical conductivity σ, and thermal conductivity κ, in respect, and TE figure of merit Z as well, at $T = 300$ K of the n-type polycrystalline $PbTe_{1-x}In_xTe$ films with a thickness of 1 ± 0.1 μm deposited on the thin polyimide substrate. The same parameters for n-type $Pb_{1-x}In_xTe_{1-y}I_y$ bulk (SPS) polycrystal samples are also shown there for comparison. Note that the value of parameter $Z(300\ \text{K})$ estimated for the polycrystalline $Pb_{0.99}In_{0.01}Te$ film with the electron concentration $n = 3 \times 10^{18}$ cm^{-3} is similar to, though a bit smaller, than that for the bulk polycrystal $Pb_{0.999}In_{0.001}Te$, cf. the lines 1 and 5 in the column Z of **Table 2**.

The lattice thermal conductivity κ_{lat} column in **Table 2** was calculated according to Eqs. (15), (19), and (A2). A significant decrease in the value of κ_{lat} for polycrystalline

$Pb_{1-x}In_xTe$ films in comparison with that for bulk polycrystalline samples [24, 25] may be due to the scattering of long-wavelength phonons on the grain boundaries [5].

4.2.3 Scattering of electrons in the polycrystalline n-$Pb_{1-x}In_xTe$ films

As shown in sub-subsection 4.2.2, in the polycrystalline n-$Pb_{1-x}In_xTe$ films deposited on the polyimide substrate, a significant decrease in the conductivity, seen in **Figure 22(c)**, which results in the strong μ_H decrease over the entire operating temperature range, since R_H, and hence n_H, variation is very weak. This may be due to an additional electron scattering on the grain boundaries. The increase of $|S|$ with increasing T in the $Pb_{1-x}In_xTe$ films, seen in **Figure 22(b)**, points at a larger scattering parameter λ, see Eq. (14) as compared to one specific for the scattering on the acoustic phonons ($\lambda = 0$), dominant in bulk $Pb_{1-x}In_xTe$ single crystals, as seen in Section 3. Such a difference is due to some extra electron-scattering mechanism inherent to the poly-crystallinity, and this effect was discovered for the nanocrystalline PbTe films [72, 74–76].

When λ is not known in advance, the parameter μ^* can be extracted from Eq. (A4) replacing n by n_H, which was *ad hoc* measured for the $Pb_{1-x}In_xTe$ films over the $200 - 300$ K temperature range. With the thus acquired μ^* data, by trial-and-error fitting the S data, measured at the same temperatures, to Eq. (A1), the parameter λ was extracted to be in the range of $1.5 - 2$. A possible interpretation of this result is as follows. At the grain boundaries in the polycrystalline $Pb_{1-x}In_xTe$ films, some energy barriers for electrons emerge. This is sketched in **Figure 23**, which presents the speculated energy-band diagram for the films in question. Such barriers allow fast electrons, that is, those with $E > eV_B$, to pass through while delaying slow electrons $E < eV_B$. Recently, a detailed semi-phenomenological explanation of the observed decrease of σ, and hence, μ_H with increasing temperature, based on a quantitative separation of electron scattering on the grain boundaries, have been proposed [77].

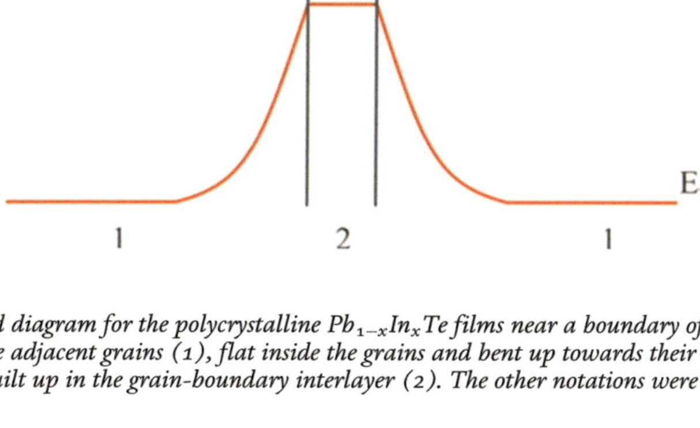

Figure 23.
Suggested energy band diagram for the polycrystalline $Pb_{1-x}In_xTe$ films near a boundary of two grains. It depicts the bands' edges in the adjacent grains (1), flat inside the grains and bent up towards their boundary due to the energy barrier eV_B built up in the grain-boundary interlayer (2). The other notations were defined previously in the text.

4.3 Other miscellaneous properties

Closing this section, we briefly touch on a few properties that are aside from the main focus of the present study but may be important for use in flexible devices. First, these films withstand well the thermal cycling without hysteresis on the scale of temperature variations allowed in the annealing process. Second, these nanocrystalline films have much larger mechanical strength than their brittle bulk poly- and single-crystal counterparts. In our experience, many of these films at our disposal have beyond 20 years longevity with preserving their physical properties. Finally, the adhesion of the polyimide layers to SiO_2 is good, with a peel strength of $50\ \text{g/mm}$ for a 12μm thick layer, and increases proportionally with the increase in the thickness. This property may be useful for merging the thin PbTe film coolers with microelectronics chips.

5. Conclusions

5.1 Brief summary

In this study, experimental and theoretical investigations on thermoelectrics of PbTe-based compounds are presented. With this, we state a complete reproducibility of all results of our measurements to within specific experimental errors. This chapter contributes to a series of investigations for a long period until nowadays related to the progress of advanced bulk and film PbTe TE materials over a wide operating temperature range. The main results of this research can be briefly summarized as follows.

1. Fundamental knowledge of the In level behavior in PbTe is reported. The electronic structure calculations confirm the formation of the In quasi-local level broadened into a narrow impurity band (IB) close to the conduction band edge. The power factor (PF) enhancement for co-doped PbTe samples was explained by the favorable chemical potential position provided by charge carriers released from the IB.

2. Significant success is achieved in increasing the TE efficiency η over a wide temperature range of $500 - 850\ \text{K}$ in In and I co-doped single- and poly-crystals of PbTe. The maximum value of the dimensionless TE figure of merit reaches the value of $(ZT)_{\text{max}} \approx 1.25$ at $T = 723$ K for $Pb_{0.999}In_{0.001}Te_{0.999}I_{0.001}$ samples. The temperature-averaged dimensionless TE figure of merit at the temperature difference of $\Delta T = 475$ K for this compound $(ZT)_{\text{av}} \approx 1.05$ is among the best reported values. For a TE-element, including the proposed n-PbTe leg and a p-type leg with the same TE properties, a value of η has been assessed. An extremely high η of up to $\sim 14.2\%$ for the above ΔT is achieved for the co-doped PbTe-based specimens with the concentration of impurities of $0.05 - 0.1\ \text{at. \%}$.

3. The TE figure of merit Z for the $Pb_{0.99}In_{0.01}Te$ films on a thin flexible polyimide substrate reaches the value $Z \approx 0.90 \times 10^{-3}\ \text{K}^{-1}$ at 300 K, which is comparable to Z of the In-doped bulk poly- and single-crystalline n-PbTe. Such a Z for these thin films opens new horizons for the fabrication of compact micro-TEGs.

5.2 Bulk and thin-film polycrystalline In-doped PbTe: Which is better for thermoelectrics?

The answer to the question in the subsection title is ambiguous and depends on the specific TE device, in which one of these two materials is to be applied. Since their figures of merit are close, seen in **Table 2**, thin-film In-doped PbTe is preferable for use in compact, lightweight, and wearable/flexible TE devices. On the contrary, since Z of the SPS fabricated n-type doped PbTe is yet, though not much, larger than Z of the FE deposited In-doped PbTe film, using the former is beneficial for TE systems, for which the main target is high power yield.

Let us emphasize that PbTe-based semiconductors with n-type conductivity have unbeatable TE performance over the mid-T range of $500 - 900$ K, from both $(ZT)_{max}$ and $(ZT)_{av}$ viewpoints. PbTe-based TEGs have been used in NASA spacecraft missions for the last decade [34].

5.3 Recent publications and future directions

In the past 5 years, a huge number of studies devoted to PbTe-based thermoelectrics, and updating the physical vapor deposition of n-type doped polycrystalline films targeted to both thermoelectrics and infrared photonic detectors. In this connection, we recommend solid review papers [78–80]. In our opinion, the future directions will be with the technology benchmark improvements of the TE devices.

Acknowledgements

The authors acknowledge the gratitude to Professor T. Parashchuk (AGH University of Science and Technology, Kraków, Poland) for fruitful discussions and for sharing the experimental research on In-doped PbTe in recent years.

Author contributions

S.N. and Z.D. contributed to conceptualization; S.N. and M.A. supervised the theoretical part; Z.D. and R.S. contributed to the experimental part supervision and critical discussion of experimental data. S.N. and Z.D. contributed to writing—original draft preparation; R.S. contributed to resources and funding acquisition; M.A. contributed to text editing and writing—final manuscript. All authors discussed the text and agreed with the final version of the chapter.

Conflict of interest

The authors declare no conflict of interest.

Dedication

S. Nemov and Z. Dashevsky dedicate this work to his late supervisor, Prof. V.I. Kaĭdanov, who pioneered the research on In-doped PbTe.

A. Appendices

Appendix A: Formulas from semiconductors theory

The equations displayed below are valid for the general parabolic band structure and an arbitrary quasi-elastic electron scattering parameter λ ($\lambda = 0$ for scattering by acoustic phonons), introduced in the main text.

The Seebeck coefficient S, and the Lorenz number L that emerges in Eq. (19), are calculated by

$$S = \mp \frac{k_B}{|e|} \left[2 \frac{F_{\lambda+1}(\mu^*)}{F_\lambda(\mu^*)} - \mu^* \right], \tag{21}$$

where signs − and + refer to n- and p-type materials, respectively, and

$$L = \left(\frac{k_B}{e} \right)^2 \left\{ \frac{3F_{\lambda+2}(\mu^*)}{F_\lambda(\mu^*)} - \left[\frac{2F_{\lambda+1}(\mu^*)}{F_\lambda(\mu^*)} \right]^2 \right\}. \tag{22}$$

In Eqs. (A1) and (A2), k_B is the Boltzmann constant, $e < 0$ is the electron charge; $\mu^* = \mp(E_\pm - \mu)/(k_B T)$ is the reduced chemical potential, in which $E_+ = E_c$ is the conduction-band bottom, $E_- = E_v$ is the valence-band top, μ is the chemical potential (Fermi energy E_F in the degenerate case), and $F_\alpha(x)$ are the Fermi integrals:

$$F_\alpha(x) = \int_0^\infty \frac{y^\alpha}{1 + \exp(y - x)} dy, \tag{23}$$

Within the same band-structure model, the carriers' concentration equals

$$n \text{ or } p = 4\pi \frac{\left(2m_d^* k_B T\right)^{3/2}}{h^3} F_{1/2}(\mu^*). \tag{24}$$

Above, m_d^* is the DOS effective mass, defined by $m_d^* = N^{2/3}(m_1 m_2 m_3)^{1/3}$, where the mass parameters and N are defined in the main text.

Appendix B: DFT simulation details and other perspectives

The electronic structure of PbTe doped with In and I was computed using the KKR method and CPA to model chemical disorder, which was implemented *via* the Munich SPR-KKR package [81]. The focus was on impurity-induced changes near the conduction band edge, which are relevant to the observed transport properties. Hence, the comparison with the known band structure of extrinsically undoped, highest-stoichiometry PbTe [34] was not performed.

Calculations employed the experimental lattice constant of $a = 6.46$ Å. A dense **k**-points grid in the irreducible Brillouin zone was applied, namely, with 4000 points for self-consistency, and 107 points for calculating the DOS and BSF to ensure accuracy near the band edge and at low impurity concentrations, down to 0.1%. The exchange-correlation (XC) effects were described *via* local density approximation (LDA) with the XC potential quantitative parameterization due to Vosko, Wilk, and Nusair [82]. Fully relativistic, full-potential calculations were performed with an angular momentum cut-off of $l_{max} = 3$. The self-consistency cycle relied on tight convergence criteria of 10^{-5} Ry accuracy for both total energy and E_F, with the latter determined using the Lloyd formula [83].

The paper [24] reported investigation of the band structure, transport and TE properties, such as σ, κ, and S, respectively, of n-type polycrystalline $Pb_{1-x}In_xTe$ samples series synthesized by SPS technique, over the entire temperature range of $298 - 773$ K.

Abbreviations

TE	thermoelectric
TEG	TE generator
SPS	spark plasma sintering
FE	flash evaporation
CB	conduction band
IB	impurity band
DOS	density of states
SPB	simple parabolic band
BSF	Bloch spectral density function
XRD	X-ray diffraction
FWHM	full width at half maximum
PXRD	powder XRD
SEM	scanning electron microscope(y)
HRSEM	high-resolution SEM
PF	power factor

Author details

Sergey Nemov[1,2†], Mark Auslender[3*], Rony Shneck[4] and Zinovi Dashevsky[4*†]

1 Department of Radiophysics and Electronics, Saint Petersburg's State Electrotechnical University, Saint Petersburg, Russia

2 Peter The Great Saint-Petersburg's Polytechnic University, Saint Petersburg, Russia

3 School of Electrical and Computer Engineering, Ben Gurion University of the Negev, Beer Sheva, Israel

4 Department of Materials Engineering, Ben-Gurion University of the Negev, Beer-Sheva, Israel

*Address all correspondence to: marka@bgu.ac.il and dashevsky45@mail.ru, zdashev@bgu.ac.il

† These authors contributed equally.

References

[1] Zoui MA, Bentouba S, Stocholm JG, Bourouis M. A review on thermoelectric generators: Progress and applications. Journal of Energies. 2020;**13**:3606

[2] Mason LS, Casani J, Elliott JO, Fleurial JP. A small fission power system for NASA planetary science missions. Journal of the British Interplanetary Society. 2011;**64**:76

[3] Oualid SE, Kosior F, Dauscher A, Candolfi C, Span G, Mehmedovic E, et al. Innovative design of bismuth-telluride-based thermoelectric micro-generators with high out power. Energy & Environmental Science. 2020;**13**:3579

[4] Goncalves AP, Godart C. Alternative strategies for thermoelectric materials development. In: Zlatic V, Hewson A, editors. New Materials for Thermoelectric Applications: Theory and Experiment. New York: Springer; 2013

[5] Goltzman BM, Dashevsky ZM, Kaĭdanov VI, Kolomoets NV. Film Thermoelements: Physics and Application. Moscow: Nauka; 1985 (in Russian)

[6] Karthikeyan V, Surjadi JU, J. Wongat at el. Wearable and flexible thin film thermoelectric module for multi-scale energy harvesting. Journal of Power Sources. 2020;**455**:227983

[7] Tappura K, Jaakkola K. A thin-film thermoelectric generator for large-area applications. Proceedings of EUROSENSORS. 2018;**2**:779

[8] Settaluri KT, Lo H, Ram RJ. Thin thermoelectric generator system for body energy harvesting. Journal of Electronic Materials. 2012;**41**:984

[9] Venkatasubramanian R, Siivola E, Colpitts T, O'Quinn B. Thin-film thermoelectric devices with high room-temperature figures of merit. Nature. 2001;**413**:597

[10] Fan P, Zheng Z, Cai Z, et al. The high performance of a thin film thermoelectric generator with heat flow running parallel to film surface. Applied Physics Letters. 2013;**102**:033904

[11] Ioffe AF, Moĭzhes BY, Stil'bans LS. Energetic applications of thermoelements. Soviet Physics: Solid State [English translation from Fizika Tverdogo Tela]. 1960;**11**:2834

[12] Dashevsky Z, Skipidarov S. Investigating the performance of bismuth– Antimony telluride. In: Skipidarov S, Nikitin M, editors. Novel Materials and Device Design Concepts. Thermoelectric Power Generation. New York: Springer; 2019. pp. 1-20

[13] Snyder GJ, Toberer ES. Complex thermoelectric materials. Nature Materials. 2008;7:2

[14] Sorell CC, Sugihara S, Nowotny J. The dimensionless figure of merit ZT for various thermoelectric materials. In: Materials for Energy Conversion Devices. Cambridge, England: Woodhead Publishing Ltd.; 2005

[15] Ben-Yehuda O, Shuker R, Gelbstein Y, Dashevsky Z, Dariel MP. High textured Bi_2Te_3-based materials for thermoelectric energy conversion. Journal of Applied Physics. 2007;**101**:25

[16] Dashevsky ZM, Konstantinov PP, Skipidarov SY. New way of application for thermoelectric energy converters. Semiconductors. 2019;**53**:861

[17] Meroz O, Elkabets N, Gelbstein Y. Enhanced thermoelectric properties of

n-type $Bi_2Te_{3-x}Se_x$ alloys following melt-spinning. ACS Applied Energy Materials. 2020;**3**:2090

[18] Fu L, Yin M, Wu D, Li W, Feng D, Huang L, et al. Large enhancement of thermoelectric properties in n-type PbTe via dual-site point defects. Energy & Environmental Science. 2017;**10**:2030

[19] Chen Z, Jian Z, Li W, Chang Y, Ge B, Hanus R, et al. Lattice dislocations enhance thermoelectric PbTe in addition to band convergence. Advanced Materials. 2017;**29**:1606768

[20] Yue L, Cui W, Zheng S, et al. Enhanced thermoelectric performance of In and Se co-doped GeTe compounds. Technology. 2020;**9**:4106

[21] Witting IT, Grovogui JA, Dravid VP, Snyder GJ. Thermoelectric transport enhancement of Te-rich bismuth antimony telluride ($Bi_{0.5}Sb_{1.5}Te_{3+x}$) through controlled porosity. Journal of Materiomics. 2020;**6**:532

[22] Witting IT, Chasapis TC, Peters FRM, Heinz NA, Hautier G, Snyder GJ. The thermoelectric properties of Bismuth telluride. Advanced Electronic Materials. 2019;**5**:1800904

[23] Gelbstein Y, Dashevsky Z, Dariel MP. High performance n-type PbTe-based materials for thermoelectric applications. Physica B: Condensed Matter. 2005;**363**:96

[24] Parashchuk T, Dashevsky Z, Wojciechowski K. Feasibility of a high stable PbTe: In semiconductor for thermoelectric energy applications. Journal of Applied Physics. 2019;**125**: 245103

[25] Wojciechowski K, Parashchuk T, Wiendlocha B, Cherniushok O, Dashevsky Z. Advanced electronic structure engineering achieved by In and I doping in PbTe for the manufacture of a high-performance n-type thermoelectric leg. Journal of Materials Chemistry C. 2020;**8**:13270

[26] Bomstein N, Spindonov G, Dashevsky Z, Gelbstein Y. Thermoelectric, structural and mechanical properties of spark-plasma-sintered sub-micro- and microstructured p-type $Bi_{0.5}Sb_{1.5}Te_3$. Journal of Electronic Materials. 2012;**41**:1546

[27] Bali A, Chetty R, Sharma A, Rogl G, Heinrich P, Suwas S, et al. Thermoelectric properties In and I-doped PbTe. Journal of Applied Physics. 2016;**120**:175101

[28] Bali A, Kim I-H, Rogl PF, Mallik RC. Thermoelectric properties of two-phase PbTe with indium inclusions. Journal of Electronic Materials. 2014;**43**:1630

[29] Zheng Y, Tan XY, Cheng X, et al. Thermal stability and mechanical response of Bi2Te3-based materials for thermoelectric applications. ACS Applied Energy Materials. 2020;**3**:2078

[30] Gelbstein Y, Dashevsky Z, Dariel MP. The search for mechanically stable PbTe based thermoelectric materials. Journal of Applied Physics. 2008;**104**:33

[31] Gelbstein Y, Gotesman G, Lishzinker Y, Dashevsky Z, Dariel MP. Mechanical properties of PbTe - based thermoelectric semiconductors. Scripta Materilia. 2008; **58**:251

[32] Dashevsky Z, Dudkin L. Generator thermoelectric materials. Journal of Thermoelectric Materials. 1993;**1**:1

[33] Vikhor LN, Anatychuk LI, Mikhailovsky VY, Strutynska LT.

Innovations in organic-fueled thermoelectric generators development. In: Dempsey WP, editor. Thermoelectric Power. Hauppauge, NY, USA: Nova Science Publishers Inc.; 2011. pp. 117-154

[34] Ravich YI, Efimova BA, Smirnov IA. Semiconducting Lead Chalcogenides. New York: Plenum Press; 1970

[35] Dashevsky Z. The applications of lead chalcogenides in thermoelectric devices. In: Khokhlov D, editor. Lead Chalcogenides: Physics and Application. New York: Taylor & Francis; 2003. pp. 555-586

[36] Lalonde AD, Pei Y, Wang H, Snyder J. Lead telluride alloy thermoelectrics. Materials Today. 2011; **14**:526

[37] Kaĭdanov VI, Ravich YI. Deep and resonance states in a^{IV} B^{VI} semiconductors. Soviet Physics Uspekhi. 1985;**28**:31

[38] Kaĭdanov VI, Nemov SA, Ravich YI. Self-compensation of electrically active impurities by intrinsic defects in IV-VI semiconductors. Semiconductors. 1993; **28**:223

[39] Kaĭdanov VI. Resonance (quasilocal) states in a^{IV} B^{VI} semiconductors. In: Proceedings of Defect & Diffusion Forum – Defects in Semiconductors I 103-105. Bäch, Switzerland: Trans Tech Publications, Ltd; 1993. pp. 387-406

[40] Akimov BA, Dmirtiev AV, Khokhlov DR, Ryabova LI. Carrier transport and non-equilibrium phenomena in doped PbTe and related materials. Physica Status Solidi A: Applications and Materials Science. 1993;**137**:9

[41] Volkov BA, Ryabova LI, Khokhlov DP. Mixed-valence impurities in lead telluride- based solid solutions. Physics-Uspekhi. 2002;**45**:819

[42] Parashchuk T, Chernyak L, Nemov S, Dashevsky Z. Influence of deformation on $Pb_{1-x}In_xTe_{1-y}I_y$ and $Pb_{1-x-y}Sn_xIn_yTe$ films. Physica Status Solidi (B). 2020;**2000304**:1-9

[43] Rosenberg AJ, Wald F. Massive heterovalent substitutions in octahedrally coordinated semiconductors. Journal of Physics and Chemistry of Solids. 1965;**26**:1079

[44] Averkin AA, Kaĭdanov VI, Melnik RB. Nature of indium impurity states in lead telluride. Soviet Physics – Semiconductors. 1971;**5**:75

[45] Koster GE, Slater JC. Wave functions for impurity levels. Physics Review. 1954;**95**:1167

[46] Koster GE, Slater JC. Simplified impurity calculation. Physics Review. 1954;**96**:1208

[47] Landau LD, Lifshitz EM. Quantum Mechanics. Oxford: Pergamon Press; 1974

[48] Anselm A. Introduction to Semiconductor Theory. Moscow: Mir; 1981

[49] Nemov SA, Ravich YI. Thallium dopant in lead chalcogenides: Investigation methods and peculiarities. Physics-Uspekhi. 1998;**168**:817

[50] Misra S, Wiendlocha B, Tobola J, Fesquet F, Dauscher A, Lenoir B, et al. Journal of Materials Chemistry C. 2020; **8**:977

[51] Wiendlocha B. Fermi surface and electron dispersion of PbTe doped with resonant Tl impurity from KKR-CPA

calculations. Physical Review B. 2013;**88**: 205205

[52] Heremans JP, Jovovic V, Toberer ES, Saramat A, Kurosaki K, Charoenphakdee A, et al. Enhancement of thermoelectric efficiency in PbTe by distortion of the electronic density of states. Science. 2008;**321**:554

[53] Wiendlocha B. Thermopower of thermoelectric materials with resonant levels: PbTe:Tl versus PbTe:Na and $Cu_{1-x}Ni_x$. Physical Review B. 2018;**97**: 205203

[54] Wiendlocha B. Localization and magnetism of the resonant impurity states in Ti-doped PbTe. Applied Physics Letters. 2014;**105**:133901

[55] Kim S, Wiendlocha B, Jin H, Tobola J, Heremans JP. Electronic structure and thermoelectric properties of p-type Ag-doped Mg_2Sn and $Mg_2Sn_{1-x}Si_x$ (x = 0.05, 0.1). Journal of Applied Physics. 2014;**116**:153706

[56] Dariel MP, Dashevsky Z, Jarashnely A, Shusterman S, Horowitz A. Carrier concentration gradient generated in p-type PbTe crystals by unidirectional solidification. Journal of Crystal Growth. 2002;**234**:164

[57] Dashevsky Z, Kasiyan V, Mogilko E, Butenko F. High-temperature PbTe diodes. Thin Solid Films. 2008;**516**:7065

[58] Dashevsky Z, Shusterman S, Dariel MP, Drabkin I. Thermoelectric efficiency in graded indium-doped PbTe crystals. Journal of Applied Physics. 2002;**92**:1425

[59] Wang XK, Veremchuk I, Bobnar M, Zhao JT, Grin Y. Solid solution $Pb_{1-x}Eu_xTe$: Constitution and thermoelectric behavior. Inorganic Chemistry Frontiers. 2016;**3**:1152

[60] Lalonde AD, Pei Y, Snyder J. Reevaluation of $PbTe_{1-x}I_x$ as high performance n-type thermoelectric material. Energy & Environmental Science. 2011;**4**:2090

[61] Kwan CCY, Basinski J, Woolley JC. Analysis of the two-band hall effect and magnetoresistance. Physica Status Solidi B-basic Solid State Physics. 1971;**45**:699

[62] Pei Y, LaLonde AD, Wang H, Snyder GJ. Low effective mass leading to high thermoelectric performance. Energy & Environmental Science. 2012; **5**:7963

[63] Cao J, Querales-Flores JD, Murphy AR, Fahy S, Savic I. Dominant electron-phonon scattering mechanisms in n-type PbTe from first principles. Physical Review B. 2018;**98**:205202

[64] Parashchuk T, Shabaldin A, Cherniushok O, Konstantinov P, Horichok I, Burkov A, et al. Origins of the enhanced thermoelectric performance for p-type $Ge_{1-x}Pb_xTe$ alloys. Physica B: Condensed Matter. 2020;**596**:412397

[65] Zhang J, Wu D, He D, Feng D, Yin M, Qin X, et al. Extraordinary thermoelectric performance realized in n-type PbTe through multiphase nanostructure engineering. Advanced Materials. 2017;**29**:1703148

[66] Wang S, Yang Z, Sun Y, Xiao Y, Zhao LD. Synergistically optimizing charge and phonon transport properties in n-type PbTe via introducing ternary compound AgSb(Se,Te)2. Journal of Alloys and Compounds. 2020;**815**:152463

[67] Tan G, Stoumpos CC, Wang S, Bailey TP, Zhao L-D, Uher C, et al. Subtle roles of Sb and S in regulating the thermoelectric properties of N-type

PbTe to high performance. Advanced Energy Materials. 2017;**7**:1700099

[68] Zhang Q, Chere EK, McEnaney K, Yao M, Cao F, Ni Y, et al. Enhancement of thermoelectric performance of N-type PbSe by Cr doping with optimized carrier concentration. Advanced Energy Materials. 2015;**5**:1401977

[69] Chere EK, Zhang Q, McEnaney K, Yao M, Cao F, Sun J, et al. Enhancement of thermoelectric performance in N-type $PbTe_{1-y}Se_y$ by doping Cr and tuning Te:Se ratio. Nano Energy. 2015; **13**:355

[70] Parashchuk T, Kostyuk O, Nykyruy L, Dashevsky Z. High thermoelectric performance of p-type $Bi_{0.5}Sb_{1.5}Te_3$ films on flexible substrate. Materials Chemistry and Physics. 2020; **253**:123427

[71] Park SJ, Cho KS, Kim SH. A study of dielectric characteristics of polyimide thin film. Journal of Colloid and Interface Science. 2004;**272**:384

[72] Dzundza B, Nykyruy L, Parashchuk T, Ivakin E, Yavorsky Y, Chernyak L, et al. Transport and thermoelectric performance of n-type PbTe films. Physica B: Condensed Matter. 2020;**588**:412178

[73] Ivakin EV, Kisialiou IG, Nykyruy LI, Yavorsky YS. Optical studies of heat transfer in PbTe:Bi(Sb) thin films. Semiconductors. 2018;**52**:1691

[74] Dashevsky Z. Influence of potential barriers on the thermoelectric properties of PbTe thin films. In: Proceedings ICT'97. 16th International Conference on Thermoelectrics. Piscataway, NJ, USA: IEEE Xplore; 1997. p. 255

[75] Dashevsky Z, Kreizman R, Dariel MP. Physical properties and inversion conductivity type in nanocrystalline PbTe films. Journal of Applied Physics. 2005;**98**:094309

[76] Dashevsky Z. High photosensitive films of lead chalcogenides. In: Balandin AA, Wang KI, editors. Handbook of Semiconductor Nanostructures and Nanodevices. Vol. 2. Valencia, CA, USA: American Scientific Publishers; 2006. Chapter 11. pp. 335-359

[77] Kovalyuk V, Sheveleva E, Auslender M, Goltsman G, Shneck R, Dashevsky Z. Polycrystalline PbTe:In films on amorphous substrate: Structure and physical properties. Materials. 2022; **15**:8383

[78] Shtern Y, Sherchenkov A, Shtern M, Rogachev M, Pepelyaev D. Challenges and perspective recent trends of enhancing the efficiency of thermoelectric materials on the basis of PbTe. Materials Today Communications. 2023;**37**:107083

[79] Dashevsky Z, Mamykin S, Dzundza B, Auslender M, Shneck RZ. A review of nanocrystalline film thermoelectrics on lead chalcogenide semiconductors: Progress and application. Energies. 2023;**16**:3774

[80] Jopp J, Kovalyuk V, Towe E, Shneck R, Dashevsky Z, Auslender M. Polycrystalline films of indium-doped PbTe on amorphous substrates: Investigation of the material based on study of its structural, transport, and optical properties. Materials. 2024;**17**:6058

[81] Ebert H, Ködderitzsch D, Minár J. Calculating condensed matter properties using the KKR-Green's function method —Recent developments and applications. Reports on Progress in Physics. 2011;**74**:096501

[82] Ebert H. et al. The Munich SPR-KKR package, version 7.7.3. 2018. Available from: http://ebert.cup.uni-muenchen.de/SPRKKR

[83] Vosko SH, Wilk L, Nusair M. Canadian Journal of Physics. 1980;**58**: 1200

Chapter 5

Nanofluidic Thermoelectric Materials

Di Wang, Xin Peng and Yahui Xue

Abstract

Thermoelectric transducers have attracted significant attention owing to their immense potential in energy harvesting and biomimetic applications, such as waste heat recovery and the design of advanced thermal sensors. Although traditional thermoelectric semiconductors exhibit excellent thermoelectric performance at room temperature, their toxicity and rarity limit their practical applications. In recent years, with the emergence of advanced materials such as graphene, MXenes, and COFs, inspirations have been provided by biological thermosensitive ion channels to construct nanofluidic thermoelectric systems using these materials as fundamental building blocks, aiming to achieve efficient thermoelectric conversion. However, the thermoelectric coefficient of current nanofluidic membranes is only 1.27 mV/K in very dilute solution, much lower than that obtained by biological ion channels, that is, 5.8 mV/K. In this chapter, a detailed analysis of nanofluidic thermoelectric materials is conducted from the perspective of theoretical background, development, and applications. It is revealed that the synthetic effects of hydrodynamic slip and surface charge at the channel wall contribute significantly to the enhancement of thermoelectric properties. Furthermore, to design nanofluidic materials with better thermoelectric transducing performance, future strategies may involve integrating various external stimuli, such as pH control, electro-gating, or novel surface treatments, to advance their use in energy harvesting and biomimetic thermal sensors.

Keywords: nanofluidic thermoelectric materials, biomimetic ion channels, confined ion transport, energy recovery, biomimetic thermal sensors

1. Introduction

The thermoelectric effect has significant potential applications in low-grade heat energy recovery and the design of biomimetic thermosensitive sensors. In human industrial and societal activities, large amounts of heat are emitted, and efficiently recovering this heat represents a new approach to addressing the energy crisis. While traditional thermoelectric semiconductors exhibit relatively considerable thermoelectric performance at room temperature, their toxicity and rarity limit their practical applications [1]. In recent years, micro-nanoscale systems based on two-dimensional materials have been shown to enable efficient energy recovery. As a result, an increasing number of researchers are focusing on ion thermoelectric in micro-nano channels, aiming to improve energy recovery efficiency and reduce energy consumption in the recycling of waste heat.

The fundamental principle of ion thermoelectric in nanofluidic thermoelectric systems is as follows: under the influence of a temperature gradient, cations and anions undergo transport across the channel (**Figure 1**). The surface charge on the channel walls causes different diffusion rates for the cations and anions, leading to an imbalance in their numbers at both ends of the nanofluidic channel, which ultimately generates a potential difference. However, the mechanisms behind the generation of transmembrane potential differences under a temperature gradient in nanofluid systems with varying ion selectivities are complex and still require further exploration.

Moreover, intelligent life relies on biological ion channels to sensitively perceive changes in external temperature. In nature, the ability of animals to sense temperature is crucial. Sensory nerve endings in the body depolarize through gated ion transport channels, allowing them to transmit electrical signals from the periphery to the central nervous system. This process enables temperature sensation and prompts quick responses to extreme cold or hot objects [2, 3]. Inspired by this, nanofluidic thermoelectric systems also offer new perspectives for the development and design of biomimetic thermosensitive sensors.

To date, the Seebeck coefficient has been used to evaluate the performance of thermoelectric materials. The Seebeck coefficient (or thermoelectric potential coefficient) is defined as the voltage generated per unit temperature difference. Specifically, it describes the relationship between the voltage induced in a material under a temperature gradient, serving as an indicator of the thermoelectric conversion capability of the material. However, the Seebeck coefficient of thermoelectric systems through artificial ion-selective nanofluidic channels under temperature gradients is only 1.27 mV/K [4], which is significantly lower than the thermoelectric coefficient of natural biological nanofluidic channels at 5.8 mV/K [5]. This indicates that designing highly thermosensitive artificial nanofluidic channels and achieving efficient thermoelectric conversion remains a challenge. In this chapter, the theoretical foundations, developments, and applications of nanofluidic thermoelectric materials will be discussed, providing a valuable reference for further research on nanofluidic thermoelectric systems.

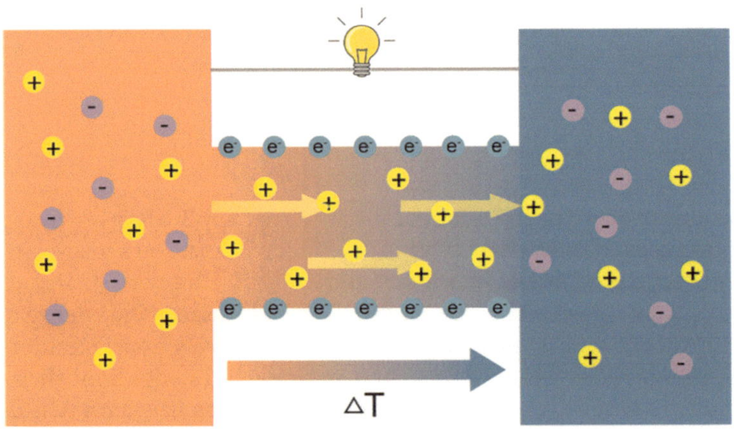

Figure 1.
Schematic diagram of nanofluidic thermoelectric system in confined space. Blue represents the low-temperature region, while red indicates the high-temperature region. Ions are transported under a thermal gradient (ΔT), resulting in the generation of electrical potential.

2. Theoretical background

The study of nanofluidic thermoelectric systems offers significant insights for efficiently recovering low-grade thermal energy and for the development and design of biomimetic thermal sensors. However, the thermoelectric conversion efficiency using water electrolytes in nanofluid channels still falls short compared to the performance of traditional thermoelectric semiconductors [6]. In addition, the ion transport driven by a thermal gradient differs significantly in bulk solutions from that in confined channels (**Figure 2**). In bulk solutions, charge separation is established due to the differing diffusivity of cations and anions, leading to the generation of electric potential. In confined spaces, the ion transport is additionally influenced by the charge density and slip length at the channel wall surface, making the underlying mechanism more comprehensive. Therefore, investigating and discussing the thermoelectric mechanisms of nanofluids is crucial for enhancing the performance of these systems. This section will briefly introduce the fundamental theory of power generation in nanofluid systems under thermal gradients, specifically focusing on the basic principles of ion and fluid transport driven by thermal gradients.

2.1 The Soret effect in the bulk electrolyte

When a thermal gradient field is applied to bulk solution, ions, molecules, suspended particles, or droplets drift toward either the cold or the warm side. The drift velocity (u) is proportional to the magnitude of the thermal gradient (∇T), that is [7],

$$u = -D_{\mathrm{T}} \nabla T. \tag{1}$$

where D_{T} is thermal diffusion coefficient (thermophoretic mobility), representing the relationship between heat and particle motion, and T is temperature. The Soret effect, also known as the thermal diffusion effect, refers to the differential diffusion of components within a mixture under the influence of a temperature gradient, leading to their spatial redistribution. Specifically, for the positive Soret effect, the particles tend to diffuse from the high-temperature region to the low-temperature region. The negative Soret effect will cause the particles to diffuse from the low-temperature region to the high-temperature region. Under the action of the Soret effect, the ion

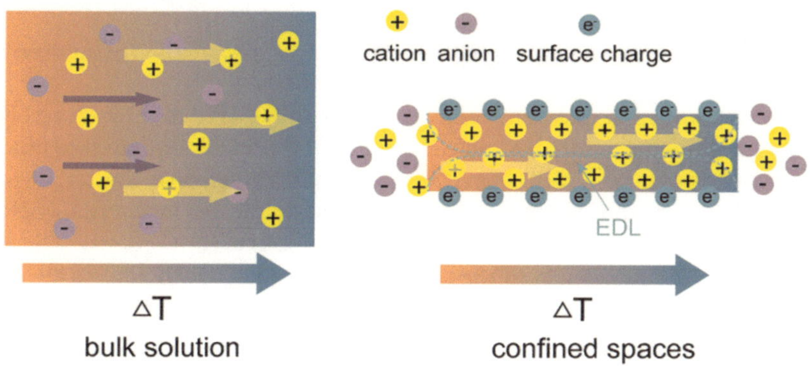

Figure 2.
Ion transport driven by a thermal gradient in bulk solution (left) and confined space (right).

migration forms a concentration gradient (∇c), and the asymmetric charge distribution due to the different electrophoretic mobility of cations and anions result in the formation of a thermoelectric field (E_T). In a static state, for any individual ion, the various driving forces, including the concentration gradient, thermal field, and thermoelectric field, achieve a state of mutual equilibrium.

For a binary electrolyte solution, the ion flux (*Ji*) of positive (i = '+') or negative (i = '-') ions induced by the concentration gradient, thermal field, and thermoelectric field can be obtained using the Nernst-Planck equation [8],

$$J_\pm = -D_\pm \left(\nabla c_\pm + c_\pm S_{\mathrm{T},\pm} \nabla T - c_\pm \frac{v_\pm e}{k_\mathrm{B} T} E_\mathrm{T} \right) \tag{2}$$

In the equation, D_i represents the diffusion coefficient, ci is the ion concentration in the solution, $S_{\mathrm{T,i}}(= D_\mathrm{T}/D_\mathrm{i})$ is the intrinsic Soret coefficient, vi denotes the ion valence, e is the elementary charge, and k_B stands for the Boltzmann constant, and $E_\mathrm{T} = -\nabla V_\mathrm{T}$ with V_T as the thermoelectric voltage. The Soret coefficient ($S_{\mathrm{T,i}}$) is related to the ionic heat of transport (Q_i^*) by

$$S_{\mathrm{T,i}} = \frac{Q_\mathrm{i}^*}{k_B T^2}. \tag{3}$$

Here, Q_i^* primarily originates from the entropy of hydration for specific ions and has typical values in the order of kJ/mol [9]. Thus, the equivalent Seebeck coefficient in a bulk solution is obtained as [8].

$$S_{\mathrm{e,bk}} = \frac{Q_+^* - Q_-^*}{2veT}, \tag{4}$$

with $S_{\mathrm{e,bk}}$ defined by $E_\mathrm{T}/\nabla T$. Given Q_i^* ~kJ/mol, the generated thermoelectric voltage per Kelvin in a bulk electrolyte, that is, $S_{\mathrm{e,bk}}$, lies in the scale of around 0.1 mV/K, which is significantly lower than the performance of conventional thermoelectric semiconductors. It indicates that the very low thermoelectric voltage generated by ion transport in bulk electrolyte solutions is insufficient for direct thermoelectric energy harvesting. Therefore, it is intriguing to see whether it is possible to enhance thermal energy conversion efficiency by using aqueous electrolyte in a physically confined space.

2.2 Thermoelectric response of liquid electrolyte in physical confinement

The assumption of ion distribution neutrality applies only to bulk electrolytes. In physically constrained micro/nanoscale channels, the presence of surface charge causes ions to redistribute within the channel, forming an electric double layer (EDL). When the EDL is thin and non-overlapping, the potential (φ) in the EDL field decays from the wall value to the bulk value (as shown in **Figure 3**).

A series of calculations leads to the equivalent Seeback coefficient in the physical confinement [8],

$$S_{\mathrm{e,cf}} = S_Q + S_\varphi \tag{5}$$

Where,

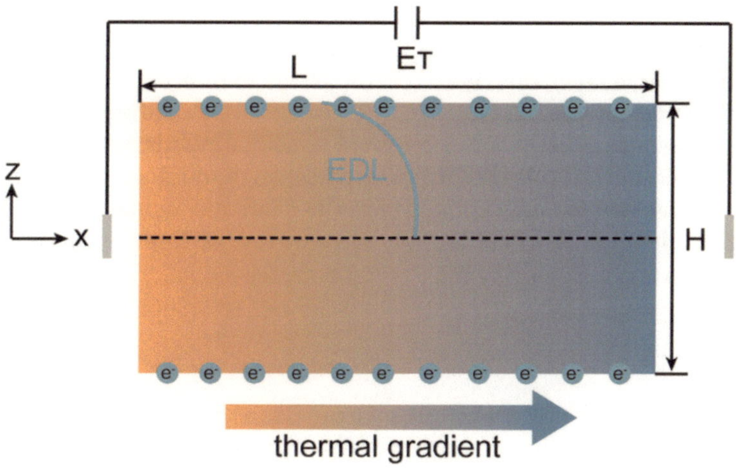

Figure 3.
Schematics of EDL in a slit channel and thermoelectric field generation by a thermal gradient.

$$S_{\mathrm{Q}} = S_{\mathrm{e,bk}} \frac{1}{1 + \frac{\overline{\zeta}^2}{4}\left[\frac{\tanh(\overline{\kappa})}{\overline{\kappa}} + \frac{1}{\cosh^2(\overline{\kappa})}\right]} \tag{6}$$

$$S_{\varphi} = \frac{\zeta}{T}\frac{\tanh(\overline{\kappa})}{\overline{\kappa}}\frac{1 + \frac{\overline{\zeta}^2}{2}\left[\frac{\tanh^2(\overline{\kappa})}{3} + \frac{1}{\cosh^2(\overline{\kappa})}\right]}{1 + \frac{\overline{\zeta}^2}{4}\left[\frac{\tanh(\overline{\kappa})}{\overline{\kappa}} + \frac{1}{\cosh^2(\overline{\kappa})}\right]} \tag{7}$$

with $S_{\mathrm{e,cf}} = E_{\mathrm{T}}/(\mathrm{d}T/\mathrm{d}x)$, $\overline{\zeta} = ev\zeta/(k_{\mathrm{B}}T)$ and $\overline{\kappa} = \kappa H/2$. Here,$S_{\mathrm{Q}}$refers to the thermoelectric contribution resulting from Soret-type thermophoretic ion motion under physical confinement. If the ionic heat transport for both positive and negative ions is the same, S_{Q} disappears, as the Seebeck coefficient becomes zero. ζ represents the zeta potential, which is the potential at the slipping plane of the wall, while κ^{-1} is the Debye length. In Eq. (7), S_{φ} represents the confinement-dominated thermoelectric effect. The Debye length (κ^{-1}) defines the thickness of the double layer. When κ^{-1} is much larger than half the channel height (H/2), that is, when $\kappa\mathrm{H} \to 0$, S_{φ} approaches ζ/T, indicating an enhancement of thermoelectric generation by increasing the surface zeta potential. Conversely, when the EDL thickness is much smaller than the channel height, that is, $\kappa\mathrm{H} \to \infty$, S_{φ} vanishes, similar to the case of bulk solutions.

In this case, the slit height is set to H = 20 nm, and the surface zeta potential is ζ =100 mV. The electrolyte used is KCl, with the heat of ionic transport difference between potassium and chloride ions given as $\Delta Q = Q_{+}^{*} - Q_{-}^{*} \approx 2$kJ/mol. The comparison of S_{Q} and S_{φ} reveals that, at low concentrations where the Debye length is large, the contribution from Soret-type thermophoretic motion is negligible, and the thermoelectric effect induced by confinement dominates. In contrast, at high concentrations where the Debye length is small, the confinement effect tends to diminish. While the thermophoretic and nanoconfined thermoelectric effects contribute additively, the total equivalent Seebeck coefficient in confined nanochannels generally stays around 0.4 mV/K, even when the zeta potential is as large as ζ = 100 mV. It is important to note that the overlap of the double layers within the slit alters the electric field distribution, but this effect does not significantly influence the total Seebeck coefficient magnitude [8].

In conclusion, the thermoelectric effects in confined spaces without convection have been modeled using the Debye-Hückel approximation. The results demonstrate that, when the double layer thickness is comparable to or greater than the channel height, the thermoelectric effect is primarily governed by confinement. Nevertheless, the equivalent Seebeck coefficient in confined nanochannels is still considerably lower than that observed in traditional thermoelectric semiconductors. It is essential to note that this model assumes no advection within the channels, which applies to channels with smooth solid walls. The next section will address the topic of thermally induced osmotic transport in confined spaces.

2.3 Thermo-osmotic flow in confined space

The confinement of liquid within nanochannels modifies its specific enthalpy. In an isothermal system, when a pressure gradient (∇p) drives liquid flow in confined nanochannels, it generates a "heat flux," which subsequently creates a thermal gradient along the channel. Conversely, when a thermal gradient (∇T) is applied across the nanochannels, it induces liquid flow, a phenomenon known as thermo-osmosis. This coupling effect is described by Onsager's linear nonequilibrium thermodynamics [10, 11].

$$\begin{bmatrix} u_{\mathrm{s}} \\ f_{\mathrm{h}} \end{bmatrix} = \begin{bmatrix} \beta_{11} & \beta_{12} \\ \beta_{21} & \beta_{22} \end{bmatrix} \begin{bmatrix} -\nabla p \\ -\dfrac{\nabla T}{T} \end{bmatrix} \tag{8}$$

where u_{s} is hydrodynamic velocity, f_{h} is heat transfer flux, and β_{ij} (i, j = 1, 2) is the phenomenological coefficient. Here, β_{11} characterizes the isothermal flow driven by a pressure gradient, β_{22} denotes heat conduction at $\nabla p = 0$, and according to the Onsager's reciprocity relations, $\beta_{12} = \beta_{21}$, which represents the thermos-osmosis coefficient. The following is intended to obtain the expression for β_{12}.

According to calculations, the presence of a slippery boundary magnifies the thermos-osmotic response by [11].

$$\beta_{12} = \beta_{12}^{\mathrm{no-slip}}(1 + b/\lambda) \tag{9}$$

Where,

$$\lambda = \frac{\int_{z_s}^{0} \mathrm{d}z(z - z_s)\delta h(z)}{\int_{z_s}^{0} \mathrm{d}z\delta h(z)} \tag{10}$$

The length scale, λ, characterizes the thickness of the interfacial liquid layer, where the enthalpy of the liquid is modified by the interaction with the nanochannel walls. This interfacial thickness is usually on the scale of a few molecules, and it can substantially improve the thermos-osmotic coupling coefficient, particularly on surfaces with ultra-low friction, such as graphite, where the slip length can extend to several tens of nanometers. Experimental studies on bare glass and Pluronic-coated substrates have observed thermos-osmotic coefficients ranging from $\beta_{12} \sim 10^{-10}$ m^2/s to

10^{-9} m^2/s [12]. Both theoretical analysis and molecular dynamics simulations demonstrate that the slip length at the liquid-solid interface is crucial for enhanced thermal-osmotic flow in confined nanofluidic channels. When the friction at the liquid-solid interface is minimal, the thermos-osmotic coefficient can increase by several orders of magnitude [11]. This suggests that thermos-osmotic flow in nanochannels with partially slippery walls should be considered in the analysis of thermoelectric coupling, as discussed in the following section.

2.4 Thermo-electric response with hydrodynamic slip

The thermos-osmotic coupling analysis above demonstrates that a temperature gradient applied across confined channels can induce substantial liquid flow, particularly when the channel walls exhibit ultra-low liquid friction. Therefore, liquid advection within these channels must be taken into account when analyzing the thermoelectric response, in contrast to the analysis in Section 2.2, where advection was not considered.

According to a series of calculations, the equivalent Seeback coefficient in slits with hydrodynamic slip boundary conditions, defined by $S_{e,hs} = -\nabla V/\nabla T$ with $j_e = 0$, is expressed by [13].

$$S_{\mathrm{e,hs}} = \frac{\alpha_{\mathrm{TE}}}{\sigma T} \tag{11}$$

where α_{TE} is the thermoelectric coupling coefficient, and σ is electrical conductivity, consisting of bulk (σ_{bulk}) and surface conductivity (σ_{surf}), of which the latter is dominated if $\kappa H < < 1$. And it is found for surface conductivity-dominated channels that [13].

$$S_{\mathrm{e,hs}} \approx \frac{-\delta h \times \lambda \times (\varepsilon\zeta/\eta)}{\sigma_{\mathrm{surf}}}. \tag{12}$$

Here, δh refers to the excess enthalpy density, while λ denotes the thickness of the solid-liquid interaction layer. The ζ potential is influenced by both the surface charge density and the boundary slip length. In summary, the above derivation suggests a direct correlation between the equivalent Seebeck coefficient and the ζ potential. Molecular dynamic simulations further confirm that simultaneously maximizing surface charge density and boundary slip length can significantly enhance the thermoelectric coupling [13]. It is important to note that the distribution of surface charges on highly slippery solid surfaces must be uniform. Any heterogeneity in the charge distribution can reduce the slip effect, which, in turn, diminishes the thermoelectric coupling.

2.5 The effect of Gibbs free energy on ions under a temperature gradient

During ion transport in confined nanochannels, the presence of a temperature gradient leads to changes in the Gibbs free energy (dG) within the system. Therefore, the change in dG caused by the temperature gradient is as follows [14],

$$\mathrm{d}G = -S\mathrm{d}T + V\mathrm{d}p + \sum_{i=1}\mu_i \mathrm{d}n_i \tag{13}$$

where G represents the Gibbs free energy, S denotes entropy, T is temperature, V is volume, and p is pressure. μ_i and n_i refer to the chemical potential and mole number of species i, respectively. In regions of high temperature, the entropy effect of the system is typically stronger, especially in confined spaces where the freedom of particle motion is restricted, leading to a higher entropy increase in the high-temperature regions. Ions generally tend to move toward regions of higher entropy, as the entropy increase in high-temperature areas helps reduce the Gibbs free energy of the system. This is particularly true in confined spaces such as nanochannels, where ion motion is significantly influenced by geometric and surface effects. As a result, the Gibbs free energy decreases with increasing temperature, while in regions where the temperature remains unchanged, the Gibbs free energy of the electrolyte remains constant. From a thermodynamic perspective, the system tends to reach a state of minimal free energy. Since Gibbs free energy determines the system's stability and direction of evolution, noticeable material diffusion can occur under the driving force of changes in Gibbs free energy.

In conclusion, hydrodynamic slip contributes to an enhancement in thermos-osmotic coupling, leading to considerable convection flow within confined slits. This flow drives ion fluxes in response to a temperature gradient, thereby enhancing the thermoelectric coupling effect. The thermoelectric coefficient can be derived using Onsager's linear nonequilibrium thermodynamics. A scaling analysis of the equivalent Seebeck coefficient reveals that enhancing the thermoelectric coupling effect can be achieved by simultaneously increasing both the surface charge density and the slip length. Additionally, the variation of Gibbs free energy under a temperature gradient also plays a crucial role in ion transport. The following section will explore the latest developments and applications of nanofluidic thermoelectric systems.

3. Development of nanofluidic thermoelectric materials

The discovery of the Seebeck effect in 1821 provided an important theoretical foundation for the conversion of thermal energy into electrical energy. Since then, due to the advantages of thermoelectric systems—such as adjustable device size, absence of additional mechanical motion, lack of acoustic or electronic noise, precise temperature control, fast response, and high reliability—extensive research has been conducted on thermoelectric systems [15].

3.1 Nanofluidic thermal transport

With the discovery of two-dimensional materials such as graphene, nanofluidic thermoelectric systems have gradually come into the spotlight of researchers. Initially inspired by biological ion channels, researchers focused on constructing two-dimensional confined nanofluidic channels, leading to in-depth studies on the electrodynamics of ions within these nanofluidic channels [16–18]. Under the influence of an electric field or concentration gradient, ions undergo directional transport in confined nanofluidic channels with surface potentials, a phenomenon known as electro-osmosis or diffusion osmosis, respectively. Moreover, fluid transport driven by pressure fields can carry ions along the charged surfaces of the channels, generating a flow of electric current.

It was only in recent years that the impact of temperature gradients on ions and fluids in confined spaces have begun to attract the attention of researchers. Dejakin and Sidorenkov were the first to study thermodiffusion in porous glass, where fluid flow is driven by a thermal gradient [19]. This phenomenon contrasts with the ion transport driven by pressure gradients, as fluid flow driven by pressure gradients arises from the excess enthalpy of the liquid near the channel walls, leading to the formation of heat flux through nanofluidic channels. Dejakin applied the diffusion-osmosis reciprocity theorem to predict thermodiffusion and the thermodiffusion slip velocity, demonstrating the existence of thermodiffusion in porous media and confirming the thermodiffusion slip effect through molecular dynamics simulations [10]. Brégère et al. [12] were the first to conduct micro-scale observations of velocity fields induced by thermodiffusion, where the reduction of friction at solid-liquid interfaces was shown to enhance the thermodiffusion response, such as at the interface between graphene and water [11].

Power generation in conductors under a thermal gradient is commonly referred to as the Soret effect [7]. In liquid electrolytes, the asymmetric thermophoretic motion of positive and negative ions leads to the accumulation of an electric field, which is the result of the Soret effect. Under physical constraints, the presence of surface charge at the wall-liquid interface allows the generation of electrical energy through the thermal gradient, even without the need for different thermophoretic mobilities of cations and anions. This effect holds promise for enhancing thermoelectric conversion efficiency through high-selectivity membranes. However, the Seebeck coefficient measured experimentally in nanofluidic channels using water electrolytes remains much lower than that of traditional thermoelectric semiconductors.

To enhance the thermoelectric performance of nanofluidic thermoelectric systems, researchers have been continuously exploring various approaches. Experimental and theoretical analyses have revealed that the surface charge density of nanofluidic channels and the slip length at the interface between the channel and the fluid-solid boundary are critical factors for improving thermoelectric performance [13]. Therefore, it is essential to design nanofluidic thermoelectric channels with ultra-high surface charge density and exceptionally low friction. Specifically, artificial ion channels with naturally high surface potentials, such as COFs-based nanofluidic channels, can be developed. Additionally, external stimuli such as pH control, voltage gating, and surface modification with functional groups can be applied to regulate the surface potential of nanofluidic channels. On the other hand, to achieve ultra-low friction in nanofluidic channels, materials with smoother surfaces, such as graphene and carbon nanotubes, can be employed to construct ion channels. Thus, integrating these optimization strategies for nanofluidic thermoelectric channels presents a promising new avenue for future research. For example, Xue et al. constructed a graphene-based, gate-controlled confined ion channel with thermoelectric response. This design not only achieved an ultra-high channel surface charge density but also utilized the slip phenomenon at the fluid-channel wall interface in the graphene channel, resulting in a Seebeck coefficient of 12.7 mV/K for the thermally responsive ion channel. This exceeds the thermoelectric performance of biological thermal TRP ion channels, offering new insights for the collection of low-grade waste heat and the development and design of biomimetic thermal sensors [20].

3.2 Nanofluidic thermoelectric materials

To enhance the thermoelectric response performance in nanofluidic thermoelectric systems, scholars have conducted extensive exploration from the perspective of both the material properties and structural design of nanofluidic thermoelectric materials. Currently, various two-dimensional materials, such as graphene, carbon nanotubes (CNTs), transition metal carbides and nitrides (MXenes), mental organic frameworks

(MOFs), and covalent organic frameworks (COFs), have been widely applied in the field of nanofluidic thermoelectric materials (**Figure 4**). By designing ion channels with distinct physicochemical characteristics for thermal response, the ion transport performance is improved, thereby boosting the thermoelectric response of nanofluid systems. For example, Sun et al. designed a nanofluid membrane based on covalent organic frameworks (COFs) [4]. Utilizing the ultra-high surface charge density and high in-plane porosity of the COF ion channels, they significantly enhanced the ion transport performance in the nanofluidic thermoelectric system, achieving a Seebeck coefficient of 1.27 mV/K. The constructed COFs-based nanofluidic thermoelectric channels can intelligently monitor temperature changes and express these changes as a continuous potential difference, providing insights for the design of biomimetic sensors. In the same year, Su et al. designed a thermosensitive ion channel with an asymmetric structure [14]. They explored the performance of the designed biomimetic thermosensitive ion channel under two conditions—presence and absence of an ion concentration gradient. The heterogeneous channel induced a highly sensitive thermoelectric response under small temperature gradients, with a sensitivity of 0.71 mV/K, which is comparable to natural thermosensation systems.

To enhance thermal energy utilization efficiency and expand the application scenarios of nanofluidic thermoelectric systems, the range of nanofluidic thermoelectric materials has become increasingly diverse. Researchers have gradually explored solar-thermal-electric energy conversion, leading to the widespread use of photothermal materials like MXenes and MOFs in nanofluidic thermoelectric systems. For example, Husam N. Alshareef and colleagues developed MXene-based nanofluid channels with

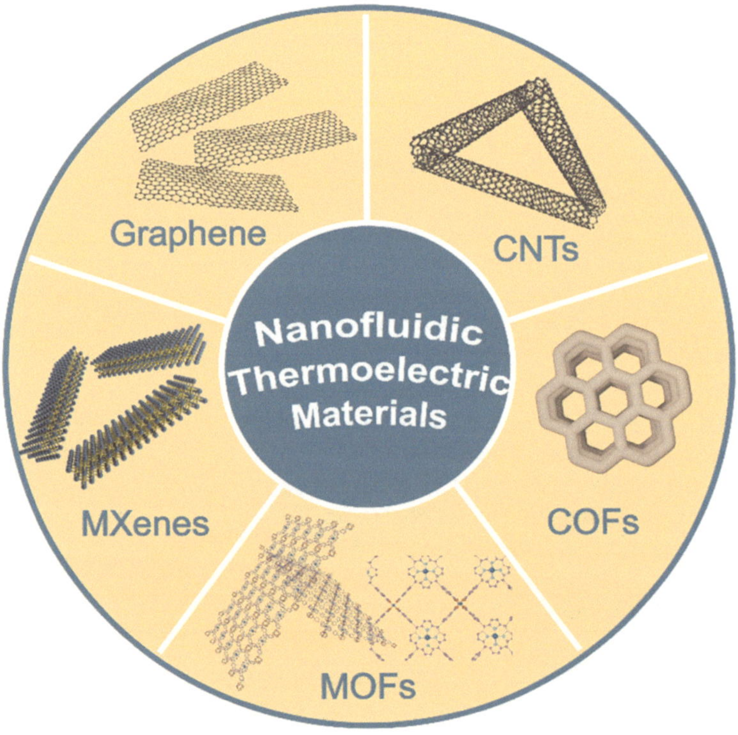

Figure 4.
Nanofluidic thermoelectric materials.

excellent photothermal conversion properties [21]. Under light irradiation, the temperature of MXenes rapidly increases, forming a temperature gradient that promotes ion transport and enables photothermal-electric conversion. The Seebeck coefficient of the constructed nanofluidic thermoelectric system reached 1 mV/K.

To further improve the thermoelectric enhancement driven by light in nanofluidic thermoelectric systems, some researchers have constructed asymmetric nanofluid channels using materials with varying photothermal properties. This approach aims to increase the temperature gradient caused by light-induced thermal fields, thereby facilitating rapid ion transport and achieving efficient thermoelectric energy conversion. For instance, Wang and colleagues designed heterogeneous photothermal-electric nanofluid materials based on MXenes and vermiculite [22]. The MXenes exhibited localized surface plasmon resonance (LSPR) effects, providing excellent photothermal conversion, while vermiculite, with its phonon scattering on its internal ordered interfaces, offered a low thermal conductivity. Under light irradiation, these heterogeneous nanofluid materials created a substantial temperature gradient, leading to superior photothermal-electric conversion performance.

Additionally, to further enhance the energy conversion efficiency of nanofluidic thermoelectric materials, researchers have even developed nanofluid materials that exhibit both photovoltaic and thermoelectric properties, offering a new platform for advancing nanofluidic thermoelectrics. For example, Wang et al. developed two-dimensional heterostructure nanofluid materials based on MXenes and g-C_3N_4, achieving the coupling of photo-electric and photothermal-electric effects, which enabled highly efficient light-responsive ion transport [23]. Under sunlight irradiation, the band structure of the channel units facilitated effective separation of photo-excited charge carriers within the hetero-structured channels, forming a built-in electric field that enhanced selective ion transport. Meanwhile, the photothermal conversion properties of the MXenes units generated a simultaneous temperature gradient within the nanofluid channels, further promoting ion transport.

To date, the exploration of nanofluidic thermoelectric materials is ongoing. Developing nanofluidic thermoelectric materials with high surface charge density and slip behavior remains a key challenge, as supported by the aforementioned theoretical findings. Additionally, broadening the application scenarios for nanofluidic thermoelectric materials is another crucial direction for advancing this field. For instance, introducing external stimuli such as light, pH, pressure, and electric control into nanofluidic thermoelectric systems offers exciting opportunities to enhance performance and enable new functionalities. In the next part, the focus will be on detailing the specific applications of nanofluidic thermoelectric materials, showcasing their potential in various fields and the innovations driving their development.

4. Applications

Nanofluidic thermoelectric systems hold immense potential for applications in energy recovery and biomimetics. For instance, they can be used in the recovery of waste heat, as well as in the design and development of biomimetic thermal sensors.

4.1 Energy recovery

With the gradual depletion of fossil fuels, the urgency for energy transition has become increasingly apparent, prompting humanity to explore renewable energy sources [24]. In recent years, the recovery and utilization of low-grade waste heat have attracted widespread attention. Low-grade heat sources typically refer to heat sources with temperatures around 200°C, which are generated in large quantities during industrial production processes and the operation of various electronic devices. Efficiently recovering and utilizing this waste heat represent a new approach to addressing the energy crisis. Currently, researchers have explored a range of materials and approaches to harness waste heat and convert it into electrical energy, including the use of thermoelectric semiconductors, ionic liquid gels, ion thermoelectric batteries, and thermally driven osmotic engines, among others [25]. However, the implementation of these methods remains limited by the rarity and toxicity of materials or the efficiency of energy conversion. There is an urgent need for cleaner and more cost-effective thermoelectric technologies for waste heat recovery.

Nanofluidic thermoelectric systems have garnered widespread attention in the field of waste heat recovery due to their excellent thermoelectric conversion efficiency, ultra-fast thermoelectric response, and simple fabrication processes. By constructing nanofluidic channels with superior ion selectivity, ion transport driven by temperature fields can be achieved, thereby converting thermal energy into electrical energy. For example, Sun et al. designed a covalent organic framework (COF) nanofluid membrane, which efficiently converts thermal energy into electrical energy, achieving a power density of 5.70 W/m^2 under a 50 K temperature gradient [26].

Additionally, introducing solar energy into nanofluidic thermoelectric systems creates a "light-heat-electric" energy conversion process, enabling efficient recovery of both solar energy and ionic energy. For example, Wen et al. proposed a bio-inspired ion pump utilizing MXenes with photothermal properties [27]. In this system, light irradiation generates a temperature gradient, which directs ion transport under the influence of the thermal field, achieving the functionality of a biological ion pump. Similarly, Xia et al. designed a Solar Thermoelectric Nanofluidic Device (STEND) to harness solar thermal energy [28]. They deposited a 3-dimensional gold nanostructure as a broadband plasmonic absorption layer on one side of an anodized aluminum oxide (AAO) membrane. This layer effectively converts light into heat through surface plasmon resonance (SPR). The resulting localized heat creates a significant temperature gradient in the nanofluidic channel, driving ionic charge separation and generating a thermoelectric potential across the membrane, enabling efficient energy recovery.

Moreover, temperature fields generated by light irradiation in nanofluidic thermoelectric systems can also promote the efficient recovery of salinity gradient energy. For instance, Wang et al. developed MOFs-based nanofluidic channels with photothermal-electric response [29]. Under the simultaneous presence of temperature and salinity gradients induced by light irradiation, they achieved a power density of 16.64 W/m^2, significantly enhancing the efficiency of salinity gradient energy recovery.

4.2 Biomimetic thermal sensors

In nature, the ability of organisms to detect changes in environmental temperature is essential for regulating vital physiological functions. In mammals, thermal signals are transduced into electrochemical signals via thermosensitive transient receptor potential (TRP) ion channels, and these ion channels are then transformed into action potentials by thermoreceptor nerve cells, such as the sensation of pain. Inspired by this biological mechanism, researchers have been dedicated to developing biomimetic thermal sensors. For instance, Sun et al. reported a biomimetic thermal sensor based on covalent organic frameworks (COFs), which can intelligently monitor temperature changes and express them as continuous potential differences [4]. This provides new insights for the development of biomimetic thermal sensors.

Furthermore, to enhance the biocompatibility of nanofluidic thermoelectric systems, some researchers have employed ion gels as thermoelectric materials. For example, Pan et al. developed an ion gel that is both thermosensitive and pressure-sensitive [30]. They designed a temperature-pressure decoupled ion sensor, mimicking the function of biomimetic skin, to capture both thermal and pressure stimuli.

In a short summary, nanofluidic thermoelectric materials have seen widespread application in energy recovery and the development of biomimetic thermal sensors. However, significant challenges remain, such as the complexity of material fabrication and the thermoelectric conversion performance, which still falls short of the efficiency seen in biological thermoelectric systems. There is still a long way to go before nanofluidic thermoelectric systems can achieve real-world practical applications. Continued research and innovation are essential to overcoming these hurdles and advancing the field.

5. Conclusion

Nanofluidic thermoelectric materials show considerable promise for applications in the recovery of low-grade heat and the development of biomimetic thermosensitive sensors, although research in this field is still in its early stages. In bulk electrolytes, the thermal gradient drives the Soret-type thermophoretic movement of cations and anions, which creates an electric field due to the differences in ion thermal transport properties. However, the resulting Seebeck coefficient is generally low, around 0.1 mV/K. In confined nanochannels, the thermoelectric effect induced by confinement dominates over the Soret effect, assuming the absence of convection flow. Nonetheless, the overall Seebeck coefficient remains relatively low. In this chapter, to improve the thermoelectric conversion efficiency of nanofluid systems, a detailed overview of the theoretical aspects, developments, and applications of nanofluidic thermoelectric systems is provided.

The results indicate that convective flow of the liquid in nanofluidic channels with partial slip boundary conditions is highly significant and must be considered in thermoelectric conversion analyses. Molecular dynamics simulations demonstrate that both high slip length and surface charge density contribute to a significant enhancement in the thermoelectric coupling coefficient. However, achieving such surfaces in nanofluidic channels is quite challenging, which is a major reason why the temperature sensitivity remains around 1 mV/K. Additionally, the uneven distribution of charges deteriorates the slip boundary conditions. Therefore, a uniform distribution of surface charges is necessary to achieve consistent low liquid friction, which would further enhance the thermoelectric performance of nanofluid systems.

With the development of nanofluidic thermoelectric materials, they are now widely applied in areas such as energy recovery and biomimetic thermal sensor design. This chapter provides a comprehensive review, offering new ideas and insights into the field of nanofluidic thermoelectrics, aiming to further promote the advancement of this field.

Acknowledgements

This work is supported by the National Program on Key Basic Research Project of China (2022YFA1203400), the National Natural Science Foundation of China (12272159), the Shenzhen Science and Technology Innovation Bureau (20231115094026001, KQTD20240729102052065), and the Guangdong S&T Programme (2023ZT10Z002, 2023A1515012592). D. Wang would like to acknowledge the technical support from SUSTech CRF.

Conflict of interest

The authors declare no conflict of interest.

Notes/thanks/other declarations

Thank you for providing us the opportunity to publishing in the chapter with you.

Author details

Di Wang, Xin Peng and Yahui Xue*
Department of Mechanics and Aerospace Engineering, Southern University of Science and Technology, Shenzhen, China

*Address all correspondence to: xueyh@sustech.edu.cn

References

[1] Barragán VM, Kristiansen KR, et al. Perspectives on thermoelectric energy conversion in ion-exchange membranes. Entropy. 2018;**20**:12-905. DOI: 10.3390/e20120905

[2] Liu S, Yang R, et al. Gated thermoelectric sensation by nanochannels grafted with thermally responsive polymers. Chemical Communications. 2020;**56**:91-14291-91-14294. DOI: 10.1039/D0CC06734B

[3] Liu B, Qin F. Single-residue molecular switch for high-temperature dependence of vanilloid receptor TRPV3. Proceedings of the National Academy of Sciences. 2017;**114**:7-1589-7-1594. DOI: 10.1073/pnas.1615304114

[4] Zhang P, Chen S, et al. Covalent organic framework nanofluidic membrane as a platform for highly sensitive bionic thermosensation. Nature Communications. 2021;**12**:1-1844. DOI: 10.1038/s41467-021-22141-z

[5] Viana F, de la Peña E, et al. Specificity of cold thermotransduction is determined by differential ionic channel expression. Nature Neuroscience. 2002; **5**(3):254-260. DOI: 10.1038/nn809

[6] Xue Y. Nanofluidics for thermoelectric energy harvesting. In: New Materials and Devices for Thermoelectric Power Generation. London: IntechOpen; 2023. DOI: 10.5772/intechopen.1002444

[7] Würger A. Thermal non-equilibrium transport in colloids. Reports on Progress in Physics. 2010;**73**:12-126601. DOI: 10.1088/0034-4885/73/12/126601

[8] Dietzel M, Hardt S. Thermoelectricity in confined liquid electrolytes. Physical Review Letters. 2016;**116**:22-225901. DOI: 10.1103/PhysRevLett.116.225901

[9] Agar J, Mou C, et al. Single-ion heat of transport in electrolyte solutions: A hydrodynamic theory. The Journal of Physical Chemistry. 1989;**93**:5-2079-2082. DOI: 10.1021/j100342a073

[10] Derjaguin BV, Churaev NV, et al. Surface Forces. New York: Springer; 1987. DOI: 10.1007/978-1-4757-6639-4

[11] Fu L, Merabia S, et al. What controls thermo-osmosis? Molecular simulations show the critical role of interfacial hydrodynamics. Physical Review Letters. 2017;**119**:21-214501. DOI: 10.1103/PhysRevLett.119.214501

[12] Bregulla AP, Würger A, et al. Thermo-osmotic flow in thin films. Physical Review Letters. 2016;**116**:18-188303. DOI: 10.1103/PhysRevLett.116.188303

[13] Fu L, Joly L, et al. Giant thermoelectric response of nanofluidic systems driven by water excess enthalpy. Physical Review Letters. 2019;**123**:13-138001. DOI: 10.1103/PhysRevLett.123.138001

[14] Chen K, Yao L, et al. Bionic thermoelectric response with nanochannels. Journal of the American Chemical Society. 2019;**141**:21-8608-21-8615. DOI: 10.1021/jacs.9b03569

[15] Qin B, Kanatzidis MG, et al. The development and impact of tin selenide on thermoelectrics. Science. 2024; **386**(6719):eadp2444. DOI: 10.1126/science.adp2444

[16] Schoch RB, Han J, et al. Transport phenomena in nanofluidics. Reviews of

Modern Physics. 2008;**80**(3):839-883. DOI: 10.1103/RevModPhys.80.839

[17] Bocquet L, Charlaix E. Nanofluidics, from bulk to interfaces. Chemical Society Reviews. 2010;**39**(3):1073-1095. DOI: 10.1039/b909366b

[18] Xue Y, Xia Y, et al. Atomic-scale ion transistor with ultrahigh diffusivity. Science. 2021;**372**(6541):501-503. DOI: 10.1126/science.abb5144

[19] Derjaguin B, Sidorenkov G. On thermo-osmosis of liquid in porous glass. Comptes Rendus de l'Académie des Sciences URSS. 1941;**32**(9):622-626

[20] Li G, Peng X, et al. Nanofluidic thermoelectric transducer with ultrahigh and tunable sensitivity. The Journal of Physical Chemistry Letters. 2024;**15**(39): 9863-9870. DOI: 10.1021/acs.jpclett.4c02370

[21] Hong S, Zou G, et al. Photothermoelectric response of Ti_3C_2Tx MXene confined ion channels. ACS Nano. 2020;**14**:9042-9049. DOI: 10.1021/acsnano.0c04099

[22] Feng Y, Li S, et al. Nanoarchitecture via synchronic stacking of metallic and nonmetallic two-dimensional nanosheets for optimal light-driven ion transport. ACS Nano. 2024;**18**(47):32793-32805. DOI: 10.1021/acsnano.4c10913

[23] Wang J, Wang D, et al. Efficient solar energy conversion via bionic sunlight-driven ion transport boosted by synergistic photo-electric/thermal effects. Energy & Environmental Science. 2023;**16**(7):3146-3157. DOI: 10.1039/D3EE00720K

[24] Elsheikh MH, Shnawah DA, et al. A review on thermoelectric renewable energy: Principle parameters that affect their performance. Renewable and Sustainable Energy Reviews. 2014;**30**: 337-355. DOI: 10.1016/j.rser.2013.10.027

[25] Kishore RA, Priya S. A review on low-grade thermal energy harvesting: Materials, methods and devices. Materials. 2018;**11**:8-1433. DOI: 10.3390/ma11081433

[26] Yin S, Li J, et al. Giant gateable thermoelectric conversion by tuning the ion linkage interactions in covalent organic framework membranes. Nature Communications. 2024;**15**:1-8137. DOI: 10.1038/s41467-024-52487-z

[27] Liu P, Zhou T, et al. Light-induced heat driving active ion transport based on 2D MXene nanofluids for enhancing osmotic energy conversion. CCS Chemistry. 2021;**3**(4):1325-1335. DOI: 10.31635/ccschem.020.202000296

[28] Li Z-Q, Wu Z-Q, et al. A solar thermoelectric nanofluidic device for solar thermal energy harvesting. CCS Chemistry. 2021;**3**(7):2174-2182. DOI: 10.31635/ccschem.020.202000366

[29] Wang J, Song Z, et al. Light-responsive and ultrapermeable two-dimensional metal-organic framework membrane for efficient ionic energy harvesting. Nature Communications. 2024;**15**:1-2125. DOI: 10.1038/s41467-024-46439-w

[30] Gao N, Huang J, et al. Biomimetic ion channel regulation for temperature-pressure decoupled tactile perception. Small. 2024;**20**(1):e2302440. DOI: 10.1002/smll.202302440

Chapter 6

Flexible Thermoelectric Films and Devices

Xiang Li and Yao Lu

Abstract

Recently, the market for portable, flexible, and wearable electronics has seen explosive growth, paralleled by the burgeoning Internet of Things (IoTs), which encompasses numerous node sensors. The reliance on traditional batteries to power these electronics and IoT node sensors not only poses environmental concerns but also significantly raises costs. Consequently, the self-powering of electronics and IoT sensors has become a necessity. Flexible thermoelectric generators (f-TEGs), assembled with flexible thermoelectric films (f-TEFs), offer a promising solution by continuously harnessing heat energy (such as body heat and sunlight) to power these devices. Consequently, f-TEFs have garnered increasing attention over the past decade, with remarkable breakthroughs occurring in the last several years. In this chapter, we review the recently reported f-TEFs, which could be categorized into freestanding films and films on flexible substrates. Strategies are proposed to improve the thermoelectric (TE) performance of these films. Additionally, we discuss the recent advancements in f-TEFs and illustrate how they can be integrated into generator designs that capitalize on their mechanical and TE properties. Furthermore, we analyze and delve into the challenges and existing problems in the study of f-TEFs and f-TEGs and provide comprehensive design guidelines pertaining to the TE properties and flexibility of the f-TEFs.

Keywords: flexible, thermoelectric, flexible devices, composite film, f-TEs

1. Introduction

Wearables and Internet of Things (IoT) markets have witnessed an explosive growth in the last decades [1]. The amount of IoT sensors has already surpassed the human population by multiple folds [2]. Wearable electronics, particularly those inspired by skin, serve as cornerstone components for personalized healthcare in the IoT ecosystem [3, 4]. However, the majority of these devices still rely on batteries with finite lifespans, necessitating frequent charging or replacement [5, 6]. This not only elevates operational costs but also contributes to environmental pollution. Furthermore, the limited lifespan of these batteries hinders their further deployment [7, 8]. Consequently, the development of self-powered flexible electronics and sensors that harness energy from the human body is imperative. Thermoelectric generators (TEGs), capable of converting low-grade

 IntechOpen

waste heat into electricity, offer a practical and feasible solution [9, 10]. TEGs offer several advantages, including the absence of moving parts or fluids, high reliability, and silence in operation. When a TEG is directly attached to the skin, heat naturally flowing from the human body and the TEG could create a voltage through the Seebeck effect based on the thermal gradient between the environment and skin. Then, the electricity produced could be stored in batteries or supercapacitors for subsequent usage, ensuring continuous running of the device [11, 12].

The efficiency of a thermoelectric (TE) generator depends on the figure of merit (ZT), defined as $ZT = S^2\sigma T/\kappa$, where S is the Seebeck coefficient, σ is the electrical conductivity, κ is the thermal conductivity, and T is the absolute temperature. It is evident that the efficiency increases with a higher ZT value. Therefore, to achieve a high-efficiency TEG, we must first identify TE materials with a high ZT. This is challenging because S, σ, and κ are intricately linked to the carrier concentration (n). S represents the average entropy carried by each charge carrier, which decreases as n increases. For metals or semiconductors, S can be calculated by the following formula:

$$S = \frac{8\pi^2 k_B^2}{3eh^2} m^* T \left(\frac{\pi}{3n}\right)^{\frac{2}{3}} \tag{1}$$

Here, m^* denotes the effective mass of the carriers, while k_B represents the Boltzmann constant and h represents Planck constant. σ is proportionate to the n and carrier mobility (μ), given by the relationship $\sigma = ne\mu$. κ primarily consists of phonon thermal conductivity (κ_L) and electronic thermal conductivity (κ_e), where κ_e is in proportion to σ based on the Wiedemann-Franz Law ($\kappa_e = L\sigma T$), albeit this relationship does not hold for organic materials. Consequently, decoupling these three variables and achieving a high ZT is exceedingly challenging.

Traditional TE materials are primarily inorganics. Numerous tactics such as nanotechnology [13], energy band engineering [14], and defect engineering [15] have increasingly been employed to enhance their TE performance. Consequently, some inorganics, such as PbTe [16] and SnSe [17], have demonstrated ZT values exceeding 2. However, nearly all inorganic TE materials are rigid and inflexible, making them unsuitable for wearable platforms. Therefore, one of the primary challenges in developing f-TEGs with high efficiency is to identify f-TEFs with excellent TE properties.

Conducting polymers (CPs), such as poly(3,4-ethylenedioxythiophene) (PEDOT) [18–20], polyaniline (PANI) [21, 22], and polypyrrole (PPy) [23, 24], exhibit remarkable bendability, lightweight, and low κ [25, 26]. Consequently, in the past decade, research on flexible TE materials has predominantly focused on CPs and their composites with inorganic TE materials. This focus aims to leverage the low thermal conductivity and excellent flexibility of CPs, while also benefiting from the high PF of inorganics [27, 28].

Recently, insulating polymers (IPs) typically including polyvinylidene fluoride (PVDF) [29], epoxy resin [30], polylactic acid (PLA) [31, 32], and polyvinyl pyrrolidone (PVP) [33] have been selected as matrices for f-TEFs. Additionally, carbon materials, such as fullerene [34], carbon nanotube (CNT) [35–37], and graphene [38], as well as carbon material-based composites [39–42], have garnered significant attention due to their impressive σ, flexibility, and light-weight. Nevertheless, inorganic alloys are most widely researched because of the superior TE performance. Consequently, inorganics-based f-TEGs have garnered more attention, and recently, notable advancements have been achieved (see **Figure 1**).

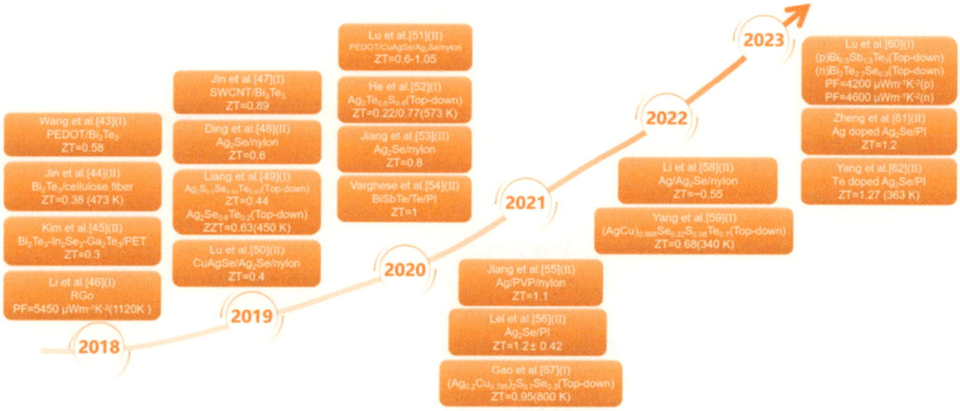

Figure 1.
Some representative works about f-TEFs in the last decades [43–62]. "I" denotes freestanding f-TEF; "II" denotes f-TEF on substrate.

The traditional method for preparing films involves utilizing rigid substrates, as these facilitate the achievement of films with superior TE performance compared to films grown on flexible substrates. However, the unbendability of these substrates inherently limits the flexibility of the resultant films. Consequently, researchers have increasingly focused on developing films on flexible substrates.

Another significant challenge in the development of f-TEGs lies in their rational design, which includes optimizing the size of f-TEGs' legs, ensuring adequate contacts, thermal interfaces, and packaging methods. This aspect of research relatively lagging behind compared to the study of f-TEFs [63, 64]. Currently, the development of f-TEGs is still at the proof-of-principle demonstration stage [65]. Based on their operation mode, f-TEGs are categorized into two types. The first type is prepared by connecting p- and n-type inorganic TE bricks electrically over a flexible substrate with patterned electrical contacts. Here, the temperature difference across the bricks is perpendicular to the substrate [66–68]. The second type is created by either directly fabricating TE legs onto a flexible substrate or by cutting strips from films, assembling them, and adhering them onto the substrate. In both cases, the TE legs harness electricity from the thermal gradient that exists in the in-plane direction of the strips [69–74]. The first form of f-TEGs always lack sufficient flexibility. This chapter will focus solely on the latter form, which is lightweight and highly flexible.

The approximate voltage ($V_{voltage}$) produced by a TEG can be expressed as:

$$V_{voltage} = N\left(S_p - S_n\right)\Delta T \tag{2}$$

where S_p and S_n represent the Seebeck coefficients of the p-type and n-type semiconductors, respectively, and *N* denotes the number of p-n thermocouples [75–79].

$$P = UI = \frac{U^2}{R_{ex} + R_{in}} \tag{3}$$

When the external load resistance (R_{ex}) is equal to the internal resistance of the device (R_{in}), the output power reaches its peak value [80, 81]. The maximum power density ($P_{density}$) is calculated by dividing the generated power by both the cross-sectional area (A) and the total number of legs (N_l), as outlined in the following equation:

$$P_{density} = \frac{P_{max}}{N_l \cdot A} = \frac{(N_l \cdot S \cdot \Delta T)^2}{N_l \cdot A} \cdot \frac{\sigma \cdot w \cdot t}{4 \cdot N_l} = \frac{S^2 \cdot \sigma}{4l} \cdot \Delta T^2 \quad (4)$$

where w, d, and l, respectively, denote the width, thickness, and length of the legs. The varying temperature differences and lengths employed in different applications necessitate the introduction of normalized power density, $P_{density} \bullet l/\Delta T^2$, which facilitates comparisons of the devices' output performance [82, 83].

In this chapter, we initiate with a systematic review of f-TEFs, which are categorized into two types: (i) free-standing f-TEFs and (ii) f-TEFs on substrates (see **Figure 2**). Considering that the output density of existing devices fails to meet practical power supply requirements, this chapter attempts to address the issue from two perspectives: materials and device design, aiming to pave a route for future researchers. To be more specific, this chapter is structured around two primary parts. Firstly, we comprehensively examine free-standing f-TEFs by analyzing various material systems. Secondly, we classify f-TEFs on substrates based

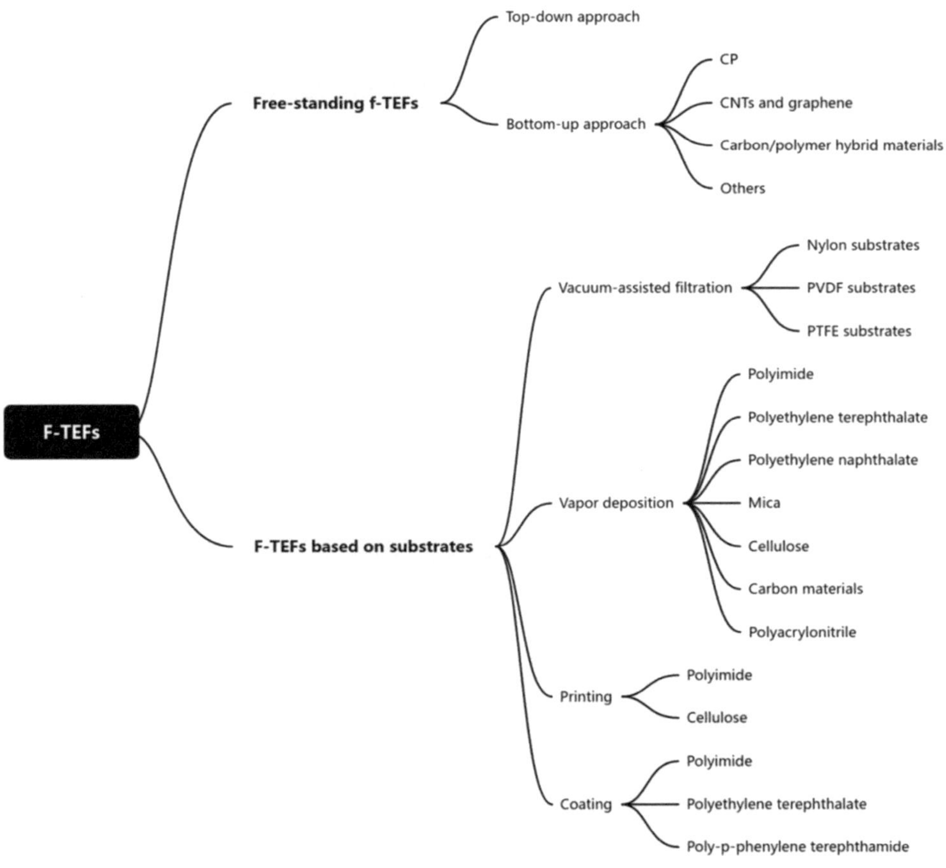

Figure 2.
The mind map of this chapter.

on the diverse flexible substrates utilized. Lastly, we highlight the main challenges in the study of f-TEFs and propose feasible solutions.

2. Freestanding film

Freestanding films could be prepared using top-down or bottom-up approaches. The top-down method involves creating large-scale templates and subsequently reducing their lateral dimensions to the nanoscale. This method has the advantage of being able to produce high-quality, high-performance materials, but it is also limited by the availability of material systems and its expensive, energy-intensive preparation process. Bottom-up method involves constructing multifunctional nano-sized materials through the self-assembly of atoms or molecules. This technique relies on the chemical reaction of precursors under specific reaction conditions, offering advantages such as low cost, simplicity of operation, and suitability for industrialization. However, it also has drawbacks, including poor controllability, the need for post-treatment, and low product quality due to poor crystallinity.

2.1 Top-down approach

The "top-down" approach to obtain inorganic f-TEFs entails the fabrication of bulks through traditional melting or annealing techniques, subsequently slicing them into thinner sections. Some progress have been made in this field (see **Figure 3**) [78].

In 2018, zone melting process was adopted by Shi et al. [79] to synthesize a range of Te/Se-doped Ag_2S-based TE materials, followed by spark plasma sintering. These bulk materials were then sliced into thin sections, which exhibited favorable mechanical properties. Notably, the ZT value reached 0.44 for $Ag_2S_{0.5}Se_{0.45}Te_{0.05}$ at RT and 0.63 for $Ag_2S_{0.8}Te_{0.2}$ at 450 K. Later on, Lu et al. [60] utilized the similar process to prepare p- and n-type Bi_2Te_3-based flexible films, achieving elastic bending (with only a 9% decrease in σ after 1000 bending tests at a 4 mm bending radius) due to the staggered-layered structure, while minimizing the degradation of carrier transport along the temperature difference, reaching high PF values of 4200 (p-type) and

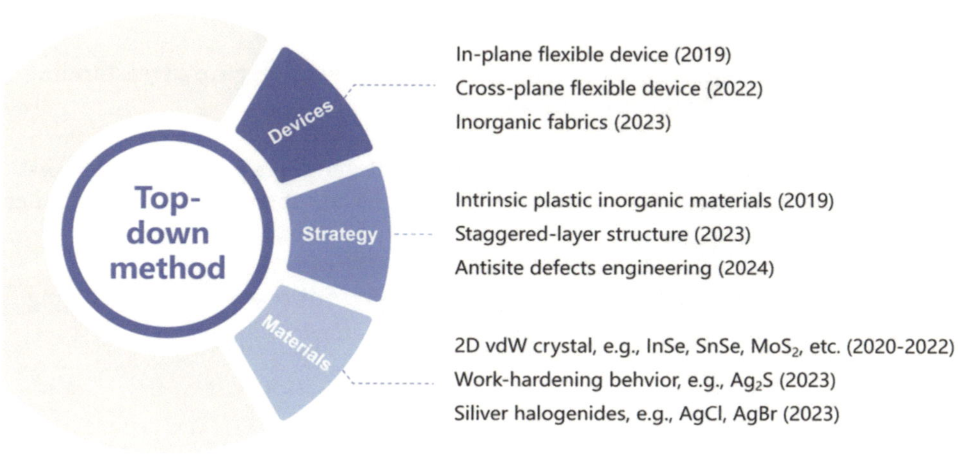

Figure 3.
Brief summary of plastic inorganic TE materials.

4600 (n-type) μW/mK^2 at RT, respectively. Very recently, Deng et al. [80] discovered that, in brittle bismuth telluride-based materials, the induction of high-density and diversified microstructures through modulation of antisite defects achieved a transformation from brittleness to plasticity. This enhancement elevated the ZT at RT to approximately 1.0, which is comparable to that of traditional brittle thermoelectric materials.

Despite the advancements, significant challenges persist within this emerging field, primarily encompassing an inadequate comprehensive grasp of the deformation mechanism, hurdles in synergistically adjusting both TE performance and deformability, as well as a scarcity of technological breakthroughs in both processing this novel class of TE materials and their seamless integration into devices.

2.2 Bottom-up approach

The bottom-up methodology entails constructing structures one atom or one molecule at a time, utilizing covalent or supramolecular interactions, analogous to the process of building a house one brick after another. This method can be realized through several approaches:

1. Initially, a precursor solution is either drop-cast or spin-coated onto a substrate, followed by peeling the resulting film. The contact angle determines the ease of preparing the film, i.e., the greater the angle of the solution droplet on the substrate, the more likely it is to detach the formed film [81].

2. Using the pressure difference between the two ends of the filter membrane via vacuum-assisted filtration, solid particles in the liquid are retained on the membrane and self-assembled to form a thin film, which is subsequently peeled off [82].

3. A self-supporting CP film or carbon-based substrate, upon which the TE inorganic is deposited through deposition process, such as electrodeposition [83].

4. Bacterial action in a petri dish could facilitate the in situ formation of composite films with bacteria cellulose as a matrix [84].

5. Interfacial polymerization method could be utilized to prepare a freestanding film at the liquid-liquid or liquid-air interface.

6. After mixing various materials using the centrifugal force generated by high-speed centrifugation, a film could be formed on the substrate material and then peeled off [85].

   Below, we provide a detailed review of the progress of related self-supporting films, categorized by material systems.

2.2.1 CP

Since the late 1970s, when doped polyacetylene was first discovered, CPs have garnered significant interest as a crucial polymer. Typically, CPs behave as semiconductors and exhibit electronic conductivity once doped with appropriate dopant materials and their high σ combining with excellent plasticity makes it attractive as TE matrices. PEDOT is one of the most extensively studied CPs in terms of excellent σ and stability. But the insolubility pose challenges for processing. To date, the mainstream method is to utilize poly(styrenesulfonate) (PSS) to decorate PEDOT. To mitigate the insulating effect of PSS, pure PEDOT film could be prepared as template [86, 87]. Apart from PSS, other dopants, like tosylate (Tos), are also used to improve the water-solubility of PEDOT [18, 41, 88]. For instance, Bubnova et al. [89] employed Tos to dope PEDOT (not in the form of a flexible film) to adjust its oxidation level, achieving a room-temperature ZT value of 0.25 with a high S of 220 μV K^{-1}. Subsequently, significant research has been conducted on post-treatment and doping of PEDOT. Post-treatment is often utilized to adjust the conformation and oxidation state of PEDOT, typically involving treated by polar solvents such as H_2SO_4, ethylene glycol, or through heat treatment. The other route designed by Fan et al. [90] to significantly improve the S involved creating a heterogeneous structure inside PEDOT:PSS via introducing ionic liquids (not a flexible film), which caused ion accumulation that scattered charge carriers with low energy, thereby significantly enhancing the S while σ almost remaining constant, leading to a PF up to 754 μW m^{-1} K^{-2}. **Figure 4** depicts the comparison of σ and S in PEDOT:PSS films doped by various agents.

2.2.2 CNTs and graphene

Up to now, numerous studies have focused on preparing CNT bucky papers with a thickness commonly between 50 to 200 μm, subsequently applying post-treatment

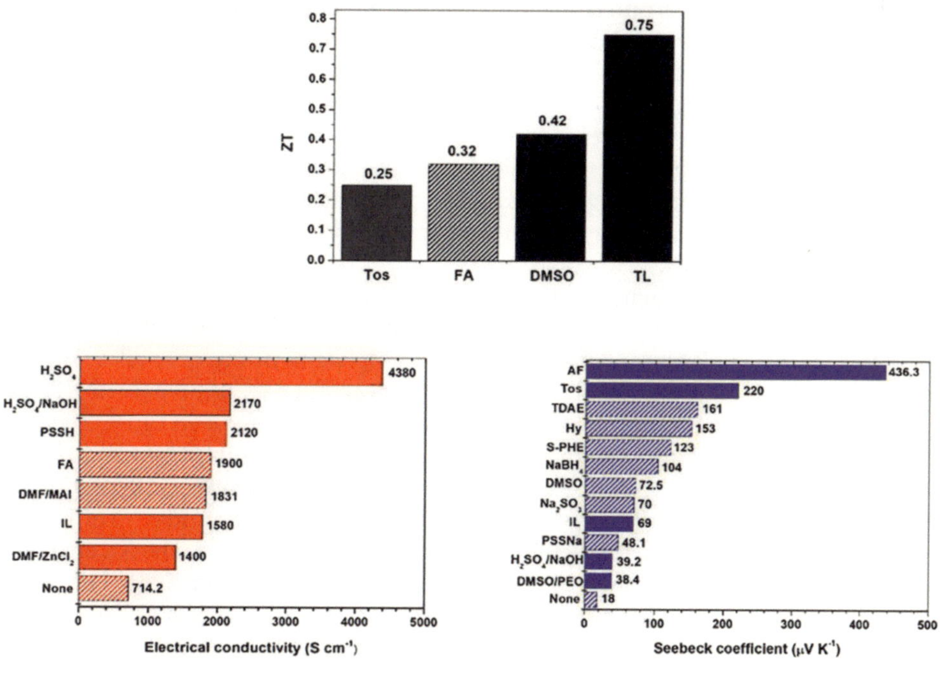

Figure 4.
Comparison of σ and S in PEDOT:PSS with different treatments, where the column filled with slashes represents the flexible film and solid column represents inflexible film. These data have been sourced from refs [91–98]. Image reproduced with permission from Ref. [72] published by Elsevier.

to enhance their inherently low TE properties. For example, An et al. [99] subjected spun CNT bucky paper to post-treatment combining heat treatment and doping with benzyl viologen, achieving a PF of approximately 3103 μW m^{-1} K^{-2} at RT. Interestingly, polyetherimide was found to transfer the CNT from p-type to n-type [100]. Meanwhile, the TE properties of graphene could be also improved by post-processing techniques. For instance, Li et al. [46] reduced graphene oxide film at high temperatures, resulting in a film that exhibited a peak PF of 5450 μW m^{-1} K^{-2} at 1120 K and remained to have excellent thermal stability, making it suitable for applications at extremely high temperatures.

2.2.3 Carbon/polymer hybrid materials

Due to their extensive π-conjugated structure and substantial specific surface areas, carbon nanoparticles could significantly enhance effective interfacial interactions with polymers. And, the high κ of these nanoparticles can be modulated through encapsulation or connection with polymers, and μ could also be substantially elevated. Researchers have endeavored to develop carbon nanoparticles/polymer f-TEFs [36, 38, 40, 42, 101, 102]. For example, Wu et al. [103] drop-casted SWCNTs dispersion onto the as-prepared PANI film, which facilitated to transfer the CP from a coiled state to a tensile state to form a highly ordered PANI interface, and thereby a high σ ~4000 S cm^{-1} was achieved.

2.2.4 Others

Given the intrinsic low S of polymers and carbon materials, incorporating inorganic TE elements as fillers into these matrices has increasingly become a prevalent approach to fabricate freestanding f-TEFs. The rationale behind this approach includes the following: (i) the relative high PF in inorganics, (ii) an improvement in the S through the energy-filtering effect, and (iii) a reduction in κ resulted from phonon scattering upon the interfaces.

However, achieving a robust combination between organic and inorganic materials presents a significant challenge due to their vastly different chemical properties. Therefore, the preparation of organic-inorganic hybrid f-TEFs with high TE performance is a critical concern.

In situ composite approach, involving the direct synthesis of reinforcing phases within the matrix, has been proved to be an effective method. For example, Wang et al. [43] employed polystyrene nanospheres as template to in situ crystallize Bi_2Te_3 nanoarrays, followed by deposited PEDOT upon the arrays to create a hybrid f-TEFs. The energy filtering effect and the strong acoustic mismatch between PEDOT and Bi_2Te_3 was considered to improve the ZT value to 0.58 at RT. Similarly, Jin et al. [47] utilized CNT as a template to deposited Bi–Te atoms, and the in situ crystalization on CNT contributes to the high orientation and low angle grain boundaries of Bi_2Te_3 nanocrystals, resulting in an excellent ZT value of ~0.89 at RT. Recently, Xu et al. [104] employed a wet-chemical method to create a flexible PVDF/Ta_4SiTe_4 film. The dimensionality/morphology matching strategy was proposed to guide the Ta_4SiTe_4 whiskers seamlessly interfaced with the polymer, resulting in a PF of 576 μW m^{-1} K^{-2} at RT.

Two-dimensional materials like Bi_2Te_3 and transition metal dichalcogenides have attracted attention for their unique in-plane electronic properties and large surface areas. Weak van der Waals interactions enable easy exfoliation into nanosheets (NSs) for composite processing. For instance, Du et al. [105] used lithium-ion intercalation to exfoliate

Bi_2Te_3-based alloys, mixing them with PEDOT:PSS to form freestanding f-TEFs. Similar approaches were applied to SnSe-based/PANI films [106, 107] and MoS_2/CNTs films [108]. Apart from physical (e.g., mechanical or liquid-phase exfoliation using solvents with matching surface tension [109, 110]), chemical exfoliation approach is also effective. For example, Dun et al. [111] intercalated Cu into Bi_2Se_3 nanoplatelets, dispersing them in PVDF to create a freestanding f-TEF with a ZT of 0.1 near RT.

Intercalation is also effective for tuning carrier transport. Wan et al. [112] intercalated organic molecules into TiS_2 layers, selectively removing some during heat treatment to reduce n, enhancing PF to 904 $\mu W\ m^{-1}\ K^{-2}$ at RT. Organic layers in the interspace lowered κ, achieving an ZT of ~0.33 at 413 K. Overall, reducing polymer content while maintaining flexibility enhances TE performance.

3. Flexible film based on substrate

3.1 Vacuum-assisted filtration

Vacuum filtration is a commonly used method for preparing thin films, featuring simple operation and low cost. Its main principle involves depositing material molecules onto the substrate surface through vacuum adsorption technology, thereby forming a nanoscale thin film. This method boasts advantages such as producing thin films with high purity, high density, high compactness, and strong adhesion. Thanks to their high chemical stability, nylon-based, PVDF-based, and PTFE-based filter membranes are widely used in thin film preparation processes. In the following sections, we will review the relevant works in detail, one by one.

3.1.1 Nylon membrane

Cai's group [113] was the first to utilize nylon membranes as substrates for preparing TE films using vacuum-assisted filtration. They successfully integrated SWCNTs as fillers into a highly conductive PEDOT:PSS to create p-type TE films with an optimal PF of 83.9 $\mu W\ m^{-1}\ K^{-2}$. Further, they [114] introduced Te nanorods with a high S [115] into this system and significantly improved the PF to 104 $\mu W\ m^{-1}\ K^{-2}$ at RT. Similarly, PEDOT/SWCNT f-TEFs were also reported by Fan et al. [116], with a PF doubled compared with the previous work [114].

It is well-known that inorganic TE materials exhibit superior TE performance. Therefore, it is desirable to combine the high TE properties of inorganic materials with the excellent flexibility of organic substrates. Cai's group [48] developed a novel method to achieve this goal, that is, firstly preparing Ag_2Se NWs and then utilizing vacuum filtration to form the film on a nylon membrane combining with post hot-pressing treatment at 573 K and 1 MPa for 30 min. And finally, they obtained a porous Ag_2Se film on nylon with a PF of 987.4 $\mu W\ m^{-1}\ K^{-2}$ and a ZT of approximately 0.6 at RT, accompanied with excellent flexiblity (**Figure 5a-d**). To enhance the PF of the Ag_2Se film, they [53] tried to raise the synthesis temperature or incorporate IPs, like PVP [55], into the system, and the resulted PF was successfully increased by an order of magnitude, up to ~1910 $\mu W\ m^{-1}\ K^{-2}$ at 300 K (**Figure 5e-h**).

On the other hand, Lu et al. [51] successfully incorporated Cu^{2+} into the synthesis of Ag_2Se nanowires to improve the TE performance via the energy filtering effect (as illustrated in **Figure 5i**), thereby achieving an impressive PF of approximately 1593 $\mu Wm^{-1}\ K^{-2}$. Further, Lu et al. [50] introduced PEDOT:PSS in situ during the growth

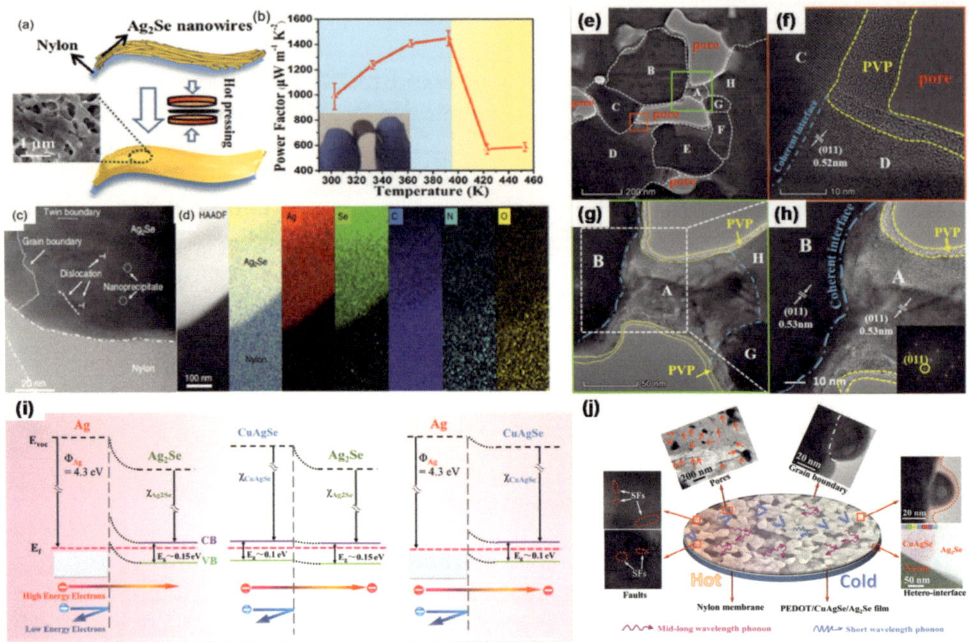

Figure 5.
(a) Schematic of Ag_2Se/nylon film. (b) PF vs. temperature plot for the film. (c) HRSTEM image showing seamless Ag_2Se-nylon integration. (d) HAADF image of Ag_2Se-nylon heterointerface, EDS mapping, and elemental images for Ag, Se, C, N, and O (Ref. [48], Springer Nature). (e) TEM cross-sectional image of the film showing pores and grains (A-G). (f) Enlarged view of (e) red square. (g) Zoomed view of (e) green square with PVP-coated Ag_2Se grains. (h) HRTEM of (g) white square (Ref. [55], Elsevier). (i) Band diagram for Ag/Ag_2Se, CuAgSe/Ag_2Se, and Ag/CuAgSe interfaces (Ref. [50], RSC). (j) Diagram of phonon scattering in PEDOT/CuAgSe/Ag2Se film (Ref. [51], Elsevier).

of Ag_2Se/CuAgSe Nws to reduce the thermal conductivity. And the composite film exhibited a remarkable PF of approximately 1603 W m^{-1} K^{-2}, corresponding to a ZT value within the range of 0.6 to 1.05 at 300 K due to the hierarchical microstructure defects inside (as illustrated in **Figure 5j**). Generally, the methodology outlined above could also be adapted to other TE components, like Ag/Ag_2Te/PVP films [33] and PEDOT:PSS/Cu_2Se films [66].

Concurrent with above work [48], Zeng et al. [117] also reported the fabrication of an Ag_2Te film on a nylon filter membrane through vacuum filtration, followed with mechanical-thermal welding. This hybrid film exhibited an impressive PF of approximately 315.1 μW m^{-1} K^{-2} around 400 K. Leveraging this film, a TE sensor was developed, boasting a swift response time of only 1.05 seconds—faster than many similar documented sensors [118–121].

The aforementioned studies underscore the potential of nylon membranes as exceptional substrates for hosting inorganic TE films or inorganic/polymer hybrid films. By integrating vacuum filtration with subsequent heat treatment, these films can achieve superior TE performance while maintaining exceptional flexibility.

3.1.2 PVDF membrane

PVDF filter membrane possesses high chemical stability, excellent mechanical strength, and heat resistance. Thanks to the porous structure of PVDF enabling the removal of excess PSS during the filtration process, a range of composite films, based on PEDOT:PSS, has been successfully fabricated onto PVDF substrates. For instance, Jiang et al. [122] employed a physical mixing method to create a PEDOT:PSS/SiC NWs composite film, which exhibited an enhanced S of 20.9 μV K^{-1} due to an energy-filtering effect. Notably, after treatment with H_2SO_4 to eliminate non-conductive PSS, σ skyrocketed to 3113 S cm^{-1}, resulting in an optimized PF of 128.3 μW m^{-1} K^{-2} at RT. See et al. [123] synthesized a PEDOT:PSS-coated Te nanorod composite film on a quartz substrate. Also, Novak et al. [124] prepared graphene on a PVDF membrane and the film showed a PF of 673 μW m^{-1} K^{-2} (RT).

In addition to these two filter membranes, PTFE membranes can serve as effective substrates. While PTFE membranes may not offer the same level of flexibility of these two membranes, their temperature tolerance, which is at least 320°C, surpasses that of other filter papers. Therefore, for applications requiring higher temperature treatment (>400°C), such as with Bi_2Te_3, PTFE membranes would be a more suitable choice.

3.2 Vapor deposition

Vapor deposition is a well-established technique for depositing films onto substrates, encompassing both chemical and physical vapor deposition. This process typically necessitates high temperatures, making it imperative that the substrate maintains stability under such conditions.

3.2.1 Polyimide (PI)

Zhu et al. [125] deposited Bi_2Te_3-based alloy films on PI substrates via magnetron sputtering, followed by annealing at 573 K, achieving PFs of 769 and 400 μW m^{-1} K^{-2} for p- and n-type films, respectively. RF magnetron sputtering was later used to deposit Te/Bi_2Te_3 on PI, with annealing at 673 K enhancing crystallinity and raising PF to 1145 μW m^{-1} K^{-2} [126]. Nuthongkum et al. [127] optimized PF to 1200 μW m^{-1} K^{-2} at 195°C by adjusting sputtering pressure. Further increases in pressure yielded PFs of ~2170 μW m^{-1} K^{-2} for Bi_2Te_3 and 1750 μW m^{-1} K^{-2} for Sb_2Te_3 [128]. Improvements were attributed to reduced grain size, diminished gaps, and preferential orientation.

Similarly, Shen et al. [129] found enhanced crystallization and reduced porosity at 4 Pa, achieving a PF of ~518 μW m^{-1} K^{-2} for Sb_2Te_3. Khumtong et al. [130] reported that lower pressure improved grain size and crystallinity, while higher pressure enhanced film formation, achieving a PF of 1060 μW m^{-1} K^{-2} at 1 Pa.

Jitthamapirom et al. [131] compared DC and RF sputtering, with DC achieving a PF of ~3500 μW m^{-1} K^{-2} at 558 K, significantly outperforming RF sputtering. Additional strategies, such as preventing element re-evaporation [132], doping [133, 134], and substrate selection [135], further enhanced TE performance.

Beyond Bi-Te alloys, Zn-Sb films have gained attention due to their non-toxic composition. Fan's group [136, 137] incorporated Ti into Zn-Sb alloys using multi-step sputtering, reducing defects and achieving a PF of ~3590 μW m^{-1} K^{-2} at 250°C. Toko et al. [138] applied this technique to Ge-based films, achieving a PF of 190 μW m^{-1} K^{-2}. Transparent TE films with ZT ~0.29 were also developed using CuI or Ga-doped ZnO on PI substrates [139].

3.2.2 Polyethylene terephthalate (PET)

PET stands as another exceptional polymer substrate for the fabrication of fully transparent films. Yang and his team [140], for instance, employed reactive sputtering at room temperature to deposit γ-CuI films on both PET and glass substrates. The film that deposited on the glass substrate showed a ZT of approximately 0.21 at RT, which surpassed the performance of other transparent materials during that period. Furthermore, it was observed that compressive stress enhanced σ, whereas tensile stress led to a reduction in σ. This phenomenon can be attributed to changes in the compactness of the grains. Notably, a P of 8.2 nW at a thermal gradient of 11 K was generated through a transparent f-TEGs with a single leg.

Additionally, Seo and colleagues [141] designed removable f-TEGs by depositing Bi_2Te_3 and Cu electrodes directly onto a PET substrate using RF and DC magnetron sputtering techniques, respectively. The device, which comprises 16 modules, is then encapsulated with a cellulose film. These f-TEGs can generate a power density of approximately 42 μW cm^{-2} at a thermal gradient of 14 K, and this performance remains consistent even when the device is bent. Furthermore, its high-resolution sensing capabilities for temperature less than 0.19 K and velocity less than 0.03 cm s^{-1} underscore its significant application prospect in the field of sensors.

3.2.3 Polyethylene naphthalate (PEN)

PEN film also emerges as an excellent substrate with great transparency and flexibility. As an example, Umeda and his team [142] designed f-TEGs by depositing an amorphous indium-gallium-zinc-oxide thin film onto a PEN substrate via magnetron sputtering technique. The f-TEGs features two legs connected in series through Au/Mo electrodes and Cu pads. An output voltage of ~6.7 mV and an output power of about 0.12 nW at a thermal gradient of 53 K could be obtained.

3.2.4 Mica

$Ca_3Co_4O_9$ is a non-toxic p-type TE material. Eklund's group [143] grew $Ca_3Co_4O_9$ films on mica substrates via multi-step sputtering, achieving a PF of ~1180 μW m^{-1} K^{-2} at 573 K. The films remained intact under tensile or compressive stress (r = 14 mm). Later, they deposited $Zn_{0.99}Ga_{0.01}O$ films on flexible mica using CVD, achieving a PF of ~100 μW m^{-1} K^{-2} at RT [144].

Recently, Zhong et al. [145] used one-step CVD to deposit polycrystalline SnSe on fluorphlogopite mica, obtaining highly oriented SnSe (400) films with a ZT of 0.15 at 550 K. A single-leg f-TEG produced 17 mV at a 50.5 K thermal gradient, and devices with varying units were designed for diverse applications.

3.2.5 Cellulose

Because of the biosafety and exceptional adaptability of cellulose, the TE textile-based generator which could maintain direct contact with human body garners significant focus. This innovative concept was recently proposed in a pioneering study [146]. Motivated by Kato's research [147], which reported the self-assembly of a porous structure, Jin and his team [44] utilized an unbalanced magnetron sputtering technique to prepare Bi_2Te_3 onto cellulose fiber (CF) paper. The strong scattering effect significantly enhanced S while reducing κ, resulting in a ZT of approximately

0.38 at 473 K. Notably, the Bi_2Te_3/CF film exhibited superior flexibility compared to the Bi_2Te_3/PI film. Furthermore, the researchers fabricated a novel TE device where one end of the strip was clamped with an alligator clip and the other end was held by fingers (as illustrated in **Figure 5n-o**). This generator, which featured 12 units of p-n legs, showed a stable voltage of 0.14 V with a thermal gradient of 50 K. Similar work was also reported by Rojas and his team [148]. The atomic layer deposition technique was employed by Karttunen and colleagues [149] to create ZnO-based films on cotton textile substrates. Among these, the $(Al_2O_3)_{50}(Zn_{0.99}Al_{0.01}O)_{600}$ film exhibited the PF of 137 μW m^{-1} K^{-2} at 300 K, with great flexibility.

3.2.6 Carbon materials

Till now, graphene is the most common carbon materials used as substrates for deposition due to its smaller lattice mismatch with TE materials [150]. It could greatly reduce residual stress during deposition process and prevents deformation, including wrinkling and warping. For instance, Lee and his team [151] deposited n-type (Bi_2Te_3) and p-type (Sb_2Te_3) films onto graphene using plasma-enhanced CVD. The deposited films exhibited excellent PFs at RT of 3520 μW m^{-1} K^{-2} (n-type) and 3590 μW m^{-1} K^{-2} (p-type) due to its high orientation along the (00 l) plane.

3.2.7 Polyacrylonitrile (PAN)

PAN is a polymer compound obtained by free radical polymerization of acrylonitrile. It possesses good resistance to chemical reagents, especially inorganic acids, bleaching powder, hydrogen peroxide, and general organic reagents. For instance, Lee and his colleagues [152] were the first to utilize an electrospinning method to create highly orientated PAN sheets, followed by depositing Bi_2Te_3 and Sb_2Te_3 onto these as-prepared PAN substrates to obtain a nanofiber structure. The resulting film demonstrated remarkable flexibility and exceptional tensile strength, with minimal changes in electrical resistance during stretching up to approximately 10.5% strain. These twist yarns were then fabricated into f-TEGs, with an output power density of 8.56 W m^{-2} at a thermal gradient of 200 K.

3.3 Printing

Thanks to its streamlined processing steps, minimized material waste, low costs, and straightforward, printing technology has become exceptionally appealing and has advanced to enable the patterning of a diverse array of electronic materials onto various substrates. It mainly involves dispenser printing, screen printing, inkjet printing and 3D printing. The former two printing methods are widely used due to their convenience, are the most prevalent printing processes; however, they frequently lack precision and controllability. The latter two approaches address these deficiencies, albeit with relatively costly equipment.

3.3.1 PI

Zhang's research group has published several studies on the preparation of TE films using printing processes [54, 153–156]. Some of these studies focused on rigid substrates [154], while others were conducted on PI substrates. For example, in one study, they [153] first synthesized $Bi_2Te_{2.8}Se_{0.2}$ through microwave-stimulated wet-chemical synthesis and then use screen printing technique to prepare it on a PI, followed by thermal treatment at 703 K to remove the surfactant and enhance film density. The high density (up to 85% relative density) and phonon scattering caused by nanoscale grains and pores resulted in a ZT value of 0.43 at 448 K. Later on, they tried to use different printing method to prepare Bi-Te alloys on PI substrate, like 3D conformal aerosol jet printing [156]. However, it seems that printing technique has little effect on TE performance while the processing technology of Bi-Te alloy raw materials and post-sintering process has more impact on the resulted film [54].

Additionally, Hou et al. [157] employed brush-printing to deposit $Bi_{0.5}Sb_{1.5}Te_3$/epoxy flexible films on a PI, with a hot-pressing treatment. Thanks to the preferential orientation of (000 l), the film showed an impressive PF of approximately 840 $\mu W\ m^{-1}\ K^{-2}$ at RT, with excellent flexibility. When the current was set at 60 mA, an f-TEG with eight elements generated a thermal gradient of 6.2 K, surpassing previously reported Bi_2Te_3-based cooling devices [158–160]. Similar work was also reported in Chen et al. [161]. Most recently, Du et al. [162] fabricated several types of chalcogenide nanowires based-inks and then utilized ink printing method to prepared Ag_2Te-based f-TEFs on glass fiber, which exhibited a PF of 493.8 $\mu W\ m^{-1}\ K^{-2}$ at 400 K, and the resulted f-TEGs showed a maximum power density of 0.9 $\mu W\ cm^{-2}\ K^{-2}$.

3.3.2 Cellulose

Ferhat et al. [163] utilized inkjet printing to prepare a TiS_2-based component on a flexible paper substrate, with a post heat-treatment at a temperature range of 383–388 K. By optimizing the hexylammonium content, a maximum PF of 210 $\mu W\ m^{-1}\ K^{-2}$ could be obtained. f-TEGs with 7 units could generate a power density of 17.6 $\mu W\ cm^{-2}$ at a thermal gradient of 20 K.

More recently, Zhao et al. [164] developed TE ink containing Bi_2Te_3 and bacterial cellulose nanofibers, enhanced with an alkyl ketene dimer emulsion to improve adhesion and water resistance. The resulted honeycomb-like generator with 96 pair of p-n legs could generate a $V_{voltage}$ of 70.5 mV and a maximum P of 596 nW at a thermal gradient of 55 K. In Ref. [165], untreated paper served as the substrate for p-type ($Bi_2Se_{0.3}Te_{2.7}$) and n-type ($Bi_{0.5}Sb_{1.5}Te_3$) legs, fabricated via modified dispenser printing to form a f-TEG. With ten legs, the f-TEG produced 8 mV and 10 nW of power under a 35 K thermal gradient. Similar work was also reported in Shin et al. [166].

When considering printing techniques for TE materials, the following suggestions are offered: (a) When preparing the slurry, it is advisable to select a binder with a lower melting point and gasification temperature, to ensure its positive role in recrystallization annealing and minimize residual binder during heat treatment. (b) The ink's contact angle should be greater than 40° but less than 90° to ensure accurate pattern reproduction during the printing process. (c) For substrates with a rough surface, such as glass fiber, a pre-coating of a polymer like chitosan may be necessary to facilitate proper ink adhesion and pattern formation.

3.4 Coating

Coating is also a convenient technique utilized for forming films, encompassing spin coating, roll coating, drop casting, spray coating, and dip coating (**Figure 6**).

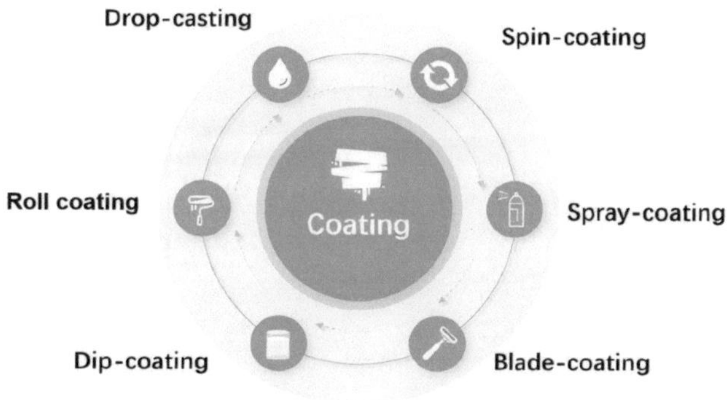

Figure 6.
The classification of coating technology.

3.4.1 PI

Takashiri's team [167] synthesized hexagonal Bi_2Te_3 nanoplates via solvothermal method and drop-cast them onto a PI substrate. After refining thermal annealing, they [168] achieved a maximum PF of ~350 μW m^{-1} K^{-2} at RT. They then applied the same method to prepare n-type $Bi_2(Se_xTe_{1-x})_3$ films, reaching a PF of ~410 μW m^{-1} K^{-2} at RT for $Bi_2(Se_{0.75}Te_{0.25})_3$.

In addition to inorganics, numerous works have tried to drop-cast CPs on PI substrates. For example, Bharti et al. [169] drop-cast PEDOT:PSS-based solutions onto PI substrates, followed by annealing. A full-organic f-TEGs with 30 legs generated an U of approximately 17.6 mV and an P of approximately 207 nW at a thermal gradient of 80 K.

Bulk Cu_2Se also exhibits excellent TE performance [170]. Consequently, Lin et al. [171] prepared Cu_2Se-based ink and then deposited onto a flexible PI substrate through spin coating, with a post-annealing at a temperature around 573–773 K. The resulted film exhibited a PF of approximately 460 μW m^{-1} K^{-2} at 664 K, which represents the highest value reported for Cu_2Se-based f-TEFs.

3.4.2 PET

Layer-by-layer self-assembly is a process by which composite membranes are constructed through alternating deposition of specific polymers, quantum dots, nanoparticles, biomolecules, and other substances, under the action of complementary interactions such as electrostatic interaction, hydrogen bonding, coordination bonding, and covalent bonding. Cho's group utilized this approach to fabricate organic TE films on PET substrates using dip-coating. They [172] successfully prepared a multiple-layered CP-carbon composite (PANi/graphene-PEDOT:PSS/PANi/DWCNT-PEDOT:PSS) film on a PET substrate. The composite film exhibited a PF of ~2710 μW m^{-1} K^{-2} at RT.

Tian et al. [173] used NMP to exfoliate TiS_2 powder into nanosheets, intercalated hexylamine to form a hybrid superlattice film, and dip-coated it onto a UV-treated PET substrate, achieving a PF of ~210 μW m^{-1} K^{-2} at RT. Choi et al. [174] synthesized SnSe nanosheets via hydrothermal synthesis, drop-casting them onto a PET substrate. The film showed an out-of-plane S of ~1250 μV K^{-1}), much higher than SnSe single crystals (809 μV K^{-1}), likely due to carrier filtering, despite a low σ of 2.4 S cm^{-1}.

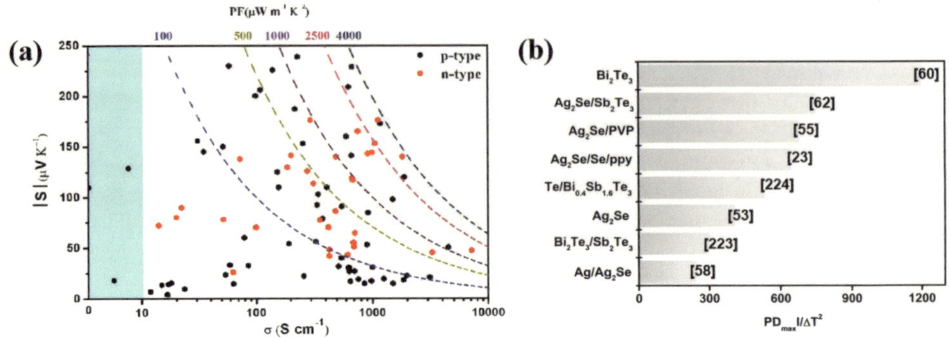

Figure 7.
(a) Selected PFs and corresponding σ and S of the f-TEFs reported in literatures. The blue region represents areas with very low PF values, indicating material systems that are not worth exploring in the future. The dashed lines in different colors correspond to the PF values at their respective peaks. Black dots represent p-type TE materials, while red dots represent n-type TE materials. Image reproduced with permission from Ref. [72] published by Elsevier. (b) Selected normalized maximum powder density ($PD_{max}l/\Delta T^2$) in high-performance f-TEGs [23, 53, 55, 58, 60, 62, 176–179].

3.4.3 Poly-p-phenylene terephthamide (PPTA)

Jang et al. [175] employed the mechanical alloying technique to synthesize BiSbTe alloy particles and subsequently incorporated nanofiber cellulose as a binder to formulate a mixed paste. This paste was then drop-cast onto chitosan-precoated PPTA, followed by cold pressing (**Figure 7e**). By refining the composition, a sample containing $Bi_{0.46}Sb_{1.34}Te_{3.2}$ after pressing treatment, showed a peak PF of approximately 611 μW m^{-1} K^{-2} at RT.

Apart from the aforementioned substrates, other materials, such as glass fiber, have been utilized for coating purposes. The coating technique is advantageous in terms of convenience, speed, and environmentally friendly method. However, the TE performance of the resulting films typically remains modest.

4. Challenge and prospective

4.1 TE property measurement

4.1.1 Electrical conductivity

Electrical conductivity, a well-established property that might initially seem redundant to mention, presents unique challenges when measured in TE films, beyond those encountered in bulks. Conventionally, the four-probe method is employed to test σ of films, necessitating an accurate determination of film thickness. For freestanding TE films, thickness measurement is straightforward; however, for films deposited on substrates, this task becomes more intricate, particularly if the substrate has an uneven surface or a porous structure, like glass fiber, which can significantly impact the film's thickness.

To address this issue, enhancing the substrate's surface smoothness, such as through pre-coating with chitosan, can be beneficial [175]. Furthermore, ensuring uniform film deposition, for instance, by using a homogeneously dispersed precursor solution during vacuum filtration, is crucial. Selecting an appropriate method for thickness measurement is also vital. Observing the sample's cross-section under scanning electron microscopy (SEM) is a common and intuitive approach, where preparing a clear cross-section is pivotal. Brittle fracture of samples in liquid nitrogen is a standard technique for obtaining such cross-sections. However, when cellulose-based materials, like nylon, serve as substrates, even this method may fail to produce a clean break. In such cases, employing a knife-assisted liquid nitrogen brittle fracture method, as detailed in the supporting information of refs. [50, 53], can be helpful.

Lastly, to enhance measurement accuracy, it is advisable to observe as wide a field as possible under SEM, measure numerous points, and average the data.

4.1.2 In-plane thermal conductivity

Measuring the in-plane κ of a film remains a significant challenge to this day. Three primary methods have been reported for this purpose [180]. The first involves using the LINSEIS TFA apparatus that operates based on the 3ω-method. Nevertheless, this approach necessitates depositing the film on a specialized substrate, thereby limiting its broad applicability. The second method employs a suspended microdevice that consists of two adjacent SiN_x membranes. Each membrane is adorned with a serpentine platinum resistance thermometer, accompanied by two electrodes and six supporting beams, enabling the conduction of four-probe measurements of TE properties [181]. While sophisticated, this method also presents its own set of challenges. The third method employs the Laser-Flash technique to measure in-plane thermal diffusivity (D). By combining this with the film's density (ρ) and specific heat capacity at constant pressure (Cp), the thermal conductivity (κ) can be calculated using the equation: $\kappa = \rho C_p D$. Details of this calculation are provided in the supporting information of reference [53]. However, a limitation of this method is that it requires a relatively large film size. For freestanding f-TEFs, this method can directly yield the in-plane thermal conductivity. The fourth method is called transient electro-thermal (TET) technique, which is an approach with high accuracy to measure the thermal diffusivity of various solid materials. In a typical TET measurement, the sample is suspended between two electrodes, where the electrodes also serve as heat sink during the thermal transport characterization and this method was proved to be accurate by the measurement of some freestanding films [60]. Yet, for f-TEFs deposited on substrates, the measured in-plane thermal conductivity inevitably includes contributions from the substrate, posing a challenge in isolating the thermal conductivity of the TE film itself.

4.1.3 Seebeck coefficient

While measuring the Seebeck coefficient may seem straightforward, it is crucial to remember that this coefficient is determined by the slope of the linear relationship between the thermal electromotive force and the temperature difference (ranging from approximately 2–5 K) between the two ends of a film. Given that the value of ZT is in proportional to the square of the S, any inaccuracies in measuring S can pose significant issues, particularly for materials with low S, like CP and carbon materials.

4.2 f-TEGs

Currently, the research on f-TEGs is still in the nascent stages of principle demonstration, lagging significantly behind the advancements made in the field of f-TEFs. Hence, it is imperative to intensify efforts in studying f-TEGs.

In the case of f-TEGs, the legs are constructed from thin films, distinguishing them vastly from devices made of bulk TE materials. For identical materials and TE leg lengths, the resistance of thin films is several hundred times greater than that of bulk materials. Furthermore, the electrical power generated by f-TEGs typically falls within the micro-watt range, necessitating amplification via a chip equipped with DC/DC amplification and capacitor storage for practical use. Consequently, if the R_{in} of f-TEGs is excessively high, the V drop across the f-TEGs will be substantial, resulting in an insufficient voltage supplied to the chip, thereby hindering power generation or even preventing the chip from triggering. To address this, researchers should endeavor to produce thicker TE films and select material systems with higher electrical conductivity.

To minimize contact resistance, the ends of each leg should be sputtered with electrodes, such as gold. It is worth noting that silver electrodes, while commonly used, are prone to aging in air at room temperature. Therefore, if silver electrodes are utilized, the device must be adequately packaged to prevent degradation.

4.3 Summary

To sum up, we have provided an in-depth examination of f-TEFs, focusing on two distinct aspects: freestanding f-TEFs and substrate-based f-TEFs, and based on our review, we propose the following outlooks:

1. In case of freestanding TE films, the integration of room-temperature metal-like plasticity with adjustable bandgaps and high carrier mobility to create plastic inorganic semiconductors remains a promising direction for future exploration. For instance, utilizing anti-situ defects to transform brittle materials into plastic materials has been proven to be an effective strategy. Meanwhile, it is essential to explore more cost-effective methods for slicing bulk materials into films, as this is crucial for future large-scale production. In terms of f-TEFs based on substrates, post-treatment often significantly affects the TE performance of thin films, while the physical properties of the substrate often constrain the overall post-treatment process of the samples. For instance, in the post-treatment of certain thin films, higher heat treatment temperatures can significantly enhance film performance, necessitating substrates with higher melting points. Consequently, substrates with specific properties, such as high-temperature-resistant PDMS or highly flexible PPTA, deserve greater attention (see **Table 1**).

2. As shown in Eq. 4, the PF of f-TEFs is directly proportional to the maximum power density of f-TEGs. Consequently, achieving superior output performance in f-TEGs requires a higher PF.

   Figure 7a presents typical data on PF, along with corresponding σ and S values, for f-TEFs reported in various literature sources. As observed in **Figure 7a**, high PF values are generally achieved when σ exceeds 1000 S/cm, and the value of S varies between 100 and 250 μV/K.

3. Early research on TE generators primarily focused on applications in power generation through temperature difference. In recent years, TE devices have also been explored in various other fields, such as temperature sensors [225], fire alarms [226], and gesture recognition [227]. Furthermore, TE conversion tech-

Type of substrate	Molecular structure	Electrical conductivity ($S\ cm^{-1}$)	Thermal conductivity ($W{\cdot}m^{-1}{\cdot}K^{-1}$)	Melting point (°C)	Tensile strength (MPa)
Nylon membrane	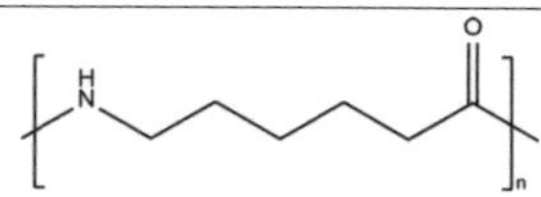	10^{-15} [180]	0.24 [181]	268.46 [182]	304 [183]
PVDF membrane	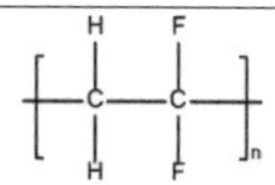	$\approx 10^{-12}$ [184]	0.25 [185]	156.5–162.3 [186]	50 [187]
PTFE membrane	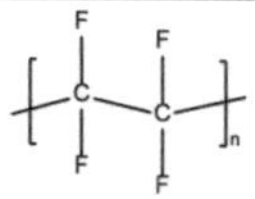	5.1×10^{-19} [188]	0.27 [189]	331 [190]	7.52 [191]
PI	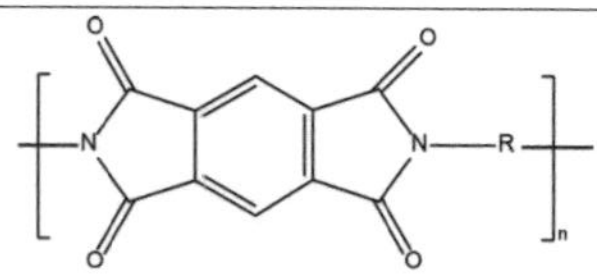	10^{-16} [192]	0.27 [193]	>400 [194]	1562 [195]
PET	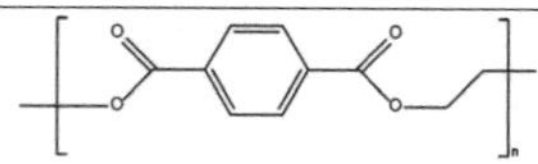	10^{-14} [196]	0.19 [181]	250–255 [197]	998 [198]
PEN	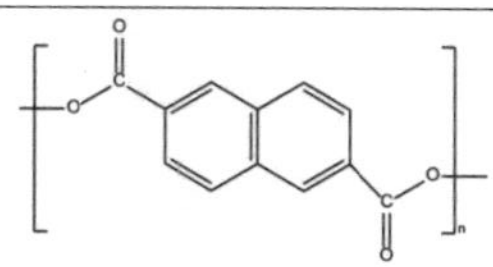	Unknown	0.26 [199]	265 [200]	200 [201]
PAN	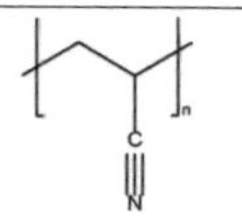	5×10^{-8} [202]	0.01–0.06 [203]	135 [204]	1800 [205]
Mica	$K(Mg,Fe,Al)_3(Si,Al)_4O_{10}(OH,F)_2$	10^{-12} [206]	0.1 [207]	700 [143]	32 [208]

Type of substrate		Molecular structure	Electrical conductivity (S cm^{-1})	Thermal conductivity (W·m^{-1}·K^{-1})	Melting point (°C)	Tensile strength (MPa)
Cellulose	Paper	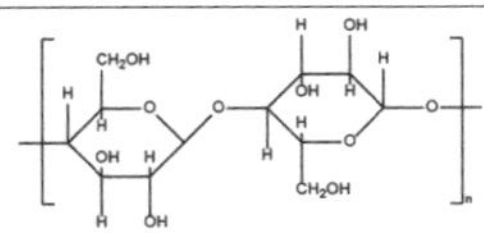	10^{-6} [209]	0.029–0.17 [210]	450 [211]	34 [212]
	Cotton			0.071 [210]	450 [211]	128–597 [213]
Carbon materials	CNT	/	5.15 [214]	2000 [215]	4527 [216]	43,000 [217]
	Graphene	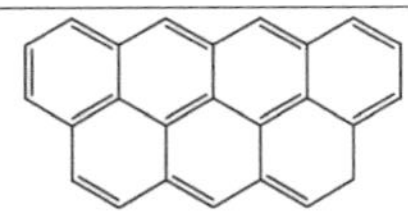	59 [218]	3080–5100 [219]	4727 [220]	150 [218]
PPTA		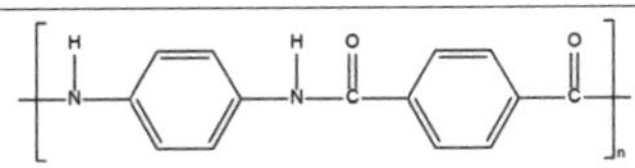	10^{-9} [221]	0.023–0.04 [222]	550 [223]	3600 [224]

Table 1.
Typical physical properties of substrates for the preparation of f-TEFs.

nology, based on the Peltier effect of semiconductor materials, enables localized heat transport and active cooling in confined spaces. With features like miniaturization, fast response, and ease of integration, it has been tested for applications such as chip cooling [228]. Future research on TE devices could place greater emphasis on these emerging applications.

4. Considering that the f-TEGs designed are intended for applications near RT, it is advisable to opt for materials that exhibit high TE performance around RT, such as Ag_2Se and Bi_2Te_3, as the target material for the inorganic component. Meanwhile, the currently reported high-performance f-TEGs are concentrated within these two material systems (**Figure 7b**). Also, more room-temperature TE materials should be given attention, such as Mg_2Sb_3-based materials.

5. Inorganic materials generally outperform organic ones in TE properties, making it essential to maximize their content in organic/inorganic composite films. However, the chemical disparity between components complicates achieving seamless interfaces. In situ synthesis of core/shell nanostructures and hybrid films offers a promising solution [50, 55].

We hope that this chapter provides valuable guidance for the preparation of high-performance f-TEFs, processing strategies, and f-TEGs design, ultimately contributing to the advancement and practical application of f-TEGs.

Acknowledgements

This chapter was supported by the National Natural Science Foundation of China (NSFC) (Grant no. 52402232), Guangdong Basic and Applied Basic Research Foundation (Grant no. 2023A1515110512), Southern University of Science and Technology (Grant no. Y01796223), and the open research fund of Songshan Lake Materials Laboratory (Grant no. 2023SLABFN15).

Conflict of interest

There are no conflicts of interest to declare.

Notes/thanks/other declarations

Special thanks are due to Prof. Kefeng Cai for his valuable comments.

Author details

Xiang Li and Yao Lu*
School of Microelectronics, Southern University of Science and Technology, Shenzhen, China

*Address all correspondence to: luy8@sustech.edu.cn

References

[1] Wortmann F, Flüchter K. Internet of things: technology and value added. Business & Information Systems Engineering. 2015;**57**:221-224. DOI: 10.1007/s12599-015-0383-3

[2] Haras M, Skotnicki T. Thermoelectricity for IoT–A review. Nano Energy. 2018;**54**:461-476. DOI: 10.1016/j.nanoen.2018.10.013

[3] Madakam S, Ramaswamy R, Tripathi S. Internet of Things (IoT): A literature review. Journal of Computer and Communications. 2015;**3**:164-173. DOI: 10.4236/jcc.2015.35021

[4] Idhalama OU, Oredo JO. Exploring the next generation Internet of Things (IoT) requirements and applications: A comprehensive overview. Information Development. 2024:02666669241267852. DOI: 10.1177/02666669241267852

[5] Nozariasbmarz A, Collins H, Dsouza K, Polash MH, Hosseini M, Hyland M, et al. Review of wearable thermoelectric energy harvesting: From body temperature to electronic systems. Applied Energy. 2020;**258**:114069. DOI: 10.1016/j.apenergy.2019.114069

[6] Kanahashi K, Pu J, Takenobu T. 2D materials for large-area flexible thermoelectric devices. Advanced Energy Materials. 2020;**10**:1902842. DOI: 10.1002/aenm.201902842

[7] Lin MC, Gong M, Lu B, Wu Y, Wang DY, Guan M, et al. An ultrafast rechargeable aluminium-ion battery. Nature. 2015;**520**:324-328. DOI: 10.1038/nature14340

[8] Ponrouch A, Frontera C, Bardé F, Palacín MR. Towards a calcium-based rechargeable battery. Nature Materials. 2016;**15**:169-172. DOI: 10.1038/nmat4462

[9] Wang Y, Yang L, Shi XL, Shi X, Chen L, Dargusch MS, et al. Flexible thermoelectric materials and generators: challenges and innovations. Advanced Materials. 2019;**31**:1807916. DOI: 10.1002/adma.201807916

[10] Siddique ARM, Mahmud S, Van B. A review of the state of the science on wearable thermoelectric power generators (TEGs) and their existing challenges. Renewable and Sustainable Energy Reviews. 2017;**73**:730-744. DOI: 10.1016/j.rser.2017.01.177

[11] Scrosati B, Hassoun J, Sun YK. Lithium-ion batteries. A look into the future. Energy & Environmental Science. 2011;**4**:3287-3295. DOI: 10.1039/C1EE01388B

[12] Lechêne BP, Cowell M, Pierre A, Evans JW, Wright PK, Arias AC. Organic solar cells and fully printed super-capacitors optimized for indoor light energy harvesting. Nano Energy. 2016;**26**:631-640. DOI: 10.1016/j.nanoen.2016.06.017

[13] Zhao W, Liu Z, Sun Z, Zhang Q, Wei P, Mu X, et al. Superparamagnetic enhancement of thermoelectric performance. Nature. 2017;**549**:247-251. DOI: 10.1038/nature23667

[14] Chen Z, Jian Z, Li W, Chang Y, Ge B, Hanus R, et al. Lattice dislocations enhancing thermoelectric PbTe in addition to band convergence. Advanced Materials. 2017;**29**:1606768. DOI: 10.1002/adma.201606768

[15] Zhu T, Liu Y, Fu C, Heremans JP, Snyder JG, Zhao X. Compromise and synergy in high-efficiency thermoelectric materials. Advanced Materials. 2017;**29**:1605884. DOI: 10.1002/adma.201605884

[16] Wu Y, Nan P, Chen Z, Zeng Z, Liu R, Dong H, et al. Thermoelectric enhancements in PbTe alloys due to dislocation-induced strains and converged bands. Advanced Science. 2020;**7**:1902628. DOI: 10.1002/advs.201902628

[17] Zhao LD, Lo SH, Zhang Y, Sun H, Tan G, Uher C, et al. Ultralow thermal conductivity and high thermoelectric figure of merit in SnSe crystals. Nature. 2014;**508**:373-377. DOI: 10.1038/nature13184

[18] Wang J, Cai K, Shen S. A facile chemical reduction approach for effectively tuning thermoelectric properties of PEDOT films. Organic Electronics. 2015;**17**:151-158. DOI: 10.1016/j.orgel.2014.12.007

[19] Wang J, Cai K, Shen S. Enhanced thermoelectric properties of poly (3, 4-ethylenedioxythiophene) thin films treated with H_2SO_4. Organic Electronics. 2014;**15**:3087-3095. DOI: 10.1016/j.orgel.2014.09.012

[20] Li Y, Du Y, Dou Y, Cai K, Xu J. PEDOT-based thermoelectric nanocomposites–A mini-review. Synthetic Metals. 2017;**226**:119-128. DOI: 10.1016/j.synthmet.2017.02.007

[21] Wu J, Sun Y, Xu W, Zhang Q. Investigating thermoelectric properties of doped polyaniline nanowires. Synthetic Metals. 2014;**189**:177-182. DOI: 10.1016/j.synthmet.2014.01.007

[22] Yao Q, Chen L, Xu X, Wang C. The high thermoelectric properties of conducting polyaniline with special submicron-fibre structure. Chemistry Letters. 2015;**34**:522-523. DOI: 10.1246/cl.2005.522

[23] Li Y, Lou Q, Yang J, Cai K, Liu Y, Lu Y, et al. Exceptionally high power factor Ag_2Se/Se/polypyrrole composite films for flexible thermoelectric generators. Advanced Functional Materials. 2022;**32**:2106902. DOI: 10.1002/adfm.202106902

[24] Tong L, Gao M, Jiang C, Cai K. Ultra-high performance and flexible polypyrrole coated CNT paper electrodes for all-solid-state supercapacitors. Journal of Materials Chemistry A. 2019;**7**:10751-10760. DOI: 10.1039/C9TA01856E

[25] Xu S, Shi XL, Dargusch M, Di C, Zou J, Chen ZG. Conducting polymer-based flexible thermoelectric materials and devices: from mechanisms to applications. Progress in Materials Science. 2021;**121**:100840. DOI: 10.1016/j.pmatsci.2021.100840

[26] Kim N, Lienemann S, Petsagkourakis I, Alemu D, Kee S, Ederth T, et al. Elastic conducting polymer composites in thermoelectric modules. Nature Communications. 2020;**11**:1424. DOI: 10.1038/s41467-020-15135-w

[27] Bharti M, Singh A, Samanta S, Aswal DK. Conductive polymers for thermoelectric power generation. Progress in Materials Science. 2018;**93**:270-310. DOI: 10.1016/j.pmatsci.2017.09.004

[28] Qin J, Du Y, Wang L, Du M, Eklund P, Ke Q. Flexible two-dimensional inorganic nanosheet/PEDOT: PSS thermoelectric composite films. Energy Material Advances. 2024;**5**:0097. DOI: 10.34133/energymatadv.0097

[29] Jiang Q, Pan D, Wang Y, Liu Y, Luo Y, Yang J, et al. High thermoelectric performance related to PVDF ferroelectric domains in p-type flexible PVDF-$Bi_{0.5}Sb_{1.5}Te_3$ composite film.

Small. 2024;**20**:2306786. DOI: 10.1002/smll.202306786

[30] Nie X, Ke S, Zhang H, Xu W, Li X, Wei P, et al. Enhancing electrical properties of flexible BiSbTe/epoxy composite films via liquid-phase extrusion. Journal of Power Sources. 2024;**623**:235405. DOI: 10.1016/j.jpowsour.2024.235405

[31] Du Y, Chen J, Meng Q, Xu J, Paul B, Eklund P. Flexible ternary carbon black/Bi_2Te_3 based alloy/polylactic acid thermoelectric composites fabricated by additive manufacturing. Journal of Materiomics. 2020;**6**:293-299. DOI: 10.1016/j.jmat.2020.02.010

[32] Du Y, Chen J, Meng Q, Dou Y, Xu J, Shen SZ. Thermoelectric materials and devices fabricated by additive manufacturing. Vacuum. 2020;**178**:109384. DOI: 10.1016/j.vacuum.2020.109384

[33] Meng Q, Qiu Y, Cai K, Ding Y, Wang M, Pu H, et al. High performance and flexible polyvinylpyrrolidone/Ag/Ag_2Te ternary composite film for thermoelectric power generator. ACS Applied Materials & Interfaces. 2019;**11**:33254-33262. DOI: 10.1021/acsami.9b11217

[34] Rincón-García L, Ismael AK, Evangeli C, Grace I, Rubio-Bollinger G, Porfyrakis K, et al. Molecular design and control of fullerene-based bi-thermoelectric materials. Nature Materials. 2016;**15**:289-293. DOI: 10.1038/nmat4487

[35] El-Shamy AG. The role of nitrogen-carbon dots (NC) nano-particles in enhancing thermoelectric power functions of PEDOT: PSS/Te nano-composite films. Chemical Engineering Journal. 2021;**417**:129212. DOI: 10.1016/j.cej.2021.129212

[36] Wei S, Zhang Y, Lv H, Deng L, Chen G. SWCNT network evolution of PEDOT: PSS/SWCNT composites for thermoelectric application. Chemical Engineering Journal. 2022;**428**:131137. DOI: 10.1016/j.cej.2021.131137

[37] Xia B, Shi XL, Zhang L, Luo J, Chen WY, Hu B, et al. Vertically designed high-performance and flexible thermoelectric generator based on optimized PEDOT: PSS/SWCNTs composite films. Chemical Engineering Journal. 2024;**486**:150305. DOI: 10.1016/j.cej.2024.150305

[38] El-Shamy AG. Review on the recent advance in PEDOT: PSS/Carbonic fillers based nanocomposite for flexible thermoelectric devices and sensors. Materials Today Physics. 2023;**35**:101101. DOI: 10.1016/j.mtphys.2023.101101

[39] Bae EJ, Kang YH, Lee C, Cho SY. Engineered nanocarbon mixing for enhancing the thermoelectric properties of a telluride-PEDOT: PSS nanocomposite. Journal of Materials Chemistry A. 2017;**5**:17867-17873. DOI: 10.1039/C7TA04280A

[40] Yoo D, Kim J, Lee SH, Cho W, Choi HH, Kim FS, et al. Effects of one-and two-dimensional carbon hybridization of PEDOT: PSS on the power factor of polymer thermoelectric energy conversion devices. Journal of Materials Chemistry A. 2015;**3**:6526-6533. DOI: 10.1039/C7TA04280A

[41] Taroni PJ, Santagiuliana G, Wan K, Calado P, Qiu M, Zhang H, et al. Toward stretchable self-powered sensors based on the thermoelectric response of PEDOT: PSS/polyurethane blends. Advanced Functional Materials. 2018;**28**:1704285. DOI: 10.1002/adfm.201704285

[42] Zhang Y, Zhang Q, Chen G. Carbon and carbon composites for

thermoelectric applications. Carbon Energy. 2020;**2**:408-436. DOI: 10.1002/cey2.68

[43] Wang L, Zhang Z, Liu Y, Wang B, Fang L, Qiu J, et al. Exceptional thermoelectric properties of flexible organic– inorganic hybrids with monodispersed and periodic nanophase. Nature Communications. 2018;**9**:3817. DOI: 10.1038/s41467-018-06251-9

[44] Jin Q, Shi W, Zhao Y, Qiao J, Qiu J, Sun C, et al. Cellulose fiber-based hierarchical porous bismuth telluride for high-performance flexible and tailorable thermoelectrics. ACS Applied Materials & Interfaces. 2018;**10**:1743-1751. DOI: 10.1021/acsami.7b16356

[45] Kim SH, Min T, Choi JW, Baek SH, Choi JP, Jr C. Ternary $Bi_2Te_3In_2Te_3Ga_2Te_3$ (n-type) thermoelectric film on a flexible PET substrate for use in wearables. Energy. 2018;**144**:607-618. DOI: 10.1016/j.energy.2017.12.063

[46] Li T, Pickel AD, Yao Y, Chen Y, Zeng Y, Lacey SD, et al. Thermoelectric properties and performance of flexible reduced graphene oxide films up to 3,000 K. Nature Energy. 2018;**3**:148-156. DOI: 10.1038/s41560-018-0086-3

[47] Jin Q, Jiang S, Zhao Y, Wang D, Qiu J, Tang DM, et al. Flexible layer-structured Bi_2Te_3 thermoelectric on a carbon nanotube scaffold. Nature Materials. 2019;**18**:62-68. DOI: 10.1038/s41563-018-0217-z

[48] Ding Y, Qiu Y, Cai K, Yao Q, Chen S, Chen L, et al. High performance n-type Ag_2Se film on nylon membrane for flexible thermoelectric power generator. Nature Communications. 2019;**10**:841. DOI: 10.1038/s41467-019-08835-5

[49] Liang J, Wang T, Qiu P, Yang S, Ming C, Chen H, et al. Flexible thermoelectrics: from silver chalcogenides to full-inorganic devices. Energy & Environmental Science. 2019;**12**:2983-2990. DOI: 10.1039/C9EE01777A

[50] Lu Y, Qiu Y, Cai K, Ding Y, Wang M, Jiang C, et al. Ultrahigh power factor and flexible silver selenide-based composite film for thermoelectric devices. Energy & Environmental Science. 2020;**13**:1240-1249. DOI: 10.1039/C9EE01609K

[51] Lu Y, Qiu Y, Cai K, Li X, Gao M, Jiang C, et al. Ultrahigh performance PEDOT/Ag_2Se/CuAgSe composite film for wearable thermoelectric power generators. Materials Today Physics. 2020;**14**:100223. DOI: 10.1016/j.mtphys.2020.100223

[52] He S, Li Y, Liu L, Jiang Y, Feng J, Zhu W, et al. Semiconductor glass with superior flexibility and high room temperature thermoelectric performance. Science Advances. 2020;**6**:eaaz8423. DOI: 10.1126/sciadv.aaz8423

[53] Jiang C, Ding Y, Cai K, Tong L, Lu Y, Zhao W, et al. Ultrahigh performance of n-type Ag_2Se films for flexible thermoelectric power generators. ACS Applied Materials & Interfaces. 2020;**12**:9646-9655. DOI: 10.1021/acsami.9b21069

[54] Varghese T, Dun C, Kempf N, Saeidi-Javash M, Karthik C, Richardson J, et al. Flexible thermoelectric devices of ultrahigh power factor by scalable printing and interface engineering. Advanced Functional Materials. 2020;**30**:1905796. DOI: 10.1002/adfm.201905796

[55] Jiang C, Wei P, Ding Y, Cai K, Tong L, Gao Q, et al. Ultrahigh performance polyvinylpyrrolidone/Ag_2Se composite thermoelectric film

for flexible energy harvesting. Nano Energy. 2021;**80**:105488. DOI: 10.1016/j.nanoen.2020.105488

[56] Lei Y, Qi R, Chen M, Chen H, Xing C, Sui F, et al. Microstructurally tailored thin β-Ag_2Se films toward commercial flexible thermoelectrics. Advanced Materials. 2022;7:2104786. DOI: 10.1002/adma.202104786

[57] Gao Z, Yang Q, Qiu P, Wei T, Yang S, Xiao J, et al. p-type plastic inorganic thermoelectric materials. Advanced Energy Materials. 2021;**11**:2100883. DOI: 10.1002/aenm.202100883

[58] Li X, Lu Y, Cai K, Gao M, Li Y, Wang Z, et al. Exceptional power factor of flexible Ag/Ag_2Se thermoelectric composite films. Chemical Engineering Journal. 2022;**434**:134739

[59] Yang Q, Yang S, Qiu P, Peng L, Wei T, Zhang Z, et al. Flexible thermoelectrics based on ductile semiconductors. Science. 2022;**377**:854-858. DOI: 10.1126/science.abq0682

[60] Lu Y, Zhou Y, Wang W, Hu M, Huang X, Mao D, et al. Staggered-layer-boosted flexible Bi_2Te_3 films with high thermoelectric performance. Nature Nanotechnology. 2023;**18**:1281-1288. DOI: 10.1038/s41565-023-01457-5

[61] Zheng ZH, Shi XL, Ao DW, Liu W, Meng L, Kou L, et al. Harvesting waste heat with flexible Bi_2Te_3 thermoelectric thin film. Nature Sustainability. 2023;**6**:180-191. DOI: 10.1038/s41893-022-01003-6

[62] Yang D, Shi XL, Li M, Nisar M, Mansoor A, Chen S, et al. Flexible power generators by Ag_2Se thin films with record-high thermoelectric performance. Nature Communications. 2024;**15**:923. DOI: 10.1038/s41467-024-45092-7

[63] Suarez F, Nozariasbmarz A, Vashaee D, Öztürk MC. Designing thermoelectric generators for self-powered wearable electronics. Energy & Environmental Science. 2016;**9**:2099-2113. DOI: 10.1039/C6EE00456C

[64] Feng M, Lv S, Deng J, Guo Y, Wu Y, Shi G, et al. An overview of environmental energy harvesting by thermoelectric generators. Renewable and Sustainable Energy Reviews. 2023;**187**:113723. DOI: 10.1016/j.rser.2023.113723

[65] He J, Li K, Jia L, Zhu Y, Zhang H, Linghu J. Advances in the applications of thermoelectric generators. Applied Thermal Engineering. 2024;**236**:121813. DOI: 10.1016/j.applthermaleng.2023.121813

[66] Soleimani Z, Zoras S, Ceranic B, Cui Y, Shahzad S. A comprehensive review on the output voltage/power of wearable thermoelectric generators concerning their geometry and thermoelectric materials. Nano Energy. 2021;**89**:106325. DOI: 10.1016/j.nanoen.2021.106325

[67] Freer R, Powell AV. Realising the potential of thermoelectric technology: A Roadmap. Journal of Materials Chemistry C. 2020;**8**:441-463. DOI: 10.1039/C9TC05710B

[68] Gürkan K, Karaman H, Ballikaya S. Optimization of high-performance flexible thermoelectric generator from material synthesis to simulation and device application. Energy Conversion and Management. 2023;**291**:117335. DOI: 10.1016/j.enconman.2023.117335

[69] Masoumi S, O'Shaughnessy S, Pakdel A. Organic-based flexible thermoelectric generators: From materials to devices. Nano Energy. 2022;**92**:106774. DOI: 10.1016/j.nanoen.2021.106774

[70] Zhou Q, Li H, Du C, Ye Z, Liang L, Chen G. High-performance flexible thermoelectric generators with tunable in-plane and out-of-plane architectures. Nano Energy. 2023;**118**:109007. DOI: 10.1016/j.nanoen.2023.109007

[71] Dong Z, Liu H, Yang X, Fan J, Bi H, Wang C, et al. Facile fabrication of paper-based flexible thermoelectric generator. npj Flexible Electronics. 2021;**5**:6. DOI: 10.1038/s41528-021-00103-1

[72] Li X, Cai K, Gao M, Du Y, Shen S. Recent advances in flexible thermoelectric films and devices. Nano Energy. 2021;**89**:106309. DOI: 10.1016/j.nanoen.2021.106309

[73] Liu Y, Li Y, Wu M, Lu Y, Wang Z, Wei P, et al. Nanoengineering approach toward high power factor Ag_2Se/Se composite films for flexible thermoelectric generators. ACS Applied Materials & Interfaces. 2023;**15**:36587-36593. DOI: 10.1021/acsami.3c06960

[74] Zhou Q, Zhu K, Li J, Li Q, Deng B, Zhang P, et al. Leaf-inspired flexible thermoelectric generators with high temperature difference utilization ratio and output power in ambient air. Advanced Science. 2021;**8**:2004947. DOI: 10.1002/advs.202004947

[75] Lu Y, Han X, Wei P, Liu Y, Wang Z, Zuo X, et al. Nanoengineering approach toward ultrahigh power factor Ag_2Se/polyvinylpyrrolidone composite film for flexible thermoelectric generator. Chemical Engineering Journal. 2024;**485**:149793. DOI: 10.1016/j.cej.2024.149793

[76] Lu Y, Li X, Cai K, Gao M, Zhao W, He J, et al. Enhanced-performance PEDOT: PSS/Cu_2Se-based composite films for wearable thermoelectric power generators. ACS Applied Materials & Interfaces. 2021;**13**:631-638. DOI: 10.1021/acsami.0c18577

[77] Zhang J, Zhang W, Wei H, Tang J, Li D, Xu D. Flexible micro thermoelectric generators with high power density and light weight. Nano Energy. 2023;**105**:108023. DOI: 10.1016/j.nanoen.2022.108023

[78] Qiu P, Deng T, Chen L, Shi X. Plastic inorganic thermoelectric materials. Joule. 2024;**8**:622-634. DOI: 10.1016/j.joule.2023.12.020

[79] Shi X, Chen H, Hao F, Liu R, Wang T, Qiu P, et al. Room-temperature ductile inorganic semiconductor. Nature Materials. 2018;**17**:421-426. DOI: 10.1038/s41563-018-0047-z

[80] Deng T, Gao Z, Li Z, Qiu P, Li Z, Yuan X, et al. Room-temperature exceptional plasticity in defective Bi_2Te_3-based bulk thermoelectric crystals. Science. 2024;**386**:1112-1117. DOI: 10.1126/science.adr8450

[81] Liu C, Lu B, Yan J, Xu J, Yue R, Zhu Z, et al. Highly conducting free-standing poly (3, 4-ethylenedioxythiophene)/poly (styrenesulfonate) films with improved thermoelectric performances. Synthetic Metals. 2010;**160**:2481-2485. DOI: 10.1016/j.synthmet.2010.09.031

[82] Mengistie DA, Chen CH, Boopathi KM, Pranoto FW, Li LJ, Chu CW. Enhanced thermoelectric performance of PEDOT: PSS flexible bulky papers by treatment with secondary dopants. ACS Applied Materials & Interfaces. 2015;7:94-100. DOI: 10.1021/am507032e

[83] Ni D, Song H, Chen Y, Cai K. Significantly enhanced thermoelectric performance of flexible PEDOT nanowire film via coating Te

nanostructures. Journal of Materiomics. 2020;**6**:364-370. DOI: 10.1016/j.jmat.2019.07.001

[84] Jia F, Wu R, Liu C, Lan J, Lin YH, Yang X. High thermoelectric and flexible PEDOT/SWCNT/BC nanoporous films derived from aerogels. ACS Sustainable Chemistry & Engineering. 2019;**7**:12591-12600. DOI: 10.1021/acssuschemeng.9b02518

[85] Du Y, Shen SZ, Yang WD, Chen S, Qin Z, Cai KF, et al. Facile preparation and characterization of poly (3-hexylthiophene)/multiwalled carbon nanotube thermoelectric composite films. Journal of Electronic Materials. 2012;**41**:1436-1441. DOI: 10.1007/s11664-012-1931-y

[86] Ni D, Song H, Chen Y, Cai K. Free-standing highly conducting PEDOT films for flexible thermoelectric generator. Energy. 2019;**170**:53-61. DOI: 10.1016/j.energy.2018.12.124

[87] Ni D, Chen Y, Song H, Liu C, Yang X, Cai K. Free-standing and highly conductive PEDOT nanowire films for high-performance all-solid-state supercapacitors. Journal of Materials Chemistry A. 2019;**7**:1323-1333. DOI: 10.1039/C8TA08814D

[88] Wang J, Cai K, Song H, Shen S. Simultaneously enhanced electrical conductivity and seebeck coefficient in Poly (3, 4-ethylenedioxythiophene) films treated with hydroiodic acid. Synthetic Metals. 2016;**220**:585-590. DOI: 10.1016/j.synthmet.2016.07.023

[89] Bubnova O, Khan ZU, Malti A, Braun S, Fahlman M, Berggren M, et al. Optimization of the thermoelectric figure of merit in the conducting polymer poly (3, 4-ethylenedioxythiophene). Nature Materials. 2011;**10**:429-433. DOI: 10.1038/nmat3012

[90] Fan Z, Du D, Guan X, Ouyang J. Polymer films with ultrahigh thermoelectric properties arising from significant seebeck coefficient enhancement by ion accumulation on surface. Nano Energy. 2018;**51**:481-488. DOI: 10.1016/j.nanoen.2018.07.002

[91] Kim N, Kee S, Lee SH, Lee BH, Kahng YH, Jo YR, et al. Highly conductive PEDOT: PSS nanofibrils induced by solution-processed crystallization. Advanced Materials. 2014;**26**:2268-2272. DOI: 10.1002/adma.201304611

[92] Fan Z, Li P, Du D, Ouyang J. Significantly enhanced thermoelectric properties of PEDOT: PSS films through sequential post-treatments with common acids and bases. Advanced Energy Materials. 2017;**7**:1602116. DOI: 10.1002/aenm.201602116

[93] Guan X, Cheng H, Ouyang J. Significant enhancement in the Seebeck coefficient and power factor of thermoelectric polymers by the Soret effect of polyelectrolytes. Journal of Materials Chemistry A. 2018;**6**:19347-19352. DOI: 10.1039/C8TA08387H

[94] Zhang S, Fan Z, Wang X, Zhang Z, Ouyang J. Enhancement of the thermoelectric properties of PEDOT: PSS via one-step treatment with cosolvents or their solutions of organic salts. Journal of Materials Chemistry A. 2018;**6**:7080-7087. DOI: 10.1039/C7TA11148G

[95] Fan Z, Du D, Yu Z, Li P, Xia Y, Ouyang J. Significant enhancement in the thermoelectric properties of PEDOT: PSS films through a treatment with organic solutions of inorganic salts. ACS Applied Materials & Interfaces. 2016;**8**:23204-23211. DOI: 10.1021/acsami.6b07234

[96] Tsai TC, Chang HC, Chen CH, Whang WT. Widely variable Seebeck

coefficient and enhanced thermoelectric power of PEDOT: PSS films by blending thermal decomposable ammonium formate. Organic Electronics. 2011;**12**:2159-2164. DOI: 10.1016/j.orgel.2011.09.004

[97] Zhang L, Goto T, Imae I, Sakurai Y, Harima Y. Thermoelectric properties of PEDOT films prepared by electrochemical polymerization. Journal of Polymer Science Part B: Polymer Physics. 2017;**55**:524-531. DOI: 10.1002/polb.24299

[98] Yi C, Wilhite A, Zhang L, Hu R, Chuang SS, Zheng J, et al. Enhanced thermoelectric properties of poly (3, 4-ethylenedioxythiophene): poly (styrenesulfonate) by binary secondary dopants. ACS Applied Materials & Interfaces. 2015;**2015**(7):8984-8989. DOI: 10.1021/acsami.5b01960

[99] An CJ, Kang YH, Song H, Jeong Y, Cho SY. High-performance flexible thermoelectric generator by control of electronic structure of directly spun carbon nanotube webs with various molecular dopants. Journal of Materials Chemistry A. 2017;**5**:15631-15639. DOI: 10.1039/C7TA02509B

[100] Zhou W, Fan Q, Zhang Q, Cai L, Li K, Gu X, et al. High-performance and compact-designed flexible thermoelectric modules enabled by a reticulate carbon nanotube architecture. Nature Communications. 2017;**8**:14886. DOI: 10.1038/ncomms14886

[101] Du Y, Cai KF, Shen SZ, Casey PS. Preparation and characterization of graphene nanosheets/poly (3-hexylthiophene) thermoelectric composite materials. Synthetic Metals. 2012;**162**:2102-2106. DOI: 10.1016/j.synthmet.2012.09.011

[102] Du Y, Shen SZ, Yang WD, Cai KF, Casey PS. Preparation and characterization of multiwalled carbon nanotube/poly (3-hexylthiophene) thermoelectric composite materials. Synthetic Metals. 2012;**162**:375-380. DOI: 10.1016/j.synthmet.2011.12.023

[103] Wu R, Yuan H, Liu C, Lan JL, Yang X, Lin YH. Flexible PANI/SWCNT thermoelectric films with ultrahigh electrical conductivity. RSC Advances. 2018;**8**:26011-26019. DOI: 10.1039/C8RA04863K

[104] Xu Q, Qu S, Ming C, Qiu P, Yao Q, Zhu C, et al. Conformal organic–inorganic semiconductor composites for flexible thermoelectrics. Energy & Environmental Science. 2020;**13**:511-518. DOI: 10.1039/C9EE03776D

[105] Du Y, Cai KF, Chen S, Cizek P, Lin T. Facile preparation and thermoelectric properties of Bi_2Te_3 based alloy nanosheet/PEDOT:PSS composite films. ACS Applied Materials & Interfaces. 2014;**6**:5735-5743. DOI: 10.1021/am5002772

[106] Ju H, Kim J. Chemically exfoliated SnSe nanosheets and their SnSe/poly (3, 4-ethylenedioxythiophene): poly (styrenesulfonate) composite films for polymer based thermoelectric applications. ACS Nano. 2016;**10**:5730-5739. DOI: 10.1021/acsnano.5b07355

[107] Ju H, Park D, Kim J. Solution-processable flexible thermoelectric composite films based on conductive polymer/$SnSe_{0.8}$ $S_{0.2}$ nanosheets/carbon nanotubes for wearable electronic applications. Journal of Materials Chemistry A. 2018;**6**:5627-5634. DOI: 10.1039/C7TA11285H

[108] Li J, Shi Q, Röhr JA, Wu H, Wu B, Guo Y, et al. Flexible 3D porous MoS_2/CNTs architectures with ZT of 0.17 at room temperature for wearable thermoelectric applications. Advanced

Functional Materials. 2020;**30**:2002508. DOI: 10.1002/adfm.202002508

[109] Hernandez Y, Nicolosi V, Lotya M, Blighe FM, Sun Z, De S, et al. High-yield production of graphene by liquid-phase exfoliation of graphite. Nature Nanotechnology. 2008;**3**:563-568. DOI: 10.1038/nnano.2008.215

[110] Wang L, Zhang Z, Geng L, Yuan T, Liu Y, Guo J, et al. Solution-printable fullerene/TiS_2 organic/inorganic hybrids for high-performance flexible n-type thermoelectrics. Energy & Environmental Science. 2018;**11**:1307-1317. DOI: 10.1039/C7EE03617E

[111] Dun C, Hewitt CA, Huang H, Xu J, Zhou C, Huang W, et al. Flexible n-type thermoelectric films based on Cu-doped Bi_2Se_3 nanoplate and polyvinylidene fluoride composite with decoupled Seebeck coefficient and electrical conductivity. Nano Energy. 2015;**18**:306-314

[112] Wan C, Tian R, Kondou M, Yang R, Zong P, Koumoto K. Ultrahigh thermoelectric power factor in flexible hybrid inorganic-organic superlattice. Nature Communications. 2017;**8**:1024. DOI: 10.1038/s41467-017-01149-4

[113] Song H, Qiu Y, Wang Y, Cai K, Li D, Deng Y, et al. Polymer/carbon nanotube composite materials for flexible thermoelectric power generator. Composites Science and Technology. 2017;**153**:71-83. DOI: 10.1016/j.compscitech.2017.10.006

[114] Meng Q, Cai K, Du Y, Chen L. Preparation and thermoelectric properties of SWCNT/PEDOT:PSS coated tellurium nanorod composite films. Journal of Alloys and Compounds. 2018;**778**:163-169. DOI: 10.1016/j.jallcom.2018.10.381

[115] Song H, Cai K. Preparation and properties of PEDOT:PSS/Te nanorod composite films for flexible thermoelectric power generator. Energy. 2017;**125**:519-525. DOI: 10.1016/j.energy.2017.01.037

[116] Fan W, Guo CY, Chen G. Flexible films of poly (3, 4-ethylenedioxythiophene)/carbon nanotube thermoelectric composites prepared by dynamic 3-phase interfacial electropolymerization and subsequent physical mixing. Journal of Materials Chemistry A. 2018;**6**:12275-12280. DOI: 10.1039/C8TA04838J

[117] Zeng X, Yan C, Ren L, Zhang T, Zhou F, Liang X, et al. Silver telluride nanowire assembly for high-performance flexible thermoelectric film and its application in self-powered temperature sensor. Advanced Electronic Materials. 2019;**5**:1800612. DOI: 10.1002/aelm.201800612

[118] Abdullah SM, Ahmad Z, Sulaiman K. A solution-based temperature sensor using the organic compound CuTsPc. Sensors. 2014;**14**:9878-9888. DOI: 10.3390/s140609878

[119] Nichols WW. Clinical measurement of arterial stiffness obtained from noninvasive pressure waveforms. American Journal of Hypertension. 2005;**18**:3S-10S. DOI: 10.1016/j.amjhyper.2004.10.009

[120] Trung TQ, Dang TML, Ramasundaram S, Toi PT, Park SY, Lee NE. A stretchable strain-insensitive temperature sensor based on free-standing elastomeric composite fibers for on-body monitoring of skin temperature. ACS Applied Materials & Interfaces. 2018;**11**:2317-2327. DOI: 10.1021/acsami.8b19425

[121] Huang Y, Zeng X, Wang W, Guo X, Hao C, Pan W, et al. High-resolution flexible temperature sensor based graphite-filled polyethylene oxide and polyvinylidene fluoride composites for body temperature monitoring. Sensors and Actuators A: Physical. 2018;**278**:1-10. DOI: 10.1016/j.sna.2018.05.024

[122] Wang X, Meng F, Wang T, Li C, Tang H, Gao Z, et al. High performance of PEDOT:PSS/SiC-NWs hybrid thermoelectric thin film for energy harvesting. Journal of Alloys and Compounds. 2018;**734**:121-129. DOI: 10.1016/j.jallcom.2017.11.013

[123] See KC, Feser JP, Chen CE, Majumdar A, Urban JJ, Segalman RA. Water-processable polymer–nanocrystal hybrids for thermoelectrics. Nano Letters. 2010;**10**:4664-4667. DOI: 10.1021/nl102880k

[124] Novak TG, Kim J, Kim J, Shin H, Tiwari AP, Song JY, et al. Flexible thermoelectric films with high power factor made of non-oxidized graphene flakes. 2D Materials. 2019;**6**:045019. DOI: 10.1088/2053-1583/ab2c32

[125] Zhu W, Deng Y, Cao L. Light-concentrated solar generator and sensor based on flexible thin-film thermoelectric device. Nano Energy. 2017;**34**:463-471. DOI: 10.1016/j.nanoen.2017.03.020

[126] Singkaselit K, Sakulkalavek A, Sakdanuphab R. Effects of annealing temperature on the structural, mechanical and electrical properties of flexible bismuth telluride thin films prepared by high-pressure RF magnetron sputtering. Advances in Natural Sciences: Nanoscience and Nanotechnology. 2017;**8**:035002. DOI: 10.1088/2043-6254/aa7222

[127] Nuthongkum P, Sakdanuphab R, Horprathum M, Sakulkalavek A. [Bi]:[Te] control, structural and thermoelectric properties of flexible Bi_xTe_y thin films prepared by RF magnetron sputtering at different sputtering pressures. Journal of Electronic Materials. 2017;**46**:6444-6450. DOI: 10.1007/s11664-017-5671-x

[128] Kong D, Zhu W, Guo Z, Deng Y. High-performance flexible Bi_2Te_3 films based wearable thermoelectric generator for energy harvesting. Energy. 2019;**175**:292-299. DOI: 10.1016/j.energy.2019.03.060

[129] Shen S, Zhu W, Deng Y, Zhao H, Peng Y, Wang C. Enhancing thermoelectric properties of Sb_2Te_3 flexible thin film through microstructure control and crystal preferential orientation engineering. Applied Surface Science. 2017;**414**:197-204. DOI: 10.1016/j.apsusc.2017.04.074

[130] Khumtong T, Sukwisute P, Sakulkalavek A, Sakdanuphab R. Microstructure and electrical properties of antimony telluride thin films deposited by RF magnetron sputtering on flexible substrate using different sputtering pressures. Journal of Electronic Materials. 2017;**46**:3166-3171. DOI: 10.1007/s11664-017-5303-5

[131] Jitthamapirom P, Wanarattikan P, Nuthongkum P, Sakdanuphab R, Sakulkalavek A. Comparison of thermoelectric properties of flexible bismuth telluride thin films deposited via DC and RF magnetron sputtering. Ferroelectrics. 2019;**552**:64-72. DOI: 10.1080/00150193.2019.1653083

[132] Fan P, Zhang PC, Liang GX, Li F, Chen YX, Luo JT, et al. High-performance bismuth telluride thermoelectric thin films fabricated by using the two-step single-source thermal evaporation. Journal of Alloys and Compounds. 2020;**819**:153027. DOI: 10.1016/j.jallcom.2019.153027

[133] Shang H, Li T, Luo D, Yu L, Zou Q, Huang D, et al. High-performance Ag-modified $Bi_{0.5}Sb_{1.5}Te_3$ films for the flexible thermoelectric generator. ACS Applied Materials & Interfaces. 2020;**12**:7358-7365. DOI: 10.1021/acsami.9b21771

[134] Kim BG, Bae SH, Byeon J, Lee C, Choi SM. Stress-induced change of Cu-doped Bi_2Te_3 thin films for flexible thermoelectric applications. Materials Letters. 2020;**270**:127697. DOI: 10.1016/j.matlet.2020.127697

[135] Vieira E, Figueira J, Pires AL, Grilo J, Silva MF, Pereira AM, et al. Bi_2Te_3 and Sb_2Te_3 thin films with enhanced thermoelectric properties for flexible thermal sensors. Proceedings. 2018;**2**:815. DOI: 10.3390/proceedings2130815

[136] Fan P, Fan WF, Zheng ZH, Zhang Y, Luo JT, Liang GX, et al. Thermoelectric properties of zinc antimonide thin film deposited on flexible polyimide substrate by RF magnetron sputtering. Journal of Materials Science: Materials in Electronics. 2014;**25**:5060-5065. DOI: 10.1007/s10854-014-2271-x

[137] Zheng ZH, Li F, Luo JT, Liang GX, Ma HL, Zhang XH, et al. Enhancement of power factor in zinc antimonide thermoelectric thin film doped with titanium. Materials Letters. 2017;**209**:455-458. DOI: 10.1016/j.matlet.2017.08.063

[138] Kusano K, Yamamoto A, Nakata M, Suemasu T, Toko K. Thermoelectric inorganic SiGe film synthesized on flexible plastic substrate. ACS Applied Energy Materials. 2018;**1**:5280-5285. DOI: 10.1021/acsaem.8b00899

[139] Coroa J, Faustino BM, Marques A, Bianchi C, Koskinen T, Juntunen T, et al. Highly transparent copper iodide thin film thermoelectric generator on a flexible substrate. RSC Advances. 2019;**9**:35384-35391. DOI: 10.1039/C9RA07309D

[140] Yang C, Souchay D, Kneiß M, Bogner M, Wei HM, Lorenz M, et al. Transparent flexible thermoelectric material based on non-toxic earth-abundant p-type copper iodide thin film. Nature Communications. 2017;**8**:16076. DOI: 10.1038/ncomms16076

[141] Seo B, Hwang H, Kang S, Cha Y, Choi W. Flexible-detachable dual-output sensors of fluid temperature and dynamics based on structural design of thermoelectric materials. Nano Energy. 2018;**50**:733-743. DOI: 10.1016/j.nanoen.2018.06.027

[142] Umeda K, Uenuma M, Senaha D, Felizco JC, Uraoka Y, Adachi H. Amorphous thin film for thermoelectric application. Journal of Physics: Conference Series. 2018;**1052**:012016

[143] Paul B, Lu J, Eklund P. Nanostructural tailoring to induce flexibility in thermoelectric $Ca_3Co_4O_9$ thin films. ACS Applied Materials & Interfaces. 2017;**9**:25308-25316. DOI: 10.1021/acsami.7b06301

[144] Paul B, Khranovskyy V, Yakimova R, Eklund P. Donor-doped ZnO thin films on mica for fully-inorganic flexible thermoelectrics. Materials Research Letters. 2019;**7**:239-243. DOI: 10.1080/21663831.2019.1594427

[145] Zhong Y, Zhang L, Linseis V, Qin B, Chen W, Zhao LD, et al. High-quality textured SnSe thin films for self-powered, rapid-response photothermoelectric application. Nano Energy. 2020;**2020**(72):104742. DOI: 10.1016/j.nanoen.2020.104742

[146] Cheng H, Du Y, Wang B, Mao Z, Xu H, Zhang L, et al. Flexible

cellulose-based thermoelectric sponge towards wearable pressure sensor and energy harvesting. Chemical Engineering Journal. 2018;**338**:1-7. DOI: 10.1016/j.cej.2017.12.134

[147] Kato K, Hatasako Y, Uchino M, Nakata Y, Suzuki Y, Hayakawa T, et al. Flexible porous bismuth telluride thin films with enhanced figure of merit using micro-phase separation of block copolymer. Advanced Materials Interfaces. 2014;**1**:1300015. DOI: 10.1002/admi.201300015

[148] Rojas JP, Conchouso D, Arevalo A, Singh D, Foulds IG, Hussain MM. Based origami flexible and foldable thermoelectric nanogenerator. Nano Energy. 2017;**31**:296-301

[149] Karttunen AJ, Sarnes L, Townsend R, Mikkonen J, Karppinen M. Flexible thermoelectric ZnO–organic superlattices on cotton textile substrates by ALD/MLD. Advanced Electronic Materials. 2017;**3**:1600459. DOI: 10.1002/aelm.201600459

[150] Kou L, Yan B, Hu F, Wu SC, Wehling TO, Felser C, et al. Graphene-based topological insulator with an intrinsic bulk band gap above room temperature. Nano Letters. 2013;**13**:6251-6255. DOI: 10.1021/nl4037214

[151] Lee CW, Kim GH, Choi JW, An KS, Kim JS, Kim H, et al. Improvement of thermoelectric properties of Bi_2Te_3 and Sb_2Te_3 films grown on graphene substrate. Physica status solidi (RRL)–Rapid Research Letters. 2017;**11**:1700029. DOI: 10.1002/pssr.201700029

[152] Lee JA, Aliev AE, Bykova JS, Andrade MJ, Kim D, Sim HJ, et al. Woven-yarn thermoelectric textiles. Advanced Materials. 2016;**28**:5038-5044. DOI: 10.1002/adma.201600709

[153] Varghese T, Hollar C, Richardson J, Kempf N, Han C, Gamarachchi P, et al. High-performance and flexible thermoelectric films by screen printing solution-processed nanoplate crystals. Scientific Reports. 2016;**6**:33135. DOI: 10.1038/srep33135

[154] Han C, Tan G, Varghese T, Kanatzidis MG, Zhang Y. High-performance PbTe thermoelectric films by scalable and low-cost printing. ACS Energy Letters. 2018;**3**:818-822. DOI: 10.1021/acsenergylett.8b00041

[155] Danaei R, Varghese T, Ahmadzadeh M, McCloy J, Hollar C, Sadeq Saleh M, et al. Ultrafast fabrication of thermoelectric films by pulsed light sintering of colloidal nanoparticles on flexible and rigid substrates. Advanced Engineering Materials. 2019;**21**:1800800. DOI: 10.1002/adem.201800800

[156] Saeidi-Javash M, Kuang W, Dun C, Zhang Y. 3D conformal printing and photonic sintering of high-performance flexible thermoelectric films using 2D nanoplates. Advanced Functional Materials. 2019;**29**:1901930. DOI: 10.1002/adfm.201901930

[157] Hou W, Nie X, Zhao W, Zhou H, Mu X, Zhu W, et al. Fabrication and excellent performances of $Bi_{0.5}Sb_{1.5}Te_3$/epoxy flexible thermoelectric cooling devices. Nano Energy. 2018;**50**:766-776. DOI: 10.1016/j.nanoen.2018.06.020

[158] Snyder GJ, Lim JR, Huang CK, Fleurial JP. Thermoelectric microdevice fabricated by a MEMS-like electrochemical process. Nature Materials. 2023;**2**:528-531. DOI: 10.1038/nmat943

[159] Goncalves LM, Rocha JG, Couto C, Alpuim P, Min G, Rowe DM, et al. Fabrication of flexible thermoelectric microcoolers using planar thin-film

technologies. Journal of Micromechanics and Microengineering. 2007;**17**:S168. DOI: 10.1088/0960-1317/17/7/S14

[160] Madan D, Wang Z, Chen A, Wright PK, Evans JW. High-performance dispenser printed MA p-type $Bi_{0.5}Sb_{1.5}Te_3$ flexible thermoelectric generators for powering wireless sensor networks. ACS Applied Materials & Interfaces. 2013;**5**:11872-11876. DOI: 10.1021/am403568t

[161] Chen B, Kruse M, Xu B, Tutika R, Zheng W, Bartlett MD, et al. Flexible thermoelectric generators with inkjet-printed bismuth telluride nanowires and liquid metal contacts. Nanoscale. 2019;**11**:5222-5230. DOI: 10.1039/C8NR09101C

[162] Du J, Zhang B, Jiang M, Zhang Q, Zhang K, Liu Y, et al. Inkjet printing flexible thermoelectric devices using metal chalcogenide nanowires. Advanced Functional Materials. 2023;**33**:2213564. DOI: 10.1002/adfm.202213564

[163] Ferhat S, Domain C, Vidal J, Noël D, Ratier B, Lucas B. Flexible thermoelectric device based on $TiS_2(HA)_x$ n-type nanocomposite printed on paper. Organic Electronics. 2019;**68**:256-263. DOI: 10.1016/j.orgel.2019.02.031

[164] Zhao X, Han W, Jiang Y, Zhao C, Ji X, Kong F, et al. A honeycomb-like paper-based thermoelectric generator based on a Bi_2Te_3/bacterial cellulose nanofiber coating. Nanoscale. 2019;**11**:17725-17735. DOI: 10.1039/C9NR06197E

[165] Zhao X, Han W, Zhao C, Wang S, Kong F, Ji X, et al. Fabrication of transparent paper-based flexible thermoelectric generator for wearable energy harvester using modified distributor printing technology. ACS Applied Materials & Interfaces. 2019;**11**:10301-10309. DOI: 10.1021/acsami.8b21716

[166] Shin S, Kumar R, Roh JW, Ko DS, Kim HS, Kim S I, et al. High-performance screen-printed thermoelectric films on fabrics. Scientific Reports. 2017;7:7317. DOI: 10.1038/s41598-017-07654-2

[167] Hosokawa Y, Wada K, Tanaka M, Tomita K, Takashiri M. Thermal annealing effect on structural and thermoelectric properties of hexagonal Bi_2Te_3 nanoplate thin films by drop-casting technique. Japanese Journal of Applied Physics. 2017;**57**:02CC02. DOI: 10.7567/JJAP.57.02CC02

[168] Kimura Y, Mori R, Yonezawa S, Yabuki H, Namiki H, Ota Y, et al. Solvothermal synthesis of n-type $Bi_2(Se_xTe_{1-x})_3$ nanoplates for high-performance thermoelectric thin films on flexible substrates. Scientific Reports. 2020;**10**:6315. DOI: 10.1038/s41598-020-63374-0

[169] Bharti M, Singh A, Samanta S, Debnath AK, Marumoto K, Aswal DK, et al. Elucidating the mechanisms behind thermoelectric power factor enhancement of poly (3, 4-ethylenedioxythiophene): poly (styrenesulfonate) flexible films. Vacuum. 2018;**153**:238-247. DOI: 10.1016/j.vacuum.2018.04.017

[170] Olvera AA, Moroz NA, Sahoo P, Ren P, Bailey TP, Page AA, et al. Partial indium solubility induces chemical stability and colossal thermoelectric figure of merit in Cu_2Se. Energy & Environmental Science. 2017;**10**:1668-1676. DOI: 10.1039/C7EE01193H

[171] Lin Z, Hollar C, Kang JS, Yin A, Wang Y, Shiu HY, et al. A solution processable high-performance thermoelectric copper selenide thin film. Advanced Materials. 2017;**29**:1606662. DOI: 10.1002/adma.201606662

[172] Cho C, Wallace KL, Tzeng P, Hsu JH, Yu C, Grunlan JC. Outstanding low temperature thermoelectric power factor from completely organic thin films enabled by multidimensional conjugated nanomaterials. Advanced Energy Materials. 2016;**6**:1502168. DOI: 10.1002/aenm.201502168

[173] Tian R, Wan C, Wang Y, Wei Q, Ishida T, Yamamoto A, et al. A solution-processed TiS_2/organic hybrid superlattice film towards flexible thermoelectric devices. Journal of Materials Chemistry A. 2017;**5**:564-570. DOI: 10.1039/C6TA08838D

[174] Choi J, Cho K, Yun J, Park Y, Yang S, Kim S. Large voltage generation of flexible thermoelectric nanocrystal thin films by finger contact. Advanced Energy Materials. 2017;**7**:1700972. DOI: 10.1002/aenm.201700972

[175] Jang E, Poosapati A, Jang N, Hu L, Duffy M, Zupan M, et al. Thermoelectric properties enhancement of p-type composite films using wood-based binder and mechanical pressing. Scientific Reports. 2019;**9**:7869. DOI: 10.1038/s41598-019-44225-z

[176] Tzounis L, Gravalidis C, Vassiliadou S, Logothetidis S. Fiber yarns/CNT hierarchical structures as thermoelectric generators. Materials Today: Proceedings. 2017;**4**:7070-7075. DOI: 10.1016/j.matpr.2017.07.040

[177] Weathers A, Ullah Khan Z, Brooke R, Evans D, Pettes MT, Wenzel Andreasen J, et al. Significant electronic thermal transport in the conducting polymer poly (3, 4-ethylenedioxythiophene). Advanced Materials. 2015;**27**:2101-2106. DOI: 10.1002/adma.201404738

[178] Kim SJ, We JH, Cho BJA. Wearable thermoelectric generator fabricated on a glass fabric. Energy & Environmental Science. 2014;**6**:1959-1965

[179] Chen W, Shi XL, Li M, Liu T, Mao Y, Liu Q, et al. Nanobinders advance screen-printed flexible thermoelectrics. Science. 2024;**6727**:1265-1271. DOI: 10.1126/science.ads5868

[180] King JA, Tucker KW, Vogt BD, Weber EH, Quan C. Electrically and thermally conductive nylon 6, 6. Polymer Composites. 1999;**5**:643-654. DOI: 10.1002/pc.10387

[181] Huang S, Kroschwitz J. Mark Enclyclopedia of Polymer Science and Engineering. Vol. 220. New York, USA: John Wiley & Sons; 1985. p. 244. ISBN:0471887862

[182] Zhang HT, Li SB, White CJB, Ning X, Nie HL, Zhu LM. Studies on electrospun nylon-6/chitosan complex nanofiber interactions. Electrochimica Acta. 2009;**54**:5739-5745. DOI: 10.1016/j.electacta.2009.05.021

[183] Bazbouz MB, Stylios GK. The tensile properties of electrospun nylon 6 single nanofibers. Journal of Polymer Science Part B: Polymer Physics. 2010;**15**:1719-1731. DOI: 10.1002/polb.21993

[184] Ezquerra TA, Canalda JC, Sanz A, Linares A. On the electrical conductivity of PVDF composites with different carbon-based nanoadditives. Colloid and Polymer Science. 2014;**292**:1989-1998. DOI: 10.1007/s00396-014-3252-6

[185] Zhou W, Zuo J, Ren W. Thermal conductivity and dielectric properties of Al/PVDF composites. Composites Part A: Applied Science and Manufacturing. 2012;**4**:658-664. DOI: 10.1016/j.compositesa.2011.11.024

[186] Nasir M, Matsumoto H, Danno T, Minagawa M, Irisawa T, Shioya M, et al.

Control of diameter, morphology, and structure of PVDF nanofiber fabricated by electrospray deposition. Journal of Polymer Science Part B: Polymer Physics. 2006;**44**:779-786. DOI: 10.1002/polb.20737

[187] Linares A, Acosta JL. Tensile and dynamic mechanical behaviour of polymer blends based on PVDF. European Polymer Journal. 1997;**4**:467-473. DOI: 10.1016/S0014-3057(96)00182-6

[188] Nishi Y, Iizuka S, Faudree MC, Oyama R. Electrical conductivity enhancement of PTFE (Teflon) induced by homogeneous low voltage electron beam irradiation (HLEBI). Materials Transactions. 2012;**5**:940-945. DOI: 10.2320/matertrans.M2011273

[189] Ambroski L. H-film--a new high temperature dielectric. Industrial & Engineering Chemistry Product Research and Development. 1963;**2**:189-194. DOI: 10.1021/i360007a005

[190] Wang XQ, Chen DR, Han JC, Du SY. Crystallization behavior of polytetrafluoroethylene (PTFE). Journal of Applied Polymer Science. 2002;**83**:990-996. DOI: 10.1002/app.2279

[191] Arce J, Palacios A, Alvítez-Temoche D, Mendoza-Azpur G, Romero-Tapia P, Mayta-Tovalino F. Tensile strength of novel nonabsorbable PTFE (Teflon®) versus other suture materials: an in vitro study. International Journal of Dentistry. 2019;**1**:7419708. DOI: 10.1155/2019/7419708

[192] Sessler GM, Hahn B, Yoon DY. Electrical conduction in polyimide films. Journal of Applied Physics. 1986;**1**:318-326. DOI: 10.1063/1.337646

[193] Wang T, Wang M, Fu L, Duan Z, Chen Y, Hou X, et al. Enhanced thermal conductivity of polyimide composites with boron nitride nanosheets. Scientific Reports. 2018;**1**:1557. DOI: 10.1038/s41598-018-19945-3

[194] Srinivas S, Caputo FE, Graham M, Gardner S, Davis RM, McGrath JE, et al. Semicrystalline polyimides based on controlled molecular weight phthalimide end-capped 1,3-bis (4-aminophenoxy) benzene and 3, 3, 4, 4-biphenyltetracarboxylic dianhydride: synthesis, crystallization, melting, and thermal stability. Macromolecules. 1997;**30**:1012-1022. DOI: 10.1021/ma9604597

[195] Dong J, Yin C, Zhao X, Li Y, Zhang Q. High strength polyimide fibers with functionalized graphene. Polymer. 2013;**23**:6415-6424. DOI: 10.1016/j.polymer.2013.09.035

[196] Zhang HB, Zheng WG, Yan Q, Yang Y, Wang JW, Lu ZH, et al. Electrically conductive polyethylene terephthalate/graphene nanocomposites prepared by melt compounding. Polymer. 2010;**5**:1191-1196. DOI: 10.1016/j.polymer.2010.01.027

[197] Amborski LE, Flierl DW. Physical properties of polyethylene terephthalate films. Industrial & Engineering Chemistry. 1953;**45**:2290-2295. DOI: 10.1021/ie50526a042

[198] Shaikh FUA. Tensile and flexural behaviour of recycled polyethylene terephthalate (PET) fibre reinforced geopolymer composites. Construction and Building Materials. 2020;**245**:118438. DOI: 10.1016/j.conbuildmat.2020.118438

[199] Calvert P. A Review of: "HANDBOOK OF PLASTICS, ELASTOMERS AND COMPOSITES" Second edition edited by CA Harper McGraw Hill 11 West 19th Street New York, NY10011 960 pages,

hard cover, 1992. Material and Manufacturing Process. 1993;**6**:751-752. DOI: 10.1080/10426919308934880

[200] Cygan PJ, Zheng JP, Yen SP, Jow TR. Thermal treatment of polyethylene-2, 6-naphthalate (PEN) film and its influence on the morphology and dielectric strength. In: Proceedings of the Proceedings of IEEE Conference on Electrical Insulation and Dielectric Phenomena-(CEIDP'93). Arlington, USA: IEEE; 1993. pp. 630-635. DOI: 10.1109/CEIDP.1993.378902

[201] Hine PJ, Astruc A, Ward IM. Hot compaction of polyethylene naphthalate. Journal of Applied Polymer Science. 2004;**2**:796-802. DOI: 10.1002/app.20517

[202] Teoh H, Mac Innes D, Metz PD. Electrical conductivity of doped polyacrylonitrile (PAN). Le Journal de Physique Colloques. 1983;**44**:C3-C687. DOI: 10.1051/jphyscol:19833134

[203] Sabetzadeh N, Bahrambeygi H, Rabbi A, Nasouri K. Thermal conductivity of polyacrylonitrile nanofibre web in various nanofibre diameters and surface densities. IET Micro & Nano Letters. 2012;**7**:662-666. DOI: 10.1049/mnl.2012.0375

[204] Gupta A, Singhal R, Bajaj P, Agarwal V. Effect of copolymerization with methacrylonitrile on the dielectric relaxation of polyacrylonitrile. Journal of Applied Polymer Science. 1983;**28**:1167-1177. DOI: 10.1002/app.1983.070280322

[205] Wu T, Lu C, Sun T, Li Y, Yuan S, Li D, et al. New discovery on the relationship between microstructure and tensile strength of PAN-based carbon fibers. Microporous and Mesoporous Materials. 2022;**330**:111584. DOI: 10.1016/j.micromeso.2021.111584

[206] Saltas V, Pentari D, Vallianatos F. Complex electrical conductivity of biotite and muscovite micas at elevated temperatures: A comparative study. Materials. 2020;**16**:3513. DOI: 10.3390/ma13163513

[207] Zhang Z, Feng Y, Wang D, Liang L, Wang Z, Yang K, et al. Thermal conductive network construction and enhanced thermal conductivity in mica tape composites for large generator insulation. Composites Science and Technology. 2024;**254**:110671. DOI: 10.1016/j.compscitech.2024.110671

[208] Busigin C, Lahtinen R, Martinez GM, Thomas G, Woodhams RT. The properties of mica-filled polypropylenes. Polymer Engineering & Science. 1984;**3**:169-174. DOI: 10.1002/pen.760240303

[209] Abdel-karim AM, Salama AH, Hassan ML. Electrical conductivity and dielectric properties of nanofibrillated cellulose thin films from bagasse. Journal of Physical Organic Chemistry. 2018;**9**:e3851. DOI: 10.1002/poc.3851

[210] Brown R. In: Brandrup J, Immergut EH, editors. Polymer handbook. New York: Wiley Interscience; 1989. ISBN 0-471-81244-7

[211] Nordin SB, Nyren JO, Back EL. An indication of molten cellulose produced in a laser beam. Textile Research Journal. 1974;**44**:152-154. DOI: 10.1177/004051757404400211

[212] Wanigaratne DM, Batchelor WJ, Parker IH. Comparison of fracture toughness of paper with tensile properties. Appita Journal. 2002;**5**:369-374

[213] Júnior CP, De Carvalho LH, Fonseca VM, Monteiro SN, d'Almeida JRM. Analysis of the tensile

strength of polyester/hybrid ramie–cotton fabric composites. Polymer Testing. 2004;**2**:131-135. DOI: 10.1016/S0142-9418(03)00071-0

[214] Ma PC, Tang BZ, Kim JK. Effect of CNT decoration with silver nanoparticles on electrical conductivity of CNT-polymer composites. Carbon. 2008;**11**:1497-1505. DOI: 10.1016/j.carbon.2008.06.048

[215] Fujii M, Zhang X, Xie H, Ago H, Takahashi K, Ikuta T, et al. Measuring the thermal conductivity of a single carbon nanotube. Physical Review Letters. 2005;**6**:065502. DOI: 10.1103/PhysRevLett.95.065502

[216] Zhang K, Stocks GM, Zhong J. Melting and premelting of carbon nanotubes. Nanotechnology. 2007;**18**:285703. DOI: 10.1088/0957-4484/18/28/285704

[217] Bai Y, Zhang R, Ye X, Zhu Z, Xie H, Shen B, et al. Carbon nanotube bundles with tensile strength over 80 GPa. Nature Nanotechnology. 2018;**7**:589-595. DOI: 10.1038/s41565-018-0141-z

[218] Chen H, Müller MB, Gilmore KJ, Wallace GG, Li D. Mechanically strong, electrically conductive, and biocompatible graphene paper. Advanced Materials. 2008;**18**:3557-3561

[219] Ghosh D, Calizo I, Teweldebrhan D, Pokatilov EP, Nika DL, Balandin AA, et al. Extremely high thermal conductivity of graphene: Prospects for thermal management applications in nanoelectronic circuits. Applied Physics Letters. 2008;**92**:151911. DOI: 10.1063/1.2907977

[220] Ganz E, Ganz AB, Yang L, Dornfeld M. The initial stages of melting of graphene between 4000 K and 6000 K. Physical Chemistry Chemical Physics. 2017;**19**:3756-3762. DOI: 10.1039/C6CP06940A

[221] Mazrouaa AM, Mansour NA, Abed MY, Youssif MA, Shenashen MA, Awual MR. Nano-composite multi-wall carbon nanotubes using poly (p-phenylene terephthalamide) for enhanced electric conductivity. Journal of Environmental Chemical Engineering. 2019;**2**:103002. DOI: 10.1016/j.jece.2019.103002

[222] Ventura G, Barucci M, Gottardi E, Peroni I. Low temperature thermal conductivity of Kevlar. Cryogenics. 2000;**40**:489-491. DOI: 10.1016/S0011-2275(00)00062-X

[223] Tanner D, Fitzgerald JA, Phillips BR. The kevlar story-an advanced materials case study. Angewandte Chemie International Edition. 1989;**28**:649-654. DOI: 10.1002/anie.198906491

[224] Lu Z, Si L, Zhao Y, Huang J, Yao C, Jia F, et al. Enhanced mechanical and electrical insulating properties of (poly (para-phenylene terephthamide)) PPTA-based specialty paper with nanoscale PPTA fibers. Journal of Materials Science: Materials in Electronics. 2019;**30**:14414-14423. DOI: 10.1007/s10854-019-01811-1

[225] Chen C, Liu Z, Guo L, Huo B, Sun Q, Liang L, et al. High spatiotemporal resolution biomimetic thermoreceptors realizing by jointless p-n integration thermoelectric composites. Advanced Functional Materials. 2024;**49**:2411490. DOI: 10.1002/adfm.202411490

[226] Li G, Hu Y, Chen J, Liang L, Liu Z, Fu J, et al. Thermoelectric and photoelectric dual modulated sensors for human internet of things application in accurate fire recognition and warning. Advanced Functional Materials.

2023;**41**:2303861. DOI: 10.1002/adfm.202303861

[227] Du C, Cao M, Li G, Hu Y, Zhang Y, Liang L, et al. Toward precision recognition of complex hand motions: wearable thermoelectrics by synergistic 2D nanostructure confinement and controlled reduction. Advanced Functional Materials. 2022;**36**:2206083. DOI: 10.1002/adfm.202206083

[228] Dong G, Feng J, Qiu G, Yang Y, Chen Q, Xiong Y, et al. Oriented Bi_2Te_3-based films enabled high performance planar thermoelectric cooling device for hot spot elimination. Nature Communications. 2024;**1**:9695. DOI: 10.1038/s41467-024-54017-3

Chapter 7

High-Temperature Ceramic Thermoelectric Generators: Potential Challenges and Future Prospects in Automotive Applications

Uday M. Basheer Al-Naib

Abstract

High-temperature thermoelectric generators (TEGs) have emerged as a key technology for improving fuel efficiency by recovering waste heat from automotive exhaust systems. While half-Heusler alloys are commonly used due to their excellent thermoelectric properties and mechanical stability, their susceptibility to oxidation and high production costs pose challenges for long-term application. In contrast, ceramic-based thermoelectric materials, including cobalt oxides, perovskites, and doped zinc oxides, offer superior high-temperature stability, oxidation resistance, and environmental compatibility. This chapter explores the potential of ceramic thermoelectric materials as viable replacements for half-Heusler alloys in automotive TEGs, assessing their performance, stability, fabrication challenges, and integration strategies. Furthermore, it highlights recent research advancements, material enhancements, and future prospects for optimizing ceramic-based TEG efficiency in automotive applications.

Keywords: thermoelectric generators, ceramic thermoelectric materials, automotive waste heat recovery, cobalt oxides, perovskites, high-temperature stability

1. Introduction

The growing demand for fuel efficiency and the transition toward sustainable energy solutions have intensified interest in waste heat recovery technologies, particularly in the transportation sector. The automotive industry is among the largest energy consumers, with approximately 60–70% of fuel energy lost as waste heat through exhaust and cooling systems [1]. According to the U.S. Department of Energy, nearly 60% of the energy generated by internal combustion engines is dissipated as waste heat, underscoring the urgent need for effective recovery strategies [2]. Thermoelectric generators (TEGs) present a promising solution by directly converting waste heat into electricity *via* the Seebeck effect, thereby improving overall fuel efficiency and reducing carbon emissions [3].

Traditionally, thermoelectric generators (TEGs) for automotive exhaust heat recovery have relied on Bi-Te-based materials and half-Heusler (HH) alloys due to their relatively high thermoelectric performance and mechanical robustness. Studies indicate that HH alloys can achieve figure of merit (ZT) values ranging from 1.0 to 1.5 at elevated temperatures, making them viable candidates for energy conversion applications [4]. However, these materials face significant challenges in high-temperature environments, primarily due to oxidation and thermal degradation. Additionally, the reliance on rare and expensive elements, such as hafnium and titanium, further limits their scalability for large-scale automotive applications [5]. Overcoming these limitations necessitates the development of advanced thermoelectric materials that offer enhanced stability, cost-effectiveness, and improved efficiency for practical deployment in the automotive sector.

Thermoelectric materials are categorized based on their operating temperature ranges, material composition, and environmental impact. While conventional thermoelectric materials such as Pb-Te and Bi-Te have been widely used, concerns over their toxicity and scarcity have driven research toward alternative materials such as silicides and oxides, which offer higher thermal stability and lower environmental impact. A classification of thermoelectric materials, highlighting their respective temperature ranges and sustainability considerations, is presented in **Figure 1**.

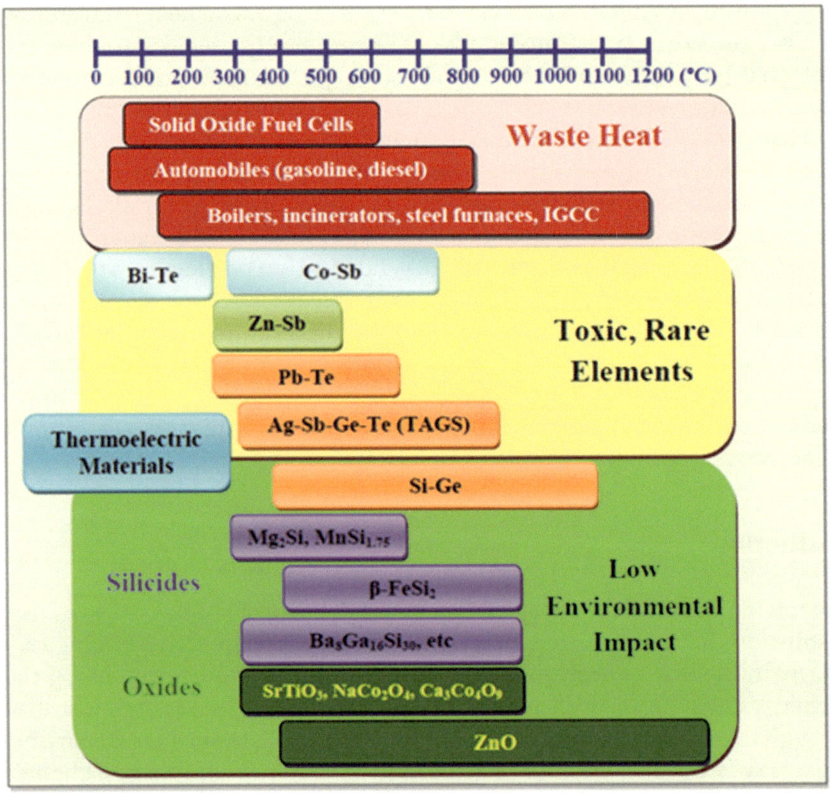

Figure 1.
Classification of thermoelectric materials based on operating temperature range, material composition, and environmental impact, highlighting their relevance for waste heat recovery applications [6].

DOI: http://dx.doi.org/10.5772/intechopen.1010921

Ceramic-based thermoelectric materials, such as perovskite oxides and layered cobaltites, present a promising alternative due to their superior thermal stability and oxidation resistance, making them well suited for high-temperature TEG applications [7]. Unlike traditional metal-based thermoelectric materials, ceramics exhibit inherently lower thermal conductivity, which helps maintain a high-temperature gradient across the thermoelectric module, thereby improving conversion efficiency [8]. Moreover, ceramics demonstrate excellent chemical stability in oxidative environments and do not require complex protective coatings, further enhancing their viability for long-term automotive applications.

Recent advances in nanostructuring and doping strategies have significantly enhanced the performance of ceramic thermoelectric materials, improving their electrical conductivity while maintaining high Seebeck coefficients [9]. For example, studies on strontium titanate ($SrTiO_3$)-based thermoelectric materials have demonstrated promising improvements in thermoelectric properties through carrier concentration optimization and grain boundary engineering [10]. Similarly, layered cobaltites, such as $Ca_3Co_4O_9$, have garnered considerable attention due to their inherently high Seebeck coefficients and excellent high-temperature stability, making them strong candidates for automotive and industrial waste heat recovery applications [11]. These advancements highlight the potential of ceramic thermoelectric materials as efficient and durable alternatives to conventional thermoelectric materials for high-temperature energy conversion [9].

Beyond material innovations, advancements in manufacturing techniques have accelerated the potential commercialization of ceramic TEGs. Spark plasma sintering (SPS), for instance, has been employed to fabricate dense, highly efficient thermoelectric ceramics with enhanced mechanical strength [12]. Additionally, the development of flexible ceramic thermoelectric films has opened new possibilities for lightweight and adaptable TEG modules [13]. However, integrating ceramic TEGs into automotive exhaust systems presents several challenges, including mechanical brittleness, thermal expansion mismatches, and high fabrication costs. To address these issues, hybrid thermoelectric modules that combine ceramic materials with metallic thermoelectrics have been proposed as a strategy to optimize both performance and durability [14]. Furthermore, advanced interface engineering techniques, such as functionally graded materials and composite architectures, are being explored to improve mechanical reliability and thermal compatibility [15].

In addition to half-Heusler alloys and oxide ceramics, other families of thermoelectric materials, such as skutterudites, clathrates, and silicides, have also garnered attention for high-temperature automotive applications due to their relatively high ZT values and tunable electronic properties. For instance, filled skutterudites exhibit low thermal conductivity and decent thermoelectric efficiency at intermediate to high temperatures, making them suitable for automotive waste heat recovery systems [5, 16]. These materials complement the ongoing efforts to identify thermoelectrics that combine thermal stability, mechanical robustness, and cost-effectiveness. Several comprehensive reviews have addressed the performance and challenges of skutterudites and related TE compounds in vehicular contexts [17].

This chapter explores the potential of ceramic thermoelectric materials in automotive applications, emphasizing recent advancements, existing challenges, and future research directions. Emerging fabrication techniques, including spark plasma sintering and nanostructuring strategies, are examined for their role in optimizing ceramic thermoelectric performance and facilitating integration within exhaust waste heat recovery systems. While thermoelectric generators (TEGs) provide a promising approach for converting waste heat into electricity, traditional metallic thermoelectric materials, such as half-Heusler (HH) alloys, have historically been the preferred choice for high-temperature automotive applications due to their high figure of merit (ZT) and mechanical robustness. However, their long-term stability is often compromised by oxidation, and their reliance on rare and expensive elements limits large-scale adoption. Given these constraints, ceramic thermoelectric materials present a compelling alternative due to their superior thermal stability, oxidation resistance, and potential for long-term durability in harsh automotive environments.

2. Comparison of Half-Heusler and ceramic-based thermoelectric materials

Oxide-based thermoelectric compounds, such as $Ca_3Co_4O_9$, $SrTiO_3$, and ZnO, have attracted significant attention due to their exceptional high-temperature stability and oxidation resistance, making them well suited for harsh environments like automotive exhaust systems [18, 19]. Unlike traditional metal-based thermoelectric materials, ceramic thermoelectrics exhibit inherent chemical stability and do not require protective coatings, ensuring long-term durability under extreme thermal conditions. However, despite these advantages, ceramic thermoelectric materials generally exhibit lower thermoelectric efficiency compared to half-Heusler (HH) compounds. This limitation primarily stems from their lower electrical conductivity and relatively higher thermal conductivity, which reduces their overall figure of merit (ZT). While advancements in nanostructuring, doping strategies, and grain .boundary engineering have led to performance improvements, ceramics remain less

Property	Half-Heusler alloys	Ceramic thermoelectric materials
Operating temperature	Typically operate efficiently between 673–1073 K [20].	Function effectively at higher temperatures, often exceeding 800 K, and can withstand operation in air [21].
Figure of merit (ZT)	Achieve ZT values around 1; optimized compositions can reach up to 1.5 [20].	Typically, exhibit lower ZT values, ranging from 0.2 to 0.8; rare cases reach ~1.2 [22].
Thermal stability	High thermal stability, retaining performance up to their temperature (~1073 K) [20, 23].	Excellent thermal stability, remaining stable up to and beyond 1000 K [24].
Oxidation resistance	Moderate resistance; require protective coatings to prevent oxidation in air [25].	Naturally resistant to oxidation, allowing direct operation in air without additional protection [11].
Mechanical conductivity	Generally possess good mechanical strength and lower brittleness compared to oxides [26].	Tend to be brittle and prone to mechanical failure under stress [27].
Electrical conductivity	High electrical conductivity due to metallic nature, enhancing thermoelectric performance [28].	Lower electrical conductivity, which can limit thermoelectric efficiency [28].
Material abundance and cost	Moderate cost: Hf-based compositions are expensive, while Zr/Ti-based alternatives are more cost-effective [29].	Made from abundant and cost-effective elements, offering a more economical solution [5].

Table 1.
Comparison between half-Heusler and ceramic thermoelectric materials.

efficient in direct energy conversion applications [18]. Nevertheless, their abundance, cost-effectiveness, and ability to operate at ultra-high temperatures (above 1000 K) make them promising candidates for applications where oxidation resistance and long-term durability are critical. A comparative analysis of half-Heusler and ceramic thermoelectric materials is presented in **Table 1**.

While ceramic-based thermoelectric materials, such as $Ca_3Co_4O_9$ and $SrTiO_3$, offer advantages like thermal stability and abundance, their performance is often constrained by low electrical conductivity and thermoelectric efficiency (ZT values typically below 1). Although isolated studies have reported ZT values exceeding 1 for certain nanostructured oxides, these results have not been widely reproduced and thus should be interpreted cautiously. Similarly, while oxides exhibit good thermal and chemical stability, their oxygen content can vary under high-temperature oxidative conditions, leading to changes in carrier concentration and degradation of thermoelectric properties [8]. Half-Heusler (HH) alloys, on the other hand, can tolerate oxidative atmospheres at moderate temperatures (~300°C) without protective coatings, although oxidation becomes significant at higher temperatures (>500–600°C) [5]. These nuances are critical when evaluating the real-world applicability of both classes of materials in automotive thermoelectric generators [17].

3. Ceramic thermoelectric materials: Properties and performance

Thermoelectric materials convert waste heat into electricity through the Seebeck effect, where a temperature gradient (ΔT) generates a voltage (ΔV) according to the relation:

$$\Delta V = S.\Delta T \tag{1}$$

where S is the Seebeck coefficient (V/K). Ceramics have emerged as promising candidates for high-temperature applications due to their stability and environmental benefits. Compared to traditional half-Heusler alloys, ceramic thermoelectrics, such as cobalt oxides, perovskites, and zinc oxides, offer enhanced oxidation resistance and mechanical strength, albeit at the expense of lower electrical conductivity.

The performance of thermoelectric materials is often quantified using the dimensionless figure of merit (*ZT*), defined as

$$ZT = \frac{S^2 \sigma T}{\kappa} \tag{2}$$

where S is the Seebeck coefficient (V/K), σ is electrical conductivity (S/m), T is absolute temperature (K), and κ is thermal conductivity (W/m.^{0}K).

Recent advancements in ceramic thermoelectric materials have concentrated on enhancing the dimensionless figure of merit (ZT) through strategies such as carrier concentration optimization, grain boundary engineering, and nanostructuring to reduce thermal conductivity. Experimental investigations have demonstrated that $Ca_3Co_4O_9$-based cobaltites typically achieve ZT values up to approximately 0.14 at 973 K. For instance, a study reported a ZT of 0.14 at 973 K for pure $Ca_3Co_4O_9$ samples. However, certain nanostructured or doped $Ca_3Co_4O_9$ samples have demonstrated higher ZT values, with some reports indicating values up to 2.7 at 900 [30]. In contrast, heavily Nb-doped $SrTiO_3$ perovskites have exhibited ZT values of approximately 0.37 at 1000 K, as evidenced by a 20% Nb-doped $SrTiO_3$ epitaxial film [31].

Another key parameter used to evaluate thermoelectric efficiency is the power factor (PF), which is given by

$$PF = S^2\sigma \tag{3}$$

where S is the Seebeck coefficient (V/K) and σ is the electrical conductivity (S/m).

Figure 2 illustrates the basic working principle of a high-temperature ceramic thermoelectric generator. The performance of these devices heavily depends on the thermoelectric properties of ceramic materials, including the Seebeck coefficient, electrical conductivity, and thermal conductivity. The use of high-temperature ceramic materials, such as perovskites and oxides, enables stable performance under extreme operating conditions. The effectiveness of heat dissipation, shown by the heat sink, is also crucial for maintaining a high-temperature gradient across the thermoelectric legs, thereby enhancing power output and efficiency.

Aluminum-doped ZnO materials have demonstrated a power factor of approximately 15×10^{-4} $W{\cdot}m^{-1}{\cdot}K^{-2}$ at 1273 K, with a dimensionless figure of merit (ZT) of 0.30 [33], making them promising candidates for hybrid thermoelectric devices. However, further research is required to enhance the electrical transport properties of ceramics while simultaneously reducing thermal conductivity. This can be achieved through defect engineering, multilayer structures, and composite material strategies. **Table 2** summarizes the key properties of commonly studied ceramic thermoelectric materials, including their Seebeck coefficient, electrical conductivity, thermal conductivity, and ZT values, which are critical for optimizing their performance in energy conversion applications. The thermoelectric performance of these materials is heavily influenced by their crystal structure, which governs charge carrier mobility and phonon-scattering behavior. For instance, layered cobalt oxides, such as $NaCo_2O_4$

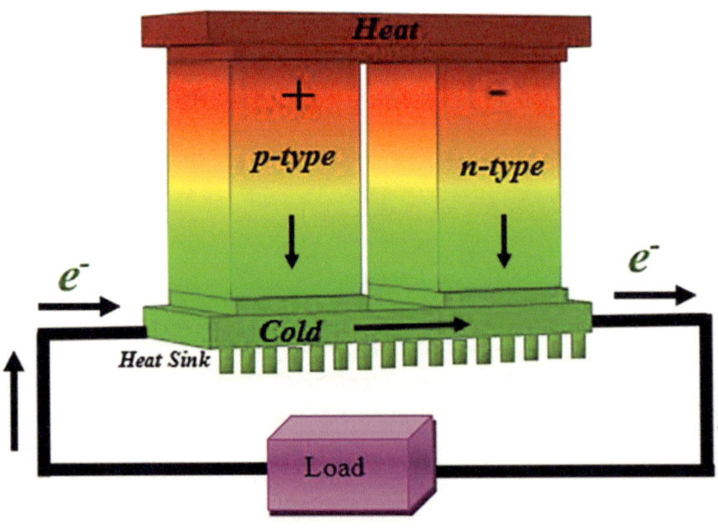

Figure 2.
Schematic of a high-temperature ceramic thermoelectric generator (TEG). The device consists of p-type and n-type ceramic-based semiconductor legs, which are connected electrically in series and thermally in parallel [32].

DOI: http://dx.doi.org/10.5772/intechopen.1010921

Material	Crystal structure	Seebeck coefficient (μV/K)	Electrical conductivity (S/cm)	Thermal conductivity (W/m·K)	Power factor (μW/cm·K^2)	Figure of merit (ZT)	Operating temperature (°C)
$NaCo_2O_4$ [31]	Layered	~100	~220	~1.6	~2.2	~0.045	Up to 723
$Ca_3Co_4O_9$ [32]	Layered	~150	~100	~2.5	~2.25	~0.3	Up to 900
$SrTiO_3$ [33, 34]	Perovskite	~100	~2000	~3.5	~20	~0.37	Up to 1000
ZnO [35]	Wurtzite	~200	~500	~10	~20	~0.3	Up to 1000
$Bi_2Sr_2Co_2O_9$ [36, 37]	Layered	~150	~50	~1.5	~1.125	~0.2	Up to 900

Table 2.
Summary of common ceramic thermoelectric materials and their properties.

and $Ca_3Co_4O_9$, exhibit moderate electrical conductivity coupled with high Seebeck coefficients, making them strong candidates for high-temperature thermoelectric applications. Meanwhile, $SrTiO_3$, a perovskite oxide, benefits from doping-induced conductivity enhancements, while ZnO and $Bi_2Sr_2Co_2O_9$ have been explored for their high thermal stability and reasonable ZT values.

3.1 Cobaltite-based thermoelectric materials

Cobalt oxides, particularly $Ca_3Co_4O_9$, are widely recognized as promising p-type thermoelectric ceramics due to their layered structure and high Seebeck coefficient. These materials exhibit exceptional thermal and chemical stability at temperatures exceeding 800°C, making them well suited for harsh environments such as automotive applications. Studies have reported that $Ca_3Co_4O_9$-based ceramics can achieve a power factor of approximately 514 μW/m·K^2 at 920 K [38]. One of the key advantages of cobaltites is their ability to maintain high thermoelectric performance under extreme temperature gradients, which is crucial for applications like automotive exhaust systems. However, their relatively low electrical conductivity compared to metallic thermoelectrics necessitates the use of doping and nanostructuring strategies to enhance performance.

To address this limitation, researchers have explored various dopant elements, such as sodium (Na) and tungsten (W), to improve charge carrier concentration and optimize electronic transport properties. For instance, dual doping with Na and W in $Ca_3Co_4O_9$ has been shown to increase the power factor to 271 μW/m·K^2 at 1000 K (727°C) [39]. The influence of different dopants on cobaltite-based thermoelectric materials is summarized in **Table 3**, highlighting how element substitution affects the Seebeck coefficient, electrical conductivity, thermal conductivity, and figure of merit (ZT).

Advancements in fabrication techniques, such as spark plasma sintering (SPS) and sol-gel processing, have significantly improved the densification and grain boundary engineering of cobaltite materials, thereby enhancing their thermoelectric

Base material	Dopant	Seebeck coefficient (μV/K)	Electrical conductivity (S/cm)	Thermal conductivity (W/m·K)	Figure of merit (ZT)	References
$Ca_3Co_4O_9$	Na	Increased	Decreased	Decreased	Improved	[40]
$Ca_3Co_4O_9$	Bi	Slightly decreased	Increased	Decreased	Improved	[41]
$Ca_3Co_4O_9$	Ag	Increased	Increased	Slightly decreased	Improved	[40, 42, 43]
Na_xCoO_2	High Na	Increased	Decreased	-	Improved	[44]
$Bi_2Sr_2Co_2O_9$	Ca	Slightly increased	Slightly decreased	Decreased	Improved	[45]
$Bi_2Sr_2Co_2O_9$	Ba	Unchanged	Decreased	-	Improved	[46]
$Ca_3Co_4O_9$	Fe	Increased	Increased	Decreased	Improved	[47]
$Ca_3Co_4O_9$	W.Na	Increased	Increased	Decreased	Improved	[39]

Table 3.
Effect of doping on cobaltite-based thermoelectric materials.

performance [48]. These methods enable precise control over microstructural features, which are essential for optimizing the thermoelectric properties of $Ca_3Co_4O_9$-based ceramics. The integration of these materials into thermoelectric modules has demonstrated promising results in laboratory-scale prototypes. However, further research is needed to address long-term durability and develop cost-effective manufacturing techniques for large-scale automotive applications. $Ca_3Co_4O_9$ ceramics exhibit exceptional thermal and chemical stability at temperatures exceeding 800°C, making them well suited for harsh automotive environments. However, their relatively low electrical conductivity compared to metallic thermoelectrics necessitates further optimization through doping and nanostructuring approaches. Overall, $Ca_3Co_4O_9$-based ceramics offer excellent thermal stability and promising thermoelectric properties for high-temperature applications. Nevertheless, continued advancements in material optimization and processing techniques are crucial to fully unlocking their potential for practical deployment in thermoelectric energy conversion systems.

3.2 Perovskite-based thermoelectric materials

Perovskite-based oxides, particularly $SrTiO_3$ doped with Nb or La, have emerged as promising n-type thermoelectric materials due to their tunable electronic structure, high-temperature stability, and mechanical resilience. Undoped $SrTiO_3$ exhibits a high Seebeck coefficient (~300–400 μV/K at elevated temperatures), attributed to the high effective mass of Ti 3d conduction electrons. However, Nb or La doping significantly enhances its electrical conductivity, albeit at the expense of a reduced Seebeck coefficient. One of the primary limitations of $SrTiO_3$ is its inherently high lattice thermal conductivity (~11–12 W/m·K at room temperature, decreasing to ~5 W/m·K at 1000 K), which restricts its overall thermoelectric efficiency [49, 50]. To mitigate this, various nanoengineering strategies, including grain boundary modifications, defect engineering, and element substitution, have been explored to reduce thermal conductivity while maintaining or enhancing electrical properties [51, 52].

Recent advancements in nanostructuring and heterostructure engineering have demonstrated that incorporating oxygen vacancies, grain refinement, and nanoinclusions can effectively enhance carrier mobility and lower lattice thermal conductivity, thereby improving overall thermoelectric performance. For instance, co-doping $SrTiO_3$ with Nb and La, combined with controlled nanostructuring, has achieved ZT values of ~0.6 at high temperatures (>800°C to 1000 K). This improvement is attributed to optimized charge carrier concentration, enhanced phonon scattering at grain boundaries, and defect-induced reductions in thermal conductivity [52, 53]. The effects of these nanoengineering strategies on $SrTiO_3$ thermoelectric performance are summarized in **Table 4**, highlighting their influence on the Seebeck coefficient, electrical conductivity, and thermal conductivity. To enable a more meaningful comparison with other thermoelectric materials presented earlier (**Table 2**), **Table 4** has been updated to include typical quantitative values derived from experimental studies. These values illustrate the measurable impact of each strategy on key thermoelectric properties and demonstrate the extent to which performance improvements, particularly in the figure of merit (ZT), can be achieved through nanostructuring, doping, and composite engineering of $SrTiO_3$-based ceramics.

Advanced composite materials, such as graphite-included Nb-doped $SrTiO_3$, have demonstrated even higher ZT values (~1.4 at ~1050 K), underscoring the potential of multi-scale engineering approaches for enhancing thermoelectric performance [53].

Strategy	Effect on Seebeck coefficient	Effect on electrical conductivity	Effect on thermal conductivity	Overall impact on ZT	References
Nano structuring	Slightly decreased	Increased	Reduced	Improved	[31]
Grain boundary engineering	Increased	Slightly increased	Reduced	Improved	[54]
Oxygen vacancy incorporation	Increased	Increased	Slightly reduced	Improved	[53]
Ruddlesden-Popper (RP) phase inclusion	Unchanged	Unchanged	Significantly reduced	Strongly improved	[55]
$SrTiO_3$-graphite composites	Slightly decreased	Strongly increased	Strongly reduced	Greatly improved	[56]

Table 4.
Effect of nanoengineering strategies on $SrTiO_3$ thermoelectric properties.

Additionally, layered perovskite structures, particularly Ruddlesden-Popper (RP) phases (e.g., Sr_2TiO_4), have emerged as promising candidates for further reducing thermal conductivity. These structures act as natural superlattices, where alternating perovskite and rock-salt layers serve as phonon-scattering interfaces, effectively minimizing lattice thermal conductivity while preserving electrical transport properties. Studies have shown that integrating RP phases within $SrTiO_3$ matrices can significantly suppress thermal conductivity, leading to enhanced ZT values [55, 57]. The incorporation of RP phases and heterostructure designs represents a viable strategy for developing high-performance thermoelectric materials suitable for energy harvesting in automotive and industrial applications [58].

3.3 Zinc oxide-based thermoelectric materials

Zinc oxide (ZnO)-based ceramics, when doped with elements such as aluminum (Al), indium (In), or gallium (Ga), exhibit notable n-type thermoelectric properties. These materials are highly valued for their high mechanical strength, thermal stability, and cost-effectiveness, making them suitable for large-scale applications. The wide bandgap of ZnO (~3.3 eV) ensures stable performance at elevated temperatures with minimal degradation. Research has primarily focused on doping strategies to enhance the electrical conductivity and Seebeck coefficient of ZnO-based materials. For instance, Al-doped ZnO (AZO) has demonstrated power factors up to 0.44 at 1000 K [59]. Similarly, Ga- and In-doped ZnO have exhibited improved charge carrier mobility and reduced lattice thermal conductivity due to enhanced phonon scattering [60].

Mitigating the inherently high thermal conductivity of ZnO (~40 W/m·K at room temperature), various nanostructuring techniques have been employed, including grain boundary engineering, defect modulation, and the incorporation of secondary phases (e.g., ZnS or Bi_2O_3). These strategies effectively suppress lattice vibrations, thereby enhancing overall thermoelectric performance [60]. Moreover, ZnO-based thermoelectric generator (TEG) modules have demonstrated strong potential for practical exhaust heat recovery applications. Studies indicate that integrating ZnO-based thermoelectric legs into automotive exhaust systems can achieve power outputs of 500–750 W under temperature gradients of several hundred degrees Celsius [60]. However, challenges such as long-term stability, interface engineering, and fabrication scalability must be addressed to enable the commercial viability of ZnO-based TEGs for automotive applications.

ZnO-based thermoelectric materials exhibit significant potential for high-temperature applications. To further enhance their efficiency, ongoing research is needed to optimize carrier transport properties and thermal management strategies. Doping ZnO ceramics with Al, In, or Ga has been shown to improve n-type thermoelectric behavior, while their intrinsic mechanical strength and cost-effectiveness make them promising for large-scale deployment. However, like perovskites, ZnO's high thermal conductivity remains a challenge that must be addressed through nanostructuring and composite approaches. For example, Al-doped ZnO nanocomposites have demonstrated significant reductions in thermal conductivity, attributed to enhanced phonon scattering, Al-induced grain refinement, and the formation of $ZnAl_2O_4$ nanoprecipitates [61].

4. Challenges in implementing ceramic TEGs for automotive applications

The implementation of ceramic thermoelectric generators (TEGs) in automotive applications presents several critical challenges, despite their inherent advantages, such as high-temperature stability and oxidation resistance. Overcoming these challenges is essential for the large-scale adoption of ceramic TEGs in efficient waste heat recovery systems for vehicles. One of the primary limitations of ceramic thermoelectric materials is their relatively low electrical conductivity compared to metallic thermoelectric materials like half-Heuslers. This deficiency significantly reduces their power factor, thereby limiting overall efficiency. To address this issue, various strategies have been explored. Doping with transition metals or rare-earth elements has been extensively studied to enhance carrier concentration and improve electrical transport properties. Research has shown that doping perovskite oxides with elements such as Nb, Ta, and La significantly increases charge carrier density and electrical conductivity [62–64].

Another effective approach involves defect engineering and grain boundary modifications to optimize charge carrier mobility. For instance, oxygen vacancy engineering in $SrTiO_3$ and $Ca_3Co_4O_9$ ceramics has been shown to enhance charge carrier concentration while minimizing scattering effects, leading to improved thermoelectric performance [65]. Additionally, nanostructuring and heterostructure engineering have been explored to create electron-conducting pathways while maintaining low thermal conductivity. Studies on nano-grained ZnO and composite thermoelectric materials have reported increased electrical conductivity and enhanced thermoelectric efficiency [66].

Furthermore, research on multi-phase ceramic composites incorporating metallic nanoparticles has shown promising results in enhancing charge transport while preserving oxidation resistance [67]. Although these methods have led to moderate improvements, further advancements are needed to develop cost-effective and scalable solutions that can significantly enhance the electrical conductivity of ceramic thermoelectric materials. Addressing these challenges will be instrumental in making ceramic TEGs viable for real-world automotive applications. The selection of ceramic thermoelectric materials depends on their stability, efficiency, and application-specific requirements. **Table 5** summarizes the performance of different ceramic thermoelectric materials in key applications, including waste heat recovery and automotive exhaust heat recovery.

Application	Preferred material	Key performance criteria	Temperature range (°C)	References
Waste heat recovery in industry	$NaCo_2O_4$, $Ca_3Co_4O_9$	High ZT, stability	600–1000	[13]
Thermoelectric generators (TEGs)	$SrTiO_3$, ZnO	High power factor	500–800	[12]
Space power systems	$Bi_2Sr_2Co_2O_9$	Thermal stability	600–900	[68]
Automotive exhaust heat recovery	$NaCo_2O_4$, ZnO	High Seebeck durability	500–900	[69]
Sensors and small-scale devices	$SrTiO_3$, ZnO	Low thermal conductivity	400–800	[69]

Table 5.
Application-specific performance of ceramic thermoelectric materials.

4.1 Thermal conductivity optimization in ceramic thermoelectric

Ceramic thermoelectric materials exhibit exceptional thermal stability, making them well suited for high-temperature applications. However, their relatively high lattice thermal conductivity reduces efficiency by facilitating excessive heat dissipation rather than maintaining a strong temperature gradient. To address this challenge, researchers have focused on reducing thermal conductivity while preserving electrical performance. One effective strategy involves incorporating secondary phases or nanoinclusions, such as Bi_2O_3 and rare-earth oxides, which enhance phonon scattering and lower lattice thermal conductivity without significantly impairing electrical transport [18, 70]. Similarly, defect and vacancy engineering, particularly oxygen vacancy introduction in perovskite-based materials, disrupts phonon propagation while improving charge carrier mobility, offering a balance between thermal and electrical optimization [71, 72]. For example, high-entropy doping in PbSe-based thermoelectrics has been reported to create lattice distortions, intensifying phonon scattering and reducing thermal conductivity while maintaining high charge carrier mobility.

Another approach to thermal conductivity reduction involves layered and composite material strategies, where ceramics are embedded within polymer or metal matrices to tailor thermal properties while preserving structural integrity. Hybrid ceramic-polymer composites, for example, optimize phonon scattering at the ceramic-polymer interface, leading to enhanced thermoelectric efficiency [12, 73]. Additionally, nanostructuring techniques, such as grain boundary engineering and superlattice structures, have demonstrated significant thermal conductivity reductions by increasing phonon-scattering sites while preserving electron transport. Studies on $SrTiO_3$ thin films confirm that reducing grain size and increasing interface density can drastically lower thermal conductivity, thereby improving overall thermoelectric performance [74, 75]. Superlattice structures, such as alternating layers of Nb-doped and undoped $SrTiO_3$, have been shown to enhance electron mobility while significantly reducing thermal conductivity due to strong phonon scattering at the interfaces.

Recent advancements in nanostructuring and hierarchical architecture have further refined thermal conductivity control in ceramic thermoelectrics. Techniques, such as spark plasma sintering and high-energy ball milling, have been successfully used to fabricate nanostructured ceramics with optimized phonon-electron interactions, enabling lower thermal conductivity without compromising electrical properties [72]. By creating fine-grained microstructures and introducing nanoscale precipitates, these processing techniques enhance phonon scattering, thereby improving thermoelectric performance.

As computational modeling and machine learning continue to advance, they are expected to accelerate the discovery of novel thermoelectric materials with finely tuned thermal transport characteristics. Future research will likely focus on integrating these strategies with emerging manufacturing technologies to further enhance efficiency and expand the applicability of ceramic thermoelectric generators for sustainable energy solutions.

4.2 Manufacturing challenges and scalability

Manufacturing high-performance ceramic thermoelectric generators (TEGs) at a commercial scale remains a significant challenge due to the complexities of material processing and system integration. One of the primary obstacles is the high-temperature sintering requirements, which leads to increased energy consumption and higher production costs. To mitigate this, advanced sintering techniques, such as spark plasma sintering (SPS) and microwave sintering, have been explored to enhance grain boundary control and reduce processing temperatures, thereby improving both efficiency and material performance [48]. Additionally, ensuring mechanical and thermal stability under cyclic thermal loads is crucial for automotive TEGs, where rapid temperature fluctuations are common. Research efforts are focused on developing thermal shock-resistant ceramic composites and flexible thermoelectric modules to enhance reliability and durability in harsh environments [76, 77].

Another major challenge in the large-scale adoption of ceramic TEGs is the cost-effectiveness of raw materials and fabrication methods. The economic feasibility of these systems depends on utilizing earth-abundant, low-cost materials while maintaining high thermoelectric performance. Recent studies have investigated alternative compositions using widely available elements to reduce costs without compromising efficiency [76]. Additionally, scalability is critical for ensuring uniform nanostructures over large areas, which is essential for reproducible and high-efficiency TEG modules. Researchers are developing scalable fabrication methods that allow for precise control over material nanostructures, ensuring performance consistency in commercial applications [78].

Integrating ceramic TEGs with automotive exhaust systems presents additional engineering challenges, particularly in maintaining robust thermal and mechanical contact between the thermoelectric elements and the heat source. Innovations in heat exchanger designs are being explored to optimize heat transfer efficiency from exhaust gases to TEGs, thereby improving overall system performance [79]. Furthermore, ensuring long-term reliability and resistance to material degradation is essential, as high-temperature applications demand stable electrical and thermal performance over extended operational periods. Recent advancements in coating technologies and encapsulation methods aim to extend the lifespan and durability of ceramic thermoelectric materials, minimizing performance loss due to oxidation and thermal fatigue [12]. Addressing these challenges through material innovations, advanced processing techniques, and system-level integration is critical for enabling the widespread adoption of ceramic TEGs in next-generation automotive waste heat recovery systems.

5. Future prospects and research directions

The future of ceramic thermoelectric generators (TEGs) in automotive applications appears highly promising, driven by advancements in material engineering, manufacturing techniques, and system-level integration. One of the key research areas is enhancing electrical conductivity, as charge carrier transport remains a significant challenge in ceramic thermoelectrics. Advanced doping strategies, including the incorporation of rare-earth elements (e.g., La, Ce, and Yb) and transition metals (e.g., Ni and Co), have demonstrated potential in optimizing carrier concentration and mobility, thereby improving power factors [80]. Additionally, defect engineering approaches, such as manipulating oxygen vacancies and grain boundary modifications, can further enhance electrical conductivity. These improvements could significantly increase the efficiency of ceramic-based TEGs, making them more viable for automotive waste heat recovery [76].

Another critical research direction is reducing thermal conductivity to enhance the thermoelectric figure of merit (ZT). Nanostructuring techniques, such as hierarchical architectures, superlattice engineering, and grain boundary engineering, have been shown to suppress lattice thermal conductivity while maintaining high electrical transport properties [81]. In particular, the introduction of multi-scale phonon-scattering mechanisms through engineered nanostructures has led to significant performance improvements in oxide-based thermoelectrics [18]. Furthermore, hybrid and composite approaches that integrate ceramic thermoelectrics with metallic or polymeric materials offer synergistic benefits, optimizing both electrical and thermal transport properties [82]. These strategies collectively contribute to the development of more efficient and durable thermoelectric modules for vehicle exhaust energy harvesting.

Scalability and long-term durability remain key challenges in the commercialization of ceramic TEGs. Cost-effective manufacturing techniques, such as spark plasma sintering (SPS), sol-gel synthesis, and chemical vapor deposition (CVD), are being explored to enable large-scale production while maintaining high material performance [83]. Additionally, integrating ceramic TEGs into automotive exhaust systems requires advancements in module design, heat exchanger integration, and thermal cycling stability [17]. The development of protective coatings, oxidation-resistant interfaces, and self-healing materials will be essential for ensuring long-term operational stability, particularly in high-temperature environments [79]. Future research should focus on dynamic power management and hybrid energy harvesting solutions to maximize the practical applicability of ceramic TEGs in next-generation energy-efficient vehicles.

6. Conclusions

Ceramic thermoelectric materials present a promising alternative to half-Heusler alloys for high-temperature automotive applications, offering excellent oxidation resistance, mechanical durability, and cost-effectiveness. Materials such as cobalt oxides, perovskites, and doped zinc oxides exhibit superior thermal stability and environmental compatibility, making them strong candidates for thermoelectric waste heat recovery in vehicles. However, challenges, such as low electrical conductivity, high thermal conductivity, and complex fabrication processes, must be addressed to achieve competitive performance. Advanced strategies, including doping, nanostructuring, and composite engineering, have shown potential for optimizing thermoelectric properties, but further research is needed to enhance efficiency to levels comparable to half-Heusler materials. Additionally, scalable and cost-effective manufacturing techniques, such as spark plasma sintering and solution-based synthesis, will be essential for large-scale production. Successful integration into automotive exhaust systems also requires improvements in thermal cycling durability, mechanical robustness, and system-level design optimization. Future advancements in hybrid material design and novel fabrication techniques will be instrumental in overcoming current limitations. With sustained interdisciplinary research and technological innovations, ceramic-based thermoelectric generators have the potential to significantly enhance vehicle fuel efficiency, reduce greenhouse gas emissions, and support sustainable energy utilization in the automotive industry.

Author details

Uday M. Basheer Al-Naib
Centre for Advanced Composite Materials (CACM), Faculty of Mechanical Engineering, Universiti Teknologi Malaysia, Malaysia

*Address all correspondence to: ummb2008@gmail.com

References

[1] Snyder GJ. Small thermoelectric generators. The Electrochemical Society Interface. 2008;**17**(3):54

[2] Dwivedi D et al. From legacy contamination to watershed systems science: A review of scientific insights and technologies developed through DOE-supported research in water and energy security. Environmental Research Letters. 2022;**17**(4):043004

[3] Hooshmand Zaferani S et al. Thermal management systems and waste heat recycling by thermoelectric generators—An overview. Energies. 2021;**14**(18):5646

[4] Zhao L-D et al. Ultralow thermal conductivity and high thermoelectric figure of merit in SnSe crystals. Nature. 2014;**508**(7496):373-377

[5] Snyder GJ, Toberer ES. Complex thermoelectric materials. Nature Materials. 2008;7(2):105-114

[6] Xing Y et al. Principles and methods for improving the thermoelectric performance of SiC: A potential high-temperature thermoelectric material. Materials. 2024;**17**(15):3636

[7] He W et al. Recent development and application of thermoelectric generator and cooler. Applied Energy. 2015;**143**:1-25

[8] Ohtaki M. Recent aspects of oxide thermoelectric materials for power generation from mid-to-high temperature heat source. Journal of the Ceramic Society of Japan. 2011;**119**(1395):770-775

[9] Yamamoto H et al. High thermal durability and thermoelectric performance with ultra-low thermal conductivity in n-type single-walled carbon nanotube films by controlling dopant concentration with cationic surfactant. Applied Physics Letters. 2025;**126**(6):063902

[10] Zhang Y et al. The electrical, thermal, and thermoelectric properties of black phosphorus. APL Materials. 2020;**8**(12):120903

[11] He J, Liu Y, Funahashi R. Oxide thermoelectrics: The challenges, progress, and outlook. Journal of Materials Research. 2011;**26**(15):1762-1772

[12] Zhang P et al. Development and applications of thermoelectric oxide ceramics and devices. Energies. 2023;**16**(11):4475

[13] Liu W-D, Chen Z-G. High-performance thermoelectric ceramics and their applications. In: Advanced Ceramics for Energy Storage, Thermoelectrics and Photonics Journal. Netherlands: Elsevier; 2023. pp. 347-362

[14] Wang W, Thesiya D, Mukhopadhyay SM. Hierarchical hybrid heat sink material for thermo-electric generators. Applied Thermal Engineering. 2024;**236**:121674

[15] Kumar A et al. Interface engineering-driven room-temperature ultralow gas sensors with elucidating sensing performance of heterostructure transition metal dichalcogenide thin films. ACS Sensors. 2023;**8**(10):3824-3835

[16] Shi X, Chen L, Uher C. Recent advances in high-performance bulk thermoelectric materials. International Materials Reviews. 2016;**61**(6):379-415

[17] Shen Z-G, Tian L-L, Liu X. Automotive exhaust thermoelectric

generators: Current status, challenges and future prospects. Energy Conversion and Management. 2019;**195**:1138-1173

[18] Liu Y et al. Oxide materials for thermoelectric conversion. Molecules. 2023;**28**(15):5894

[19] Ohta H, Sugiura K, Koumoto K. Recent progress in oxide thermoelectric materials: p-type $Ca_3Co_4O_9$ and n-type $SrTiO_3$−. Inorganic Chemistry. 2008;**47**(19):8429-8436

[20] Poon SJ. Half Heusler compounds: Promising materials for mid-to-high temperature thermoelectric conversion. Journal of Physics D: Applied Physics. 2019;**52**(49):493001

[21] Wood ND et al. A review of key properties of thermoelectric composites of polymers and inorganic materials. Materials. 2022;**15**(23):8672

[22] Jaldurgam FF, Ahmad Z, Touati F. Synthesis and performance of large-scale cost-effective environment-friendly nanostructured thermoelectric materials. Nanomaterials. 2021;**11**(5):1091

[23] He J, Tritt TM. Advances in thermoelectric materials research: Looking back and moving forward. Science. 2017;**357**(6358):eaak9997

[24] Cao P, Chen Z, Xia Z, editors. Advanced Ceramics for Energy Storage, Thermoelectrics and Photonics. Series in Advanced Ceramic Materials. Netherlands: Elsevier; 2023

[25] Gu J et al. High-temperature oxidation mechanism of ZrCoSb-based Half-Heusler thermoelectric compounds. Journal of Materials Science & Technology. 2023;**148**:242-249

[26] He R et al. Studies on mechanical properties of thermoelectric materials by nanoindentation. Physica Status Solidi (A). 2015;**212**(10):2191-2195

[27] Wu L et al. Toughening thermoelectric materials: From mechanisms to applications. International Journal of Molecular Sciences. 2023;**24**(7):6325

[28] Wunderlich W. Energy harvesting under large temperature gradient–comparison of silicides, half-heusler alloys and ceramics. Energy Harvesting and Systems. 2015;**2**(1-2):37-46

[29] Ioannou I et al. Reduction of Hf via Hf/Zr substitution in mechanically alloyed (Hf, Ti) CoSb Half-Heusler solid solutions. Inorganics. 2022;**10**(4):51

[30] Nag A. Perovskite oxide thermoelectric module-a way forward. Catalysis Research. 2023;**3**(4):1-14

[31] Ohta S et al. Large thermoelectric performance of heavily Nb-doped $SrTiO_3$ epitaxial film at high temperature. Applied Physics Letters. 2005;**87**(9):092108

[32] Tian Z, Lee S, Chen G. Comprehensive review of heat transfer in thermoelectric materials and devices. Annual Review of Heat Transfer. 2014;**17**:425-483

[33] Tsubota T et al. Thermoelectric properties of Al-doped ZnO as a promising oxidematerial for high-temperature thermoelectric conversion. Journal of Materials Chemistry. 1997;**7**(1):85-90

[34] Thong TQ, Le TTH, Tinh NT. Investigation of the influence of singly and dually doping effect on scattering mechanisms and thermoelectric properties of perovskite-type STO. Materials Transactions. 2015;**56**(9):1365-1369

[35] Kurosaki K et al. Thermoelectric properties of $NaCo_2O_4$. Journal of Alloys and Compounds. 2001;**315**(1-2):234-236

[36] Xu G et al. Thermoelectric properties of the Bi-and Na-substituted $Ca_3Co_4O_9$ system. Applied Physics Letters. 2002;**80**(20):3760-3762

[37] He Q et al. Ultra-low thermal conductivity and improved thermoelectric performance in La_2O_3-dispersed $Bi_2Sr_2Co_2Oy$ ceramics. Materials Science and Engineering: B. 2024;**299**:116976

[38] Kang M-G et al. High-temperature thermoelectric properties of nanostructured $Ca_3Co_4O_9$ thin films. Applied Physics Letters. 2011;**98**(14):142102

[39] Hira U et al. High-temperature thermoelectric properties of Na-and W-doped $Ca_3Co_4O_9$ system. RSC Advances. 2018;**8**(22):12211-12221

[40] Constantinescu G et al. Effect of Na doping on the $Ca_3Co_4O_9$ thermoelectric performance. Ceramics International. 2015;**41**(9):10897-10903

[41] Hira U et al. Improved high-temperature thermoelectric properties of dual-doped $Ca_3Co_4O_9$. ACS Omega. 2022;7(8):6579-6590

[42] Fan Y, Qi X, Zeng D. Enhanced thermoelectric properties of $Ca_{3-x}Ag_xCo_4O_9$ by the sol–gel method with spontaneous combustion and cold isostatic pressing. Materials. 2018;**11**(12):2573

[43] Wang Y et al. Comparison of the high temperature thermoelectric properties for Ag-doped and Ag-added $Ca_3Co_4O_9$. Journal of Alloys and Compounds. 2009;**477**(1-2):817-821

[44] Lee M et al. Large enhancement of the thermopower in Na_xCoO_2 at high Na doping. Nature Materials. 2006;**5**(7):537-540

[45] Chatterjee A et al. Enhanced thermoelectric properties of misfit $Bi_2Sr_{2-X}Ca_xCo_2O_y$: Isovalent substitutions and selective phonon scattering. Materials. 2023;**16**(4):1413

[46] Özkurt B et al. Enhanced thermoelectric properties in $Bi_2Sr_{2-X}Ba_xCo_2O_y$ ceramics by Ba doping. Physica B: Condensed Matter. 2022;**643**:414138

[47] Wu T et al. On the origin of enhanced thermoelectricity in Fe doped $Ca_3Co_4O_9$. Journal of Materials Chemistry C. 2013;**1**(26):4114-4121

[48] Kaszyca K et al. Using the spark plasma sintering system for fabrication of advanced semiconductor materials. Materials. 2024;**17**(6):1422

[49] Lu Z et al. High-figure-of-merit thermoelectric La-doped A-site-deficient $SrTiO_3$ ceramics. Chemistry of Materials. 2016;**28**(3):925-935

[50] Chandel YS, Khare PS. Lattice thermal conductivity of $SrTiO_3/TiO_2$ superlattice nanowires. International Journal of Scientific and Research Publications. 2015:618-621

[51] Zhang L et al. Thermal conductivity enhancement of phase change materials with 3D porous diamond foam for thermal energy storage. Applied Energy. 2019;**233**:208-219

[52] Wang Y, Zhu J, Tang W. Extracting the effective mass of electrons in transparent conductive oxide thin films using Seebeck coefficient. Applied Physics Letters. 2014;**104**(21):212103

[53] Acharya M et al. High performance (ZT> 1) n-type oxide thermoelectric composites from earth abundant materials. Nano Energy. 2021;**84**:105905

[54] Azough F et al. Self-nanostructuring in $SrTiO_3$: A novel strategy for enhancement of thermoelectric response in oxides. ACS Applied Materials & Interfaces. 2019;**11**(36):32833-32843

[55] Jerič M et al. An enhanced thermoelectric figure of merit for Sr ($Ti_{0.8}Nb_{0.2}$) O_3 based on a Ruddlesden–popper-polytype-induced microstructure. Journal of the European Ceramic Society. 2016;**36**(5):1177-1182

[56] Ekren D et al. Controlling the thermoelectric behavior of La-doped $SrTiO_3$ through processing and addition of graphene oxide. ACS Applied Materials & Interfaces. 2022;**14**(48):53711-53723

[57] Radhika T et al. Synthesis, structural characterization and thermoelectric properties of Sr_2 ($Sr_{n-1}Ti_nO_{3n+1}$) n-type ceramic materials. Journal of Ovonic Research. 2019;**15**(5):315-323

[58] Hoque MSB et al. Ruddlesden-Popper chalcogenides push the limit of mechanical stiffness and glass-like thermal conductivity in single crystals. arXiv e-prints. Dec 2023;**v1**:arXiv:2312.02534

[59] Choi M et al. High figure-of-merit for ZnO nanostructures by interfacing lowly-oxidized graphene quantum dots. Nature Communications. 2024;**15**(1):1996

[60] Sulaiman S et al. Review of the nanostructuring and doping strategies for high-performance ZnO thermoelectric materials. Crystals. 2022;**12**(8):1076

[61] Jood P et al. Al-doped zinc oxide nanocomposites with enhanced thermoelectric properties. Nano Letters. 2011;**11**(10):4337-4342

[62] Wang N et al. Enhanced thermoelectric performance of Nb-doped $SrTiO_3$ by nano-inclusion with low thermal conductivity. Scientific Reports. 2013;**3**(1):3449

[63] Okhay O et al. Thermoelectric performance of Nb-doped $SrTiO_3$ enhanced by reduced graphene oxide and Sr deficiency cooperation. Carbon. 2019;**143**:215-222

[64] Chatterjee A et al. On the thermoelectric properties of Nb-doped $SrTiO_3$ epitaxial thin films. Physical Chemistry Chemical Physics. 2022;**24**(6):3741-3748

[65] Wang HX, Wang XL, Bu TA, et al. A-site defect construction in medium-entropy $SrTiO_3$ ceramics for enhanced thermoelectric performance. Rare Metals. 2025;**44**:3324-3338. DOI: 10.1007/s12598-024-03177-8

[66] Doan UTT et al. Nanostructuring and oxygen-vacancy tuning to repress thermal transport and activate thermoelectric properties of green-synthesized ZnO. Materialia. 2024;**34**:102112

[67] Park K et al. Colloidal synthesis and thermoelectric properties of La-doped $SrTiO_3$ nanoparticles. Journal of Materials Chemistry A. 2014;**2**(12):4217-4224

[68] Kruppa K et al. Advanced thermoelectric performance of a textured ceramic composite: Encapsulation of Na_xCoO_2 into a triple-phase matrix. Journal of the American Ceramic Society. 2024;**107**(12):7951-7965

[69] d'Angelo M, Galassi C, Lecis N. Thermoelectric materials and applications: A review. Energies. 2023;**16**(17):6409

[70] Zhang S et al. Atomic-scale structure and thermoelectric properties in medium-entropy PbSnTeSe alloy. Journal of Materials Chemistry A. 2024;**12**(48):33991-34000

[71] Wei Z et al. Improving thermoelectric properties of high-entropy ($Ca_{0.27}Sr_{0.27}Ba_{0.27}La_{0.19}$) $TiO_{3-\delta}$ ceramics through defect engineering by controlling the oxygen vacancy content. Ceramics International. 2025;**51**(4):4300-4309

[72] Su L et al. High thermoelectric performance realized through manipulating layered phonon-electron decoupling. Science. 2022;**375**(6587):1385-1389

[73] Deng Q et al. Dielectric property optimization of polymer nanocomposites using $BaTiO_3$ based high-entropy ceramic filler with Dirac-cone effect. Physica B: Condensed Matter. 2024;**684**:415992

[74] Hicks LD, Dresselhaus MS. Effect of quantum-well structures on the thermoelectric figure of merit. Physical Review B. 1993;**47**(19):12727

[75] Dawley NM et al. Thermal Conductivity of the n= 1-5 and 10 Members of the $(SrTiO_3)_nSrO$ Ruddlesden-Popper Superlattices. USA: AIP Publishing; 2021

[76] Bhakta S, Kundu B. A review of thermoelectric generators in automobile waste heat recovery systems for improving energy utilization. Energies. 2024;**17**(5):1016

[77] Yang S et al. Recent developments in flexible thermoelectric devices. Small Science. 2021;**1**(7):2100005

[78] Liu Y et al. Scalable-produced 3D elastic thermoelectric network for body heat harvesting. Nature Communications. 2023;**14**(1):3058

[79] Feng X, Guo F, Luo L, et al. Engineering design of feedstock powder and relevant thermal–mechanical performance of thermal/environmental barrier coatings. Journal of Advanced Ceramics. 2025;**14**(2):9221033. DOI: 10.26599/JAC.2025.9221033

[80] Zhou JHJZJ. High-temperature thermoelectric properties with $Th_{3-x}Te_4$. arXiv preprint arXiv:2301.02763; 2023

[81] Wang Y et al. Nanostructuring and band engineering boosting thermoelectric performance of Bi-Sb-Te alloys via CsBr doping. Science China Materials. 2023;**66**(10):3991-4000

[82] Zang J et al. Printed flexible thermoelectric materials and devices. Journal of Materials Chemistry A. 2021;**9**(35):19439-19464

[83] Fiedler C et al. Solution-processed inorganic thermoelectric materials: Opportunities and challenges∇. Chemistry of Materials. 2022;**34**(19):8471-8489

Chapter 8

Recent Advances in Thermoelectric Materials for Biomedical Applications: Energy Harvesting and Wearables

Anita Ioana Visan, Irina Negut and Claudiu Hapenciuc

Abstract

Recent advances in thermoelectric materials are transforming biomedicine, particularly for energy harvesting and wearables. This chapter reviews thermoelectric generators (TEGs) utilizing the Seebeck effect to convert body heat into electricity, powering self-sufficient medical devices like implants and monitors. Performance metrics, including thermoelectric figures of merit (ZT) and power factors, are detailed, highlighting materials such as bismuth telluride (Bi_2Te_3) and carbon nanotube composites achieving ZT values over 1.0 at room temperature. The chapter covers: (1) novel nanocomposites, exploring nanostructured materials for enhanced efficiency; (2) low-temperature thermoelectrics, focusing on materials such as magnesium silicide (Mg_2Si) and organic polymers for wearables; and (3) future challenges, addressing scalability, biocompatibility, and biomedical integration. Key innovations include flexible organic polymers and hybrid nanocomposites, showing ZT values up to 0.8 and power factors exceeding 500 $\mu W/mK^2$, ideal for monitoring and therapies. Despite progress, challenges remain in achieving high ZT, material stability, and scalable production. Future efforts emphasize biocompatible materials, robust designs, and seamless integration. Thermoelectric materials can revolutionize personalized healthcare, enabling energy-autonomous devices and reducing reliance on conventional power. This review highlights the synergy between material innovation and medical technology for next-generation applications.

Keywords: thermoelectric materials, biomedicine, energy harvesting, nanostructured materials, wearable devices

1. Introduction

Thermoelectric materials convert temperature differences into electrical potential differences, a principle that has garnered significant attention in recent years due to their potential applications in biomedicine [1]. The ability to harvest waste heat from the human body and convert it into usable electrical energy presents exciting possibilities for powering medical devices. This review synthesizes recent advancements in thermoelectric materials, focusing on their applications in biomedical fields, including wearable technology and energy harvesting systems [2].

Recent studies have demonstrated the feasibility of using thermoelectric generators (TEGs) to harvest energy from the human body [2, 3]. For instance, Kim et al. [4] developed a wearable thermoelectric generator that effectively captures body heat, generating sufficient power to operate low-energy medical devices such as pacemakers and neurostimulators [4]. This aligns with findings by Remeš and Havlík, who emphasized the suitability of TEGs for powering small, portable devices by generating an electrical potential difference from a temperature drop [5]. The integration of TEGs into wearable technology not only provides a sustainable energy source but also enhances the functionality of medical devices by enabling real-time monitoring of physiological parameters.

The advancement of materials science has led to the exploration of various thermoelectric materials that exhibit improved performance characteristics. Li et al. highlighted the development of multifunctional wearable thermoelectrics that can manage personal thermal environments while effectively utilizing body heat [6]. This is complemented by the work of Zhao and You, who discussed the potential of shoe-embedded piezoelectric energy harvesters, which can be integrated with thermoelectric systems to enhance energy generation from human motion and heat [7]. The combination of these technologies could lead to more efficient energy harvesting systems that cater to the demands of modern biomedical applications. Nanostructured thermoelectric materials, such as bismuth telluride (Bi_2Te_3) and silver selenide (Ag_2Se), have shown remarkable improvements in thermoelectric performance. For example, Bi_2Te_3 exhibits a ZT value of ~1.0 at room temperature, while nanostructured Ag_2Se has achieved ZT values of up to 1.2 due to its low thermal conductivity (~0.5 W/mK) and high power factor (~20 $\mu W/cmK^2$) [8–10].

These materials are particularly suitable for biomedical applications due to their high efficiency and flexibility.

Su et al. noted that methods such as micromachining and the use of nanowires can significantly improve the efficiency of thermoelectric materials, although challenges remain in mass production [11]. Doping strategies, such as introducing copper into telluride matrices, have been shown to enhance thermoelectric properties by increasing the power factor, a key parameter for efficiency. Furthermore, Hricková's research on ductile doped Ag_2S materials revealed comparable power factor values to other thermoelectric materials, suggesting that doping strategies could enhance the thermoelectric properties of existing materials [12]. The exploration of these advanced materials is crucial for developing efficient and flexible thermoelectric devices suitable for biomedical applications.

The design of thermoelectric devices also plays a critical role in their effectiveness. For example, Kim et al. developed a hybrid generator that integrates thermoelectric and pyroelectric components, allowing for improved energy harvesting from body heat [13]. This innovative approach demonstrates the potential for creating multifunctional devices that can adapt to varying physiological activities. Additionally, the work by He et al. on stretchable and breathable thermoelectric fabrics highlights the importance of material flexibility and comfort in wearable applications, ensuring that devices can be worn continuously without discomfort [14].

The advancements in thermoelectric materials and their applications in biomedicine present a promising frontier for energy harvesting technologies. The ability to convert body heat into electrical energy not only supports the development of self-powered medical devices but also enhances the functionality of wearable technologies. Continued research into new materials, innovative device designs, and integration strategies will be essential for realizing the full potential of thermoelectric systems in healthcare.

2. Recent advances in thermoelectric materials

2.1 Thermoelectric principles and applications

Thermoelectric energy harvesting relies on the Seebeck effect, where a voltage is generated in response to a temperature gradient [15–17]. This phenomenon is fundamental to the operation of thermoelectric materials, which can convert thermal energy into electrical energy (**Figure 1**).

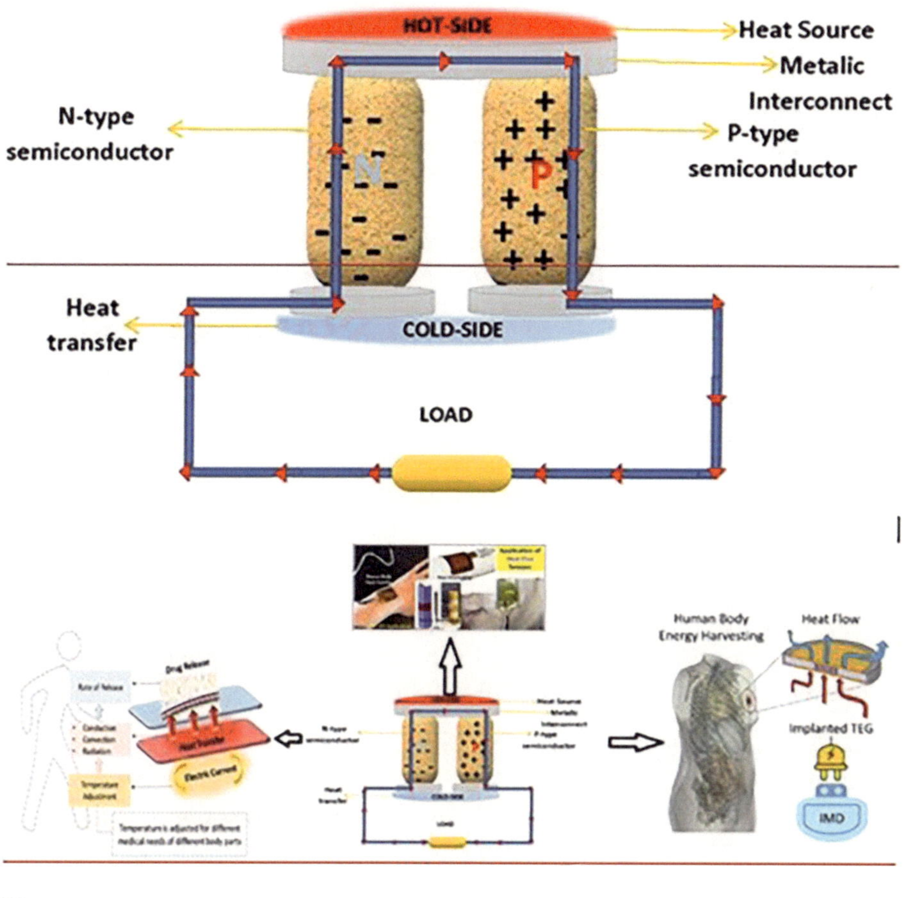

Figure 1.
Schematic representation of thermoelectric material principles in biomedical applications. This diagram illustrates the fundamental principles of thermoelectric materials, highlighting how they convert thermal energy into electrical energy and vice versa. In the context of biomedical applications, such materials can be utilized for innovative solutions such as wearable thermoelectric devices for monitoring body temperature, powering biosensors, and enabling efficient thermal management in medical implants. The arrows indicate the flow of heat and electricity, demonstrating the potential for these materials to harness body heat for energy generation or to regulate temperature in healthcare settings [18–20].

The efficiency of this conversion is characterized by the dimensionless figure of merit (ZT), which is influenced by the material's electrical conductivity, Seebeck coefficient, and thermal conductivity. The ability to harness waste heat from the human body through this mechanism opens up numerous possibilities for biomedical applications. TEGs have shown significant potential in powering implantable medical devices. For example, Kim et al. demonstrated that a wearable TEG could generate up to 40 $\mu W/cm^2$ from body heat, sufficient to power low-energy medical implants such as pacemakers [21]. This reduces the need for battery replacements and associated surgical interventions, enhancing patient quality of life. Additionally, wearable sensors utilizing thermoelectric materials can provide continuous monitoring of vital signs, such as body temperature and heart rate, by harvesting energy from the body's heat [22].

As noted by Patidar [22, 23] the potential applications of thermoelectric materials in biomedicine are vast, ranging from powering implanted devices to wearable sensors that monitor physiological parameters. The integration of thermoelectric generators (TEGs) into medical devices can facilitate self-sustaining systems that do not require external power sources, thereby enhancing the functionality and reliability of these devices. For instance, TEGs can be used to power pacemakers, neurostimulators, and other implantable devices, significantly reducing the need for battery replacements and associated surgical interventions.

Moreover, the development of wearable sensors that utilize thermoelectric materials can provide continuous monitoring of vital signs, such as body temperature, heart rate, and other physiological parameters. These sensors can be designed to operate autonomously, harvesting energy from the body's heat to power their functions. This capability is particularly beneficial for chronic disease management, where real-time monitoring is crucial for timely interventions.

Research has shown that materials such as bismuth telluride and its alloys exhibit high thermoelectric performance, making them suitable candidates for medical applications [8]. Additionally, the exploration of flexible and biocompatible thermoelectric materials has paved the way for the development of comfortable and effective wearable devices [9].

The principles of thermoelectric energy harvesting, particularly the Seebeck effect, provide a foundation for innovative applications in biomedicine. The ability to convert body heat into electrical energy not only supports the development of self-powered medical devices but also enhances the functionality of wearable health monitoring systems. As research continues to advance in this field, the integration of thermoelectric materials into biomedical applications will likely lead to significant improvements in patient care and health monitoring.

2.2 Promising materials for thermoelectric applications

Bismuth telluride (Bi_2Te_3) is one of the most widely studied thermoelectric materials due to its high efficiency at room temperature, with ZT values ranging from 0.8 to 1.2 depending on doping and nanostructuring [8, 24]. Its high power factor (~50 $\mu W/cmK^2$) and low thermal conductivity (~1.5 W/mK) make it ideal for low-temperature applications such as wearable devices and sensors [25, 26].

Cai et al. [3] highlight promising materials for thermoelectric applications, including bismuth telluride and various nanostructured composites. The versatility of bismuth telluride is further enhanced when it is used in nanostructured forms, which can significantly improve its thermoelectric properties.

Nanostructuring, which involves reducing the material's thermal conductivity while maintaining electrical conductivity, has been shown to enhance its performance [27, 28]. The reduction in thermal conductivity is crucial because it allows for a greater temperature gradient across the thermoelectric material, which is essential for maximizing the Seebeck effect.

Doping strategies, such as the introduction of copper into telluride matrices, further improve thermoelectric properties by increasing the power factor [29]. The research by Lal [29] and Jia et al. [16] on electrodeposited Cu-doped telluride films illustrates how doping can further enhance the thermoelectric properties of materials like bismuth telluride. The introduction of copper into telluride matrices has been shown to increase the power factor, a key parameter that reflects the efficiency of thermoelectric materials. This aligns with the findings of Kadel et al. [30], who also noted that nanostructuring can induce significant enhancements in thermoelectric efficiency, by increasing the power factor, through quantum confinement effects [31].

Silver selenide (Ag_2Se) has also emerged as a promising material for biomedical applications due to its flexibility and high thermoelectric performance. Recent studies have reported ZT values of up to 1.2 for Ag_2Se, with a power factor of ~20 $\mu W/cmK^2$ and thermal conductivity as low as 0.5 W/mK [9, 32]. These properties make it suitable for flexible and wearable thermoelectric devices.

The advancements in nanostructured thermoelectric materials are not limited to bismuth telluride. Other promising materials include rare-earth silicides and carbon nanotube composites, which offer unique advantages such as improved flexibility and mechanical properties, essential for developing wearable thermoelectric devices [9, 33]. The integration of these advanced materials into practical applications is crucial for the future of thermoelectric technology, particularly in the biomedical field where flexible, efficient energy harvesting solutions are needed [15].

The ability to manipulate thermal and electrical properties through nanostructuring and doping strategies is critical for enhancing the efficiency of thermoelectric devices [3]. The discovery or design of novel thermoelectric materials with a high figure of merit is the key to improving the efficiency and competitiveness of thermoelectric generators. Novel thermoelectric materials can be obtained by using advanced techniques such as band structure engineering, nanostructuring, doping, alloying, or heterostructure to manipulate the electronic and phononic behavior of the materials. Novel thermoelectric materials can also be obtained by exploring new classes of materials, such as organic, hybrid, or topological materials, that may exhibit unconventional or enhanced thermoelectric effects [17].

The development of advanced thermoelectric modules with improved performance and reliability is another important aspect of thermoelectric generator research and development. Advanced thermoelectric modules can be achieved by using novel thermoelectric materials with high figures of merit or by using novel module architectures or configurations that can increase power output or reduce thermal losses. Advanced thermoelectric modules can also be achieved by using novel fabrication methods or techniques that can reduce the cost or increase the quality of the modules [1].

The innovation of thermoelectric systems with novel designs and applications is a promising direction for thermoelectric generator research and development. Innovative thermoelectric systems can be achieved by using novel heat sources or heat sinks that can provide high or stable temperature differences or by using novel integration schemes or strategies that can optimize the system's performance or functionality. Innovative thermoelectric systems can also be achieved by exploring new application domains or scenarios that can benefit from the advantages of thermoelectric generators [15].

As research continues to evolve in this area, the potential for these materials to contribute to sustainable energy solutions and innovative biomedical applications becomes increasingly apparent.

2.3 Novel nanocomposites

Nanocomposites leverage the unique properties of nanoscale materials to optimize thermoelectric efficiency, which is critical for applications such as thermoelectric generators and coolers in medical devices [34]. The design of nanocomposites often involves the incorporation of various nanostructured materials, such as carbon nanotubes (CNTs), transition metal dichalcogenides (TMDs), and metallic nanoparticles, into a polymer or ceramic matrix. For instance, single-walled carbon nanotubes (SWCNTs) are known for their exceptional electrical and thermal conductivity, mechanical strength, and flexibility, making them ideal candidates for enhancing the thermoelectric properties of composites [35]. By integrating SWCNTs into polymer matrices, researchers have achieved significant improvements in electrical conductivity while maintaining low thermal conductivity, which is essential for high thermoelectric performance [35]. For example, single-walled carbon nanotubes (SWCNTs) exhibit exceptional electrical conductivity (~10^4 S/cm) and low thermal conductivity (~0.1 W/mK), making them ideal for enhancing the thermoelectric properties of composites [35, 36]. When integrated into polymer matrices such as PEDOT:PSS, SWCNTs have demonstrated ZT values of up to 0.4, with a power factor of ~200 μW/mK^2, making them highly effective for biomedical sensors [37, 38].

Transition metal dichalcogenides (TMDs), such as MoS_2 and WS_2, have also garnered attention due to their tunable electronic properties and high surface area.

The engineering of TMD nanosheets at the atomic scale allows for the creation of heterostructures that can exhibit enhanced thermoelectric performance [39]. Doping TMDs with different elements can further optimize their thermoelectric properties, leading to materials with improved Seebeck coefficients and reduced thermal conductivity [39]. Concretely, doping TMDs with elements like copper or nickel can further optimize their thermoelectric properties, achieving ZT values of up to 0.6 and power factors of ~15 μW/cmK^2 [39]. These materials are particularly promising for hybrid nanocomposites, where their combination with metallic nanoparticles or polymers can lead to synergistic enhancements in thermoelectric performance [40].

The effectiveness of SWCNT-PEDOT:PSS composites in biomedical sensors has been well-documented [37]. Additionally, their flexibility and biocompatibility allow for seamless integration into biomedical platforms, such as biosensors and drug delivery systems [41].

The development of hybrid nanocomposites that combine different types of nanomaterials has shown promise in achieving synergistic effects that enhance thermoelectric performance. For example, the combination of metallic nanoparticles with TMDs can lead to improved charge transport properties and enhanced thermoelectric efficiency. This hybrid approach allows for the fine-tuning of the electronic and thermal properties of the composite, making them suitable for specific biomedical applications, such as targeted drug delivery systems that utilize thermoelectric effects for localized heating [40].

The integration of additive manufacturing techniques in the fabrication of nanocomposites has enabled the precise control of material composition and structure at the nanoscale. This approach allows for the creation of complex geometries and tailored microstructures that can optimize thermoelectric performance. For instance, 3D printing techniques can be employed to fabricate nanocomposite structures with specific porosity and alignment, enhancing their thermoelectric properties [42]. Additionally, the use of advanced characterization techniques, such as scanning electron microscopy (SEM) and transmission electron microscopy (TEM), facilitates the detailed analysis of the microstructure and its correlation with thermoelectric performance [42].

The unique properties of novel nanocomposites make them particularly attractive for biomedical applications. Their ability to convert waste heat into electrical energy can be harnessed in wearable medical devices, providing a sustainable power source for sensors and monitoring devices. Furthermore, the localized heating capabilities of thermoelectric nanocomposites can be utilized in hyperthermia treatment for cancer, where controlled heating can enhance the efficacy of chemotherapy [40]. The versatility of these materials also extends to their use in biosensors, where their enhanced thermoelectric properties can improve sensitivity and response times [40]. In conclusion, the development of novel nanocomposites represents a significant advancement in the field of thermoelectric materials for biomedicine. By leveraging the unique properties of nanoscale materials and employing innovative fabrication techniques, researchers are paving the way for the next-generation of thermoelectric devices that can meet the demands of modern biomedical applications. Recent advancements in thermoelectric nanocomposites have highlighted the potential of combining carbon nanotubes (CNTs) with conductive polymers for biomedical applications. Glória et al. [43] reported that these composites exhibit enhanced thermoelectric properties, making them particularly suitable for biomedical devices due to their flexibility, biocompatibility, and ability to efficiently convert temperature gradients into electrical potential. The incorporation of CNTs into polymer matrices not only improves electrical conductivity but also maintains the mechanical integrity required for various biomedical applications, such as wearable sensors and implantable devices [44].

Furthermore, Yin et al. [37] demonstrated the promising thermoelectric performance of single-walled carbon nanotube (SWCNT) and poly(3,4-ethylenedioxythiophene) (PEDOT):PSS coated tellurium nanorod composites. This combination showed significant potential for energy harvesting in wearable devices, which is crucial for powering biomedical sensors and monitoring systems. The unique properties of these nanocomposites, including their lightweight nature and high conductivity, facilitate their integration into flexible electronic systems, enhancing the functionality of wearable health monitoring devices [41].

The biocompatibility of CNTs has been a focal point in their application within the biomedical field. Research indicates that functionalized CNTs can be effectively used in various biomedical applications, including drug delivery systems, biosensors, and tissue engineering scaffolds [45]. The ability to tailor the surface properties of CNTs through functionalization further enhances their compatibility with biological systems, allowing for improved interaction with cells and tissues [46].

2.4 Low-temperature thermoelectric generators

Low-temperature thermoelectric generators (TEGs) have gained significant attention in recent years due to their potential applications in biomedical fields, particularly for energy harvesting from body heat and other low-grade thermal sources. These devices exploit the Seebeck effect, which generates electrical voltage drop from a temperature difference across thermoelectric materials. The development of efficient low-temperature TEGs is crucial for powering wearable medical devices, sensors, and other portable electronic systems that require sustainable energy sources. **Table 1** illustrates the diversity of thermoelectric materials suitable for low-temperature applications.

The dimensionless figure of merit (ZT) is a critical parameter that quantifies the efficiency of a thermoelectric material. A higher ZT value indicates better thermoelectric performance. Bismuth telluride (Bi_2Te_3) exhibits the highest ZT values among the listed materials, making it a top choice for applications requiring high efficiency. Magnesium silicide (Mg_2Si) shows moderate ZT values of 0.7–1.0, while carbon nanotube composites and organic thermoelectric materials have lower ZT values of 0.1–0.4 and 0.1–0.5, respectively. This limits their efficiency in high-performance thermoelectric applications.

The power factor, defined as the product of the square of the Seebeck coefficient and electrical conductivity, is a crucial metric for evaluating the performance of thermoelectric materials. Bismuth telluride leads with a power factor of 40–50 $\mu W/m{\cdot}K^2$, indicating its superior ability to convert temperature gradients into electrical energy. Carbon nanotube composites exhibit a higher power factor of 100–500 $\mu W/m{\cdot}K^2$, owing to their excellent electrical conductivity, while magnesium silicide shows a moderate power factor of 20–30 $\mu W/m{\cdot}K^2$. Organic thermoelectric materials have the lowest power factor, ranging from 10 to 100 $\mu W/m{\cdot}K^2$, reflecting their lower efficiency in energy conversion.

Thermal conductivity is a measure of a material's ability to conduct heat. Lower thermal conductivity is desirable in thermoelectric materials as it helps maintain a temperature gradient, enhancing the thermoelectric effect. Bismuth telluride has a low thermal conductivity of 1.5–2.0 W/m·K, which contributes to its high ZT value. Magnesium silicide has a higher thermal conductivity of 3.0–5.0 W/m·K, which may limit its efficiency. Organic thermoelectric materials exhibit the lowest thermal conductivity of 0.1–0.5 W/m·K, making them suitable for applications where flexibility and lightweight properties are prioritized. Carbon nanotube composites have a thermal conductivity range of 0.2–1.5 W/m·K, depending on the alignment and composition of the composite.

Bismuth telluride remains the benchmark material for low-temperature applications due to its high Seebeck coefficient (~200 μV/K) and low thermal conductivity (~1.5 W/mK) [25]. However, researchers are exploring alternative materials, such as magnesium silicide (Mg_2Si), which offers a balance of performance and cost-effectiveness, with ZT values of ~0.7 and power factors of ~10 $\mu W/cmK^2$ [50].

Organic thermoelectric materials, such as conducting metallopolymers, have shown promise for low-temperature TEG applications due to their lightweight, flexibility, and ease of processing. Recent advancements in carbon-based materials have demonstrated ZT values of up to 0.3 and power factors of ~50 $\mu W/mK^2$, making them ideal for flexible and wearable devices [53]. Organic thermoelectric materials, such as conducting metallopolymers, have shown promise for low-temperature TEG applications due to their lightweight, flexibility, and ease of processing [53]. These materials can be integrated into flexible substrates, making them ideal for wearable devices that require conformability and comfort [53].

The design of low-temperature TEGs involves optimizing the thermoelectric figure of merit (ZT), which is a dimensionless quantity that indicates the efficiency of a thermoelectric material. Strategies for enhancing ZT include reducing thermal conductivity while maintaining or increasing electrical conductivity. Techniques such as nanostructuring, alloying, and the incorporation of fillers into polymer

Material	Key properties	ZT Value	Power Factor ($\mu W/m{\cdot}K^2$)	Thermal Conductivity (W/m·K)	Applications	Reference
Bismuth telluride (Bi2Te3)	High Seebeck coefficient, low thermal conductivity	~1.0–1.2 at room temperature (optimized for n-type and p-type doping)	~40–50 $\mu W/m{\cdot}K^2$	~1.5–2.0 W/m·K	Wearable devices, sensors	[25, 47, 48]
Magnesium silicide (Mg2Si)	Cost-effective, good thermoelectric performance	~0.7–1.0 (depending on doping and nanostructuring)	~20–30 $\mu W/m{\cdot}K^2$	~3.0–5.0 W/m·K.	Energy harvesting in low-temperature applications	[30, 49, 50]
Organic thermoelectric materials	Lightweight, flexible, easy to process	~0.1–0.5 (highly dependent on molecular structure and doping)	~10–100 $\mu W/m{\cdot}K^2$ (can vary widely)	~0.1–0.5 W/m·K (typically low due to disordered structure)	Wearable medical devices, sensors	[51–54]
Carbon nanotube composites	High electrical conductivity, low thermal conductivity	~0.1–0.4 (depending on composite composition and alignment)	~100–500 $\mu W/m{\cdot}K^2$ (high due to excellent electrical conductivity)	~0.2–1.5 W/m·K (varies with CNT alignment and matrix material)	Biomedical applications, energy harvesting	[35, 55]

Notes: The ZT value (figure of merit) is calculated as $ZT = S^2\sigma T/\kappa$, where S is the Seebeck coefficient, σ is electrical conductivity, T is absolute temperature, and κ is thermal conductivity. The power factor is given by $S^2\sigma$ and is a measure of the material's ability to generate electrical power from a temperature gradient. Thermal conductivity is a critical parameter for thermoelectric efficiency, as lower thermal conductivity generally improves ZT.

Table 1.
Summary of thermoelectric materials for low-temperature generators.

matrices have been employed to achieve these goals [56, 57]. Additionally, the use of segmented TEG designs, where different materials are used for the hot and cold sides, can improve overall efficiency [58].

Challenge	Description	Material comparisons	Future direction	References
Efficiency and performance	Low ZT values limit thermoelectric efficiency, especially at low temperatures	Bi_2Te_3: ZT Value: ~1.0–1.2 at room temperature. Power Factor: ~40–50 μW/m·K^2. Thermal Conductivity: ~1.5–2.0 W/m·K. [8, 48] Ag_2Se: ZT Value: ~1.2 at room temperature. Power Factor: ~20 μW/m·K^2 (not μW/cm·K^2).Thermal Conductivity: ~0.5–0.6 W/m·K. [9, 60]. Organic materials (e.g., PEDOT:PSS): ZT Value: ~0.1–0.4 (depending on doping and processing).Power Factor: ~10–100 μW/m·K^2 (not μW/cm·K^2). Thermal Conductivity: ~0.1–0.3 W/m·K [53, 61].	Focus on novel materials and composites to enhance ZT	[25, 62]
Material stability and reliability	Degradation and mechanical reliability issues hinder long-term performance	Bi_2Te_3: is known to degrade in physiological conditions due to oxidation and chemical reactions with bodily fluids. Studies have shown that exposure to moisture and elevated temperatures can lead to a reduction in thermoelectric performance over time. Degrades by ~20% in ZT after 30 days in physiological conditions [62, 63]. Ag_2Se: can release silver ions in physiological environments, which may cause cytotoxicity. Encapsulation with biocompatible polymers like PLA has been shown to reduce ion leaching and improve the material's stability and biocompatibility. [8, 64]. PEDOT:PSS: is highly sensitive to moisture, which can degrade its electrical and thermoelectric properties. Hybrid composites, such as those incorporating carbon nanotubes or inorganic fillers, have been shown to enhance stability and reduce moisture sensitivity [65, 66].	Develop robust hybrid materials with improved stability	[66, 67]
Scalability and manufacturing	Traditional methods may not support the production of complex nanostructures	Bi_2Te_3: is one of the most efficient thermoelectric materials at room temperature, but tellurium is a rare and expensive element, which limits its widespread use [68, 69]. Mg_2Si: is an environmentally friendly and cost-effective thermoelectric material with moderate performance, ZT ~0.7, power factor ~ 100 **μ**W/m·K^2 [50, 70]. CNT-PEDOT:PSS: Carbon nanotube (CNT) composites with PEDOT:PSS have been demonstrated as scalable and flexible thermoelectric materials, suitable for wearable applications [55]. Scalable *via* 3D printing, ZT ~0.4, power factor ~ 200 μW/mK^2 [37].	Explore 3D printing and additive manufacturing techniques	[71, 72]

Challenge	Description	Material comparisons	Future direction	References
Integration with biomedical systems	Need for compatibility with existing medical technologies and bio-compatibility	Bi_2Te_3: Used in implantable glucose sensors with 6-month stability [59]. Ag_2Se: Requires encapsulation to prevent silver ion toxicity [8]. CNT-PEDOT:PSS: Biocompatible, suitable for wearable sensors [37].	Research on integrating thermoelectric materials into biomedical platforms	[73, 74]
Environmental and economic considerations	Reliance on rare or toxic elements raises sustainability concerns	Bi_2Te_3: Both tellurium scarcity and the potential toxicity of bismuth limit the widespread use of Bi_2Te_3, especially in biomedical applications [48, 69]. PbTe: is a high-performance thermoelectric material, but its use is restricted due to the toxicity of lead, which poses environmental and health risks [75]. Abandoned due to lead toxicity despite high ZT (~1.5) [76]. Mg_2Si and CNT composites: are considered environmentally friendly and sustainable alternatives to traditional thermoelectric materials like Bi_2Te_3 and PbTe [70]. Abundant, nontoxic, and sustainable alternatives [50, 77].	Prioritize development of abundant, nontoxic materials	[69, 78]

Table 2.
Summary of challenges and future directions in thermoelectric materials for biomedicine.

Wearable devices that monitor vital signs can be powered by the heat generated by the body, reducing the need for external power sources or batteries [59]. Furthermore, TEGs can be integrated into implantable devices, providing a continuous power supply for sensors and drug delivery systems [59]. The ability to convert body heat into electrical energy not only enhances the functionality of these devices but also contributes to their sustainability and user convenience.

2.5 Challenges and future directions

The field of thermoelectric materials for biomedicine is rapidly evolving, yet several challenges remain that must be addressed to fully realize their potential in practical applications. **Table 2** discusses the key challenges faced in the development of thermoelectric materials and outlines future directions for research and innovation.

2.5.1 Efficiency and performance

Achieving high thermoelectric efficiency, characterized by the dimensionless figure of merit (ZT), remains a primary challenge. For example, while Bi_2Te_3 exhibits ZT values of ~1.0 at room temperature, organic thermoelectric materials typically achieve ZT values below 0.5, limiting their efficiency [79]. Future research should focus on novel materials and composites that can enhance ZT through strategies such as band engineering, nanostructuring, and the incorporation of dopants [80].

2.5.2 Material stability and reliability

Material stability and scalability are also critical challenges. The stability of thermoelectric materials under operating conditions is crucial for their long-term performance, especially in biomedical applications where devices may be subjected to varying temperatures and environmental conditions. Recent studies have highlighted the degradation of thermoelectric materials in biological environments, particularly those containing toxic elements like lead. For instance, research by Culebras et al. indicates that materials such as Bi_2Te_3, while effective, pose significant toxicity risks, necessitating the exploration of safer alternatives [81]. Furthermore, the mechanical reliability of these materials is paramount; studies have shown that under long-term fatigue loading, thermoelectric properties can degrade, leading to reduced efficiency and potential device failure [82]. Research into more robust materials, including hybrid composites that combine the advantages of organic and inorganic components, is essential for improving stability. Many thermoelectric materials, particularly organic and polymer-based systems, face challenges related to degradation and mechanical reliability [83]. Research into more robust materials, including hybrid composites that combine the advantages of organic and inorganic components, is essential for improving stability [83].

The long-term stability of thermoelectric materials in biological environments is a critical concern. For instance, bismuth telluride (Bi_2Te_3), while highly efficient, has been shown to degrade in physiological conditions over time, leading to a loss of performance. A study by Gao et al. demonstrated that Bi_2Te_3-based devices experienced a 20% reduction in ZT after 30 days of exposure to simulated body fluids, highlighting the need for protective coatings or encapsulation strategies [62]. Similarly, organic thermoelectric materials, such as PEDOT:PSS, face challenges related to moisture absorption and mechanical degradation under prolonged use [83].

To address these issues, researchers are exploring hybrid materials that combine the stability of inorganic components with the flexibility of organic polymers. For example, a recent study by Xin et al. developed a Bi_2Te_3-PEDOT:PSS composite that exhibited improved stability, with only a 10% reduction in ZT after 60 days in physiological conditions [66]. These findings underscore the importance of material hybridization in enhancing long-term reliability.

2.5.3 Scalability and manufacturing

The fabrication of thermoelectric materials and devices at scale poses significant challenges. Traditional manufacturing techniques may not be suitable for producing complex nanostructures or composites required for advanced thermoelectric applications. Innovative manufacturing methods, such as 3D printing and additive manufacturing, offer promising avenues for creating flexible and customizable thermoelectric devices [84]. Continued exploration of these techniques will be vital for the commercialization of thermoelectric technologies.

2.5.4 Integration with biomedical systems

For thermoelectric materials to be effectively utilized in biomedical applications, they must be compatible with existing technologies and systems. This includes ensuring biocompatibility and the ability to integrate with sensors, drug delivery systems, and other medical devices [85]. A notable case study involves the integration of thermoelectric materials into implantable glucose sensors. A 2023 study by Nozariasbmarz et al. demonstrated that a Bi_2Te_3-based thermoelectric generator could power a glucose sensor for up to 6 months without significant performance degradation, showcasing the potential for long-term biomedical applications [59].

However, challenges remain in ensuring the biocompatibility of thermoelectric materials. For example, silver selenide (Ag_2Se), while highly efficient, has raised concerns due to the potential release of toxic silver ions in biological environments. A study by Wei et al. investigated the cytotoxicity of Ag_2Se and found that encapsulation with biocompatible polymers, such as polylactic acid (PLA), significantly reduced silver ion leaching and improved biocompatibility [8]. Future research should focus on developing thermoelectric materials that can seamlessly integrate into biomedical platforms while maintaining their thermoelectric performance. Case studies, such as the integration of PEDOT-based thermoelectric materials in wearable devices, have shown promising results in terms of both performance and biocompatibility [86].

2.5.5 Environmental and economic considerations

The sustainability of thermoelectric materials is an increasingly important consideration. Many conventional thermoelectric materials rely on rare or toxic elements, which can pose environmental and economic challenges [76]. For instance, tellurium, a key component of Bi_2Te_3, is a rare element with limited global reserves, raising concerns about the scalability of tellurium-based thermoelectrics [69]. Similarly, the use of lead-based materials, such as PbTe, has been largely abandoned due to toxicity concerns, despite their high thermoelectric performance.

To address these issues, researchers are exploring alternative materials, such as magnesium silicide (Mg_2Si) and carbon-based composites, which are abundant and nontoxic. A 2021 study by Choudhary et al. demonstrated that Mg_2Si-based composites achieved ZT values of ~0.7, with a power factor of ~10 $\mu W/cmK^2$, making them a viable alternative for sustainable thermoelectric applications [50]. Additionally, carbon nanotube (CNT) composites have shown promise due to their high electrical conductivity and low toxicity. A 2022 study by Yin et al. reported that CNT-PEDOT:PSS composites exhibited ZT values of up to 0.4, with excellent biocompatibility, making them suitable for wearable medical devices [37]. The development of abundant, nontoxic materials, such as carbon-based thermoelectric composites, is essential for creating sustainable thermoelectric technologies [77]. Research should prioritize the exploration of environmentally friendly materials and processes. For example, Zevalkink et al. have identified Ca_3AlSb_3 as a promising nontoxic alternative for thermoelectric applications, demonstrating its potential for waste heat recovery without the associated toxicity of traditional materials [87].

2.5.6 Future research priorities

Future research should prioritize the development of materials that combine high thermoelectric performance with long-term stability and biocompatibility. For example, the use of advanced encapsulation techniques, such as atomic layer deposition (ALD), could improve the durability of thermoelectric materials in biological environments. Additionally, the exploration of novel materials, such as topological insulators and organic–inorganic hybrids, could lead to breakthroughs in both efficiency and sustainability.

Real-world validation through clinical trials and long-term studies will be essential to demonstrate the feasibility of thermoelectric materials in biomedical applications. For instance, a 2024 pilot study by Gao et al. evaluated the performance of a wearable thermoelectric device in monitoring patients with chronic conditions, showing promising results in terms of energy harvesting and device longevity [62]. Such studies provide valuable insights into the practical challenges and opportunities for thermoelectric materials in healthcare.

3. Conclusion

This chapter has elucidated the transformative potential of advanced thermoelectric materials in biomedical applications, driven by critical performance metrics such as thermoelectric figures of merit (ZT) and power factors. Materials like nanostructured bismuth telluride (Bi_2Te_3) and silver selenide (Ag_2Se) achieve ZT values exceeding 1.2 at room temperature, with Bi_2Te_3 demonstrating power factors of ~50 μW/cmK^2 and Ag_2Se exhibiting ultralow thermal conductivity (~0.5 W/mK). Flexible carbon nanotube (CNT)-polymer composites, such as SWCNT-PEDOT:PSS, further showcase ZT values of ~0.4 and power factors up to 500 μW/mK^2, enabling seamless integration into wearable biosensors and energy-autonomous health monitors. Hybrid nanocomposites, including doped transition metal dichalcogenides (TMDs), have reached ZT values of ~0.6, while organic thermoelectric polymers achieve ZT ~0.8, underscoring their viability for low-temperature, implantable systems.

Despite these advances, challenges persist in balancing efficiency with material stability and scalability. For instance, Bi_2Te_3 degrades by ~20% in ZT after 30 days in physiological conditions, and Ag_2Se requires encapsulation to mitigate silver ion toxicity. Scalable manufacturing techniques, such as 3D printing, have enabled CNT composites to achieve ZT ~0.4 and power factors of ~200 μW/mK^2, yet cost-effective alternatives like magnesium silicide (Mg_2Si, ZT ~0.7–1.0) remain critical for sustainable deployment.

Near-term research must prioritize optimizing biocompatibility and stability—for example, hybrid Bi_2Te_3-PEDOT:PSS composites show only 10% ZT loss over 60 days in simulated body fluids. Long-term efforts should focus on advancing nontoxic materials (e.g., Mg_2Si, CNT-organic hybrids) and scalable fabrication methods to bridge laboratory innovations with clinical applications. The integration of thermoelectric systems into implantable glucose sensors (6-month stability) and wearable monitors highlights their potential, provided material toxicity and thermal management challenges are resolved.

By addressing these priorities through interdisciplinary collaboration, thermoelectric materials can revolutionize personalized healthcare, enabling energy-efficient, self-powered devices that enhance patient outcomes while reducing reliance on conventional power sources. This synergy between material innovation and biomedical engineering promises to unlock next-generation technologies for real-time monitoring, targeted therapies, and sustainable medical solutions.

Funding

AV, IN, CH acknowledge a grant of the Ministry of Research, Innovation, and Digitalization, CNCS—UEFISCDI, project number PN-IV-P2–2.1-TE-2023-0993 within PNCDI IV. This research was also founded by the Romanian Ministry of Research, Innovation, and Digitalization under the Romanian National Nucleu Program LAPLAS VII—contract No. 30 N/2023.

Author details

Anita Ioana Visan, Irina Negut and Claudiu Hapenciuc*
National Institute for Lasers, Plasma and Radiation Physics, Măgurele, Romania

*Address all correspondence to: claudiu.hapenciuc@inflpr.ro

References

[1] Jaziri N, Boughamoura A, Müller J, Mezghani B, Tounsi F, Ismail M. A comprehensive review of thermoelectric generators: Technologies and common applications. Energy Reports. 2020;**6**(Supplement 7):264-287. ISSN 2352-4847. DOI: 10.1016/j.egyr.2019.12.011

[2] Zhang X, Zhao L-D. Thermoelectric materials: Energy conversion between heat and electricity. Journal of Materiomics. 2015;**1**(2):92-105

[3] Cai B et al. Promising materials for thermoelectric applications. Journal of Alloys and Compounds. 2019;**806**:471-486

[4] Kim SJ, We JH, Cho BJ. A wearable thermoelectric generator fabricated on a glass fabric. Energy & Environmental Science. 2014;7(6):1959-1965

[5] Remeš Š, Havlík J. Possibilities of using thermoelectric generators (TEG) for producing electrical energy by living subjects. In: 2020 International Conference on Applied Electronics (AE). Piscataway, New Jersey, USA: IEEE; 2020

[6] Li L et al. Multifunctional wearable thermoelectrics for personal thermal management. Advanced Functional Materials. 2022;**32**(22):2200548

[7] Zhao J, You Z. A shoe-embedded piezoelectric energy harvester for wearable sensors. Sensors. 2014;**14**(7):12497-12510

[8] Wei TR et al. Ag2Q-based (Q= S, Se, Te) silver chalcogenide thermoelectric materials. Advanced Materials. 2023;**35**(1):2110236

[9] Chen J et al. Hierarchical structures advance thermoelectric properties of porous n-type β-Ag_2Se. ACS Applied Materials & Interfaces. 2020;**12**(46):51523-51529

[10] Cervino-Solana P et al. Thermoelectric bismuth telluride nanostructures fabricated by electrodeposition within flexible templates. Heliyon. 2024;**10**(16):e36114. ISSN 2405-8440. DOI: 10.1016/j.heliyon.2024.e36114

[11] Su J et al. A batch process micromachined thermoelectric energy harvester: Fabrication and characterization. Journal of Micromechanics and Microengineering. 2010;**20**(10):104005

[12] Hricková G et al. Thermoelectric properties of ductile doped AgS for low power factor application. Acta Electrotechnica et Informatica. 2024;**24**(1):9-15

[13] Kim M-S et al. Design of wearable hybrid generator for harvesting heat energy from human body depending on physiological activity. Smart Materials and Structures. 2017;**26**(9):095046

[14] He X et al. Highly stretchable, durable, and breathable thermoelectric fabrics for human body energy harvesting and sensing. Carbon Energy. 2022;**4**(4):621-632

[15] Enescu D. Thermoelectric energy harvesting: Basic principles and applications. Green Energy Advances. 2019;**1**:38

[16] Jia Y et al. Wearable thermoelectric materials and devices for self-powered electronic systems. Advanced Materials. 2021;**33**(42):2102990

[17] Beretta D et al. Thermoelectrics: From history, a window to the future.

Materials Science and Engineering: R: Reports. 2019;**138**:100501

[18] Rao Y et al. Fabrication and characterization of a thermoelectric generator with high aspect ratio thermolegs for electrically active implants. Advanced Materials Technologies. 2024;**9**(1):2301157

[19] Li L et al. Highly sensitive flexible heat flux sensor based on a microhole array for ultralow to high temperatures. Microsystems & Nanoengineering. 2023;**9**(1):133

[20] Wai Cheung T, Li L. Development of self-care textile wearables with thermally stimulated drug delivery function via biological and physical investigations. Textile Research Journal. 2021;**91**(7-8):820-827

[21] Kim M-K et al. Wearable thermoelectric generator for harvesting human body heat energy. Smart Materials and Structures. 2014;**23**(10):105002

[22] Patidar S. Applications of thermoelectric energy: A review. International Journal for Research in Applied Science and Engineering Technology. 2018;**6**(5):1992-1996

[23] Ma H et al. Flexible Ag_2Se thermoelectric films enable the multifunctional thermal perception in electronic skins. ACS Applied Materials & Interfaces. 2024;**16**(6):7453-7462

[24] Mamur H et al. A review on bismuth telluride (Bi_2Te_3) nanostructure for thermoelectric applications. Renewable and Sustainable Energy Reviews. 2018;**82**:4159-4169

[25] Chen K et al. Flexible thermoelectrics based on plastic inorganic semiconductors. Advanced Materials Technologies. 2023;**8**(16):2300189

[26] Nozariasbmarz A et al. Review of wearable thermoelectric energy harvesting: From body temperature to electronic systems. Applied Energy. 2020;**258**:114069

[27] Chen Z-G et al. Nanostructured thermoelectric materials: Current research and future challenge. Progress in Natural Science: Materials International. 2012;**22**(6):535-549

[28] Gao W, Wang Y, Lai F. Thermoelectric energy harvesting for personalized healthcare. Smart Medicine. 2022;**1**(1):e20220016

[29] Lal S, Razeeb KM, Gautam D. Enhanced thermoelectric properties of electrodeposited Cu-doped Te films. ACS Applied Energy Materials. 2020;**3**(4):3262-3268

[30] Kadel K et al. Synthesis and thermoelectric properties of Bi_2Se_3 nanostructures. Nanoscale Research Letters. 2011;**6**:1-7

[31] Kim SL et al. Flexible power fabrics made of carbon nanotubes for harvesting thermoelectricity. ACS Nano. 2014;**8**(3):2377-2386

[32] Kwon C et al. Stretchable Ag_2Se thermoelectric fabric with simple and nonthermal fabrication for wearable electronics. Small Science. 2024;**4**(11):2400230

[33] Wang Y et al. Flexible thermoelectric materials and generators: Challenges and innovations. Advanced Materials. 2019;**31**(29):1807916

[34] Liu W et al. Recent advances in thermoelectric nanocomposites. Nano Energy. 2012;**1**(1):42-56

[35] Kharlamova MV, Kramberger C. Applications of filled single-walled

carbon nanotubes: Progress, challenges, and perspectives. Nanomaterials. 2021;**11**(11):2863

[36] Choi J et al. High-performance, wearable thermoelectric generator based on a highly aligned carbon nanotube sheet. ACS Applied Energy Materials. 2019;**3**(1):1199-1206

[37] Yin S et al. Poly (3, 4-ethylenedioxythiophene)/Te/single-walled carbon nanotube composites with high thermoelectric performance promoted by electropolymerization. ACS Applied Materials & Interfaces. 2019;**12**(3):3547-3553

[38] Lee W et al. Organic thermoelectric devices with PEDOT: PSS/ZnO hybrid composites. Chemical Engineering Journal. 2021;**415**:128935

[39] Zhang X et al. Novel structured transition metal dichalcogenide nanosheets. Chemical Society Reviews. 2018;**47**(9):3301-3338

[40] Song X et al. Catalytic biomaterials. Accounts of Materials Research. 2024;**5**(3):271-285

[41] Badawi NM et al. SWCNTs/PEDOT: PSS coated cotton for wearable clothes and supercapacitor applications. Sustainability. 2023;**15**(1):889

[42] Wu X, Su Y, Shi J. Perspective of additive manufacturing for metamaterials development. Smart Materials and Structures. 2019;**28**(9):093001

[43] Gloria J et al. Solubilization, characterization, and protein coupling analysis to multiwalled carbon nanotubes. High Performance Polymers. 2021;**33**(3):338-344

[44] Mamidi N. Cytotoxicity evaluation of carbon nanotubes for biomedical and tissue engineering applications. In: Saleh HM, El-Sheikh SMM, editors. Perspective of Carbon Nanotubes. Vol. 1. London, UK: Intech Open; 2019. pp. 197-208

[45] Uzair M et al. Review: Biomedical applications of carbon nanotubes. Nano Biomedicine And Engineering. 2021;**13**(1):82-93

[46] Sun Y et al. Silk fibroin biomaterial-functionalized carbon nanotubes for high water dispersibility and promising biomedical applications. Textile Research Journal. 2019;**89**(7):1144-1152

[47] Bark H, Lee W, Lee H. Enhanced thermoelectric performance of CNT thin film p/n junctions doped with N-containing organic molecules. Macromolecular Research. 2015;**23**:795-801

[48] Snyder GJ, Toberer ES. Complex thermoelectric materials. Nature Materials. 2008;**7**(2):105-114

[49] Hotěk P, Fiala L, Černý R. Thermoelectric properties of metals and the graphene model. In: Journal of Physics: Conference Series. Bristol, United Kingdom: IOP Publishing; 2023

[50] Choudhary S, Muthiah S, Dhakate SR. Enhancement of power factor and mechanical properties in low cost Mg_2Si_{1-x} Sn x employing a composite approach. ACS Applied Energy Materials. 2021;**5**(1):549-556

[51] Yao H et al. Recent development of thermoelectric polymers and composites. Macromolecular Rapid Communications. 2018;**39**(6):1700727

[52] Campoy-Quiles M. Will organic thermoelectrics get hot? Philosophical Transactions of the Royal Society A. 2019;**377**(2152):20180352

[53] Sun Z, Li J, Wong WY. Emerging organic thermoelectric applications from conducting metallopolymers. Macromolecular Chemistry and Physics. 2020;**221**(12):2000115

[54] Russ B et al. Organic thermoelectric materials for energy harvesting and temperature control. Nature Reviews Materials. 2016;**1**(10):1-14

[55] Kim D et al. Improved thermoelectric behavior of nanotube-filled polymer composites with poly (3, 4-ethylenedioxythiophene) poly (styrenesulfonate). ACS Nano. 2010;**4**(1):513-523

[56] Sankhla A et al. Mechanical alloying of optimized Mg_2 (Si, Sn) solid solutions: Understanding phase evolution and tuning synthesis parameters for thermoelectric applications. ACS Applied Energy Materials. 2018;**1**(2):531-542

[57] Zhou X et al. Enhanced figure of merit of poly (9, 9-di-n-octylfluorene-alt-benzothiadiazole) and SWCNT thermoelectric composites by doping with $FeCl_3$. Journal of Applied Polymer Science. 2019;**136**(5):47011

[58] Cheong K, Lim J. Numerical simulation of segmented ratio in bismuth telluride and skutterudites for waste heat recovery. In: Journal of Physics: Conference Series. Bristol, United Kingdom: IOP Publishing; 2021

[59] Nozariasbmarz A, Krasinski JS, Vashaee D. N-type bismuth telluride nanocomposite materials optimization for thermoelectric generators in wearable applications. Materials. 2019;**12**(9):1529

[60] Liu W et al. Convergence of conduction bands as a means of enhancing thermoelectric performance<? Format?> of n-type Mg 2 Si 1-x Sn x solid solutions. Physical Review Letters. 2012;**108**(16):166601

[61] Bubnova O et al. Optimization of the thermoelectric figure of merit in the conducting polymer poly (3, 4-ethylenedioxythiophene). Nature Materials. 2011;**10**(6):429-433

[62] Gao M et al. In vitro and In vivo biocompatibility assessment of chalcogenide thermoelectrics as implants. Journal of Materials Chemistry B. 2024;**12**:6847-6855

[63] Hu B et al. Thermoelectrics for medical applications: Progress, challenges, and perspectives. Chemical Engineering Journal. 2022;**437**:135268

[64] Wan Y, Qiu Z, Guo CF. Soft and flexible materials. In: Sun S, Tan W, Wei H, editors. Emergent Micro-and Nanomaterials for Optical, Infrared, and Terahertz Applications. Vol. 1. CRC Press; 2022. pp. 285-322

[65] Kim J et al. Organic-inorganic hybrid thermoelectric material synthesis and properties. Journal of the Korean Ceramic Society. 2017;**54**(4):272-277

[66] Xin J et al. Inorganic thermoelectric fibers: A review of materials, fabrication methods, and applications. Sensors. 2021;**21**(10):3437

[67] Wu X et al. Thermoelectric converters based on ionic conductors. Chemistry–An Asian Journal. 2021;**16**(2):129-141

[68] Snyder G, Toberer E. Complex thermoelectric materials. Nature Materials. 2008;**v**:7

[69] de Lourdes Gonzalez-Juarez M et al. Electrochemical deposition and thermoelectric characterisation of a semiconducting 2-D metal–organic

framework thin film. Journal of Materials Chemistry A. 2020;**8**(26):13197-13206

[70] Yakuphanoglu F et al. The dielectric spectroscopy and surface morphology studies in a new conjugated polymer poly (benzobisoxazole-2, 6-diylvinylene). Physica B: Condensed Matter. 2005;**365**(1-4):13-19

[71] Jia F et al. Bi8Se7: Delocalized interlayer π-bond interactions enhancing carrier mobility and thermoelectric performance near room temperature. Journal of the American Chemical Society. 2020;**142**(28):12536-12543

[72] Ebrahimi M et al. High-performance nanoscale metallic multilayer composites: Techniques, mechanical properties and applications. Materials. 2024;**17**(9):2124

[73] Zhang D et al. XTe (X= Ge, Sn, Pb) monolayers: Promising thermoelectric materials with ultralow lattice thermal conductivity and high-power factor. ES Energy & Environment. 2020;**10**(26):59-65

[74] Ma Z et al. Heavily boron-doped silicon layer for the fabrication of nanoscale thermoelectric devices. Nanomaterials. 2018;**8**(2):77

[75] Heremans JP et al. When thermoelectrics reached the nanoscale. Nature Nanotechnology. 2013;**8**(7):471-473

[76] Wang X, Dai X, Chen Y. Sonopiezoelectric nanomedicine and materdicine. Small. 2023;**19**(29):2301693

[77] Han T et al. 3D printed sensors for biomedical applications: A review. Sensors. 2019;**19**(7):1706

[78] Wu L et al. Toughening thermoelectric materials: From mechanisms to applications. International Journal of Molecular Sciences. 2023;**24**(7):6325

[79] Liu Y et al. Effects of preparation methods on the thermoelectric performance of SWCNT/Bi_2Te_3 bulk composites. Materials. 2020;**13**(11):2636

[80] Massetti M et al. Unconventional thermoelectric materials for energy harvesting and sensing applications. Chemical Reviews. 2021;**121**(20):12465-12547

[81] Culebras M, Gómez C, Cantarero A. Enhanced thermoelectric performance of PEDOT with different counter-ions optimized by chemical reduction. Journal of Materials Chemistry A. 2014;**2**(26):10109-10115

[82] Li W et al. Influence of nanopores on the tensile/compressive mechanical behavior of crystalline Co_Sb 3: A molecular dynamics study. Journal of Electronic Materials. 2015;**44**:1477-1482

[83] Srinivasan B, Berthebaud D, Mori T. Is LiI a potential dopant candidate to enhance the thermoelectric performance in Sb-free GeTe systems? A prelusive study. Energies. 2020;**13**(3):643

[84] Adekoya GJ et al. Recent advancements in biomedical application of polylactic acid/graphene nanocomposites: An overview. BMEMat. 2023;**1**(4):e12042

[85] Zindani D, Kumar K. Graphene-based polymeric nano-composites: An introspection into functionalization, processing techniques and biomedical applications. Biointerface Research in Applied Chemistry. 2019;**9**:3926-3933

[86] Zhang Q et al. Organic thermoelectric materials: Emerging green energy materials converting

heat to electricity directly and efficiently. Advanced Materials. 2014;**26**(40):6829-6851

[87] Zevalkink A et al. Ca_3AlSb_3: An inexpensive, non-toxic thermoelectric material for waste heat recovery. Energy & Environmental Science. 2011;**4**(2):510-518